THE OXFORD AUTHORS

General Editor: Frank Kermode

WILLIAM BLAKE was born in 1757, the son of a West End shopkeeper. His own trade was that of craftsman engraver, and until he was in his thirties there was little to distinguish him outwardly from the ordinary man of his profession. In 1789 he produced *Songs of Innocence*, the first in a beautiful series of small hand-made volumes of illustrated poetry in lyric and blank-verse—sometimes obscure, but always arresting—issued by Blake over the next six years: among them *The Book of Thel*, *The Marriage of Heaven and Hell* (chiefly prose), *Visions of the Daughters of Albion*, *America*, *Songs of Experience*, *Europe*, and *The Book of Urizen*. In 1800 Blake tried to break free from London and commercial engraving by accepting an invitation to work under the patronage of the poet and critic William Hayley near his home in Sussex. Blake's mind was already turning to mythological narrative on a much larger scale than hitherto, and his new situation was not congenial. He returned to London in 1803, and worked heroically on the etching of his grand narratives *Milton* and *Jerusalem*, and on ambitious and highly personal painting and illustrating projects. For years he was a derided or ignored figure but towards the end of his life he attracted around him a deeply admiring group of young artists of the next generation. To them is largely due the survival, alongside Blake's etched and printed works, of a body of manuscript remains full of pungent aphorisms and haunting lyrics. Blake died in 1827.

MICHAEL MASON is Senior Lecturer in English at University College London. He has published many studies of nineteenth-century poetry and fiction, and is preparing a new edition of Wordsworth's *Lyrical Ballads*.

THE OXFORD AUTHORS

WILLIAM BLAKE

EDITED BY

MICHAEL MASON

Oxford   New York

OXFORD UNIVERSITY PRESS

Oxford University Press, Walton Street, Oxford OX2 6DP

Oxford New York Toronto
Delhi Bombay Calcutta Madras Karachi
Kuala Lumpur Singapore Hong Kong Tokyo
Nairobi Dar es Salaam Cape Town
Melbourne Auckland Madrid
and associated companies in
Berlin Ibadan

Oxford is a trade mark of Oxford University Press

Introduction, edited text, and editorial matter © Michael Mason 1988

First published by Oxford University Press 1988
First issued as an Oxford University Press paperback 1988

British Library Cataloguing in Publication Data
Data available

Library of Congress Cataloging in Publication Data
Blake, William, 1757–1827
William Blake.
(The Oxford authors).
bibliography: p.
Includes indexes.
I. Mason, Michael, 1941–   II. Title.
III. Series.
PR4142.M37 1988 821'.7 87–22104
ISBN 0–19–282001–X

3 5 7 9 10 8 6 4

Printed in Great Britain by
Biddles Ltd
Guildford and King's Lynn

CONTENTS

Italics indicate a prose work

Introduction	xiii
Chronology	xxiii
Note on the Text	xxvi

BLAKE ON RELIGION AND KNOWLEDGE

From the annotations to John Caspar Lavater's Aphorisms on
  Man *(1788)*   3
From the annotations to Emanuel Swedenborg's Wisdom of
  Angels Concerning Divine Love and Divine Wisdom
  (c.1789)   5
All Religions are One (c.1788)   5
There is No Natural Religion (c.1788) (1)   6
There is No Natural Religion (c.1788) (2)   7
The Marriage of Heaven and Hell (1790)   8
From the annotations to Richard Watson's An Apology for the
  Bible *(1798)*   20
'To my friend Butts' *(1800)*   23
'With happiness stretched across the hills' *(1802)*   25
Auguries of Innocence (c.1804)   27
Prefaces to Chapters II, III and IV of Jerusalem *(1804–18)*   31
From 'Vision of the Last Judgement' (c.1810)   35
Annotation to J. G. Spurzheim's Observations on . . . Insanity
  (c.1818)   45
The Everlasting Gospel *(c.1818)*   45
From the annotations to Bishop Berkeley's Siris *(c.1820)*   54
From the annotations to R. J. Thornton's The Lord's Prayer,
  Newly Translated *(1827)*   55

BLAKE ON ART AND LITERATURE

Letter of 16 August 1799   59
Letter of 23 August 1799   60
Letter of 26 August 1799   61
Letter of 2 July 1800   63
To my Dearest Friend, John Flaxman (1800)   64

From the annotations to Henry Boyd's A Translation of the
    Inferno of Dante Alighieri (c.*1800*)   65
Letter of 11 September 1801   65
From a letter of 10 January 1802   67
Letter of 22 November 1802   68
From a letter of 25 April 1803   70
Preface to Milton (*1804*)   71
Preface to Chapter 1 of Jerusalem (*1804–18*)   72
From the annotations to volume 1 of Sir Joshua Reynolds's
    Works (mainly c.*1808*)   73
From A Descriptive Catalogue of Pictures (*1809*)   81
From 'Public Address' (c.*1811*)   102
On Homer's Poetry (c.*1821*)   115
On Virgil (c.*1821*)   116
From the annotations to Wordsworth's 1815 Poems (*1826*)   116

EARLY VISIONARY AND NARRATIVE WRITINGS

To Spring   121
To Summer   121
To Autumn   122
To Winter   122
To the Evening Star   123
To Morning   124
Fair Elenor   124
To the Muses   126
Gwin, King of Norway   127
An Imitation of Spenser   131
King Edward the Third   133
Prologue, Intended for a Dramatic Piece of *King Edward the*
    Fourth   150
Prologue to King John   150
A War Song to Englishmen   151
The Couch of Death   152
Contemplation   153
Samson   154
[*The Vision of Pride*]   157
'"*Woe!*" *cried the Muse*'   160

SEPTENARY VERSE OF THE FRENCH REVOLUTIONARY PERIOD

Tiriel (*c*.1789)   163
The Book of Thel (*c*.1789)   175
The French Revolution (1791)   180
Visions of the Daughters of Albion (1793)   196
Africa (1795)   203
America, a Prophecy (1793)   205
Europe, a Prophecy (1794)   214
Asia (1795)   222

THE LYRICS

Song ['How sweet I roamed'] (before 1783)   227
Song ['My silks and fine array'] (before 1783)   227
Song ['Love and harmony combine'] (before 1783)   228
Song ['I love the jocund dance'] (before 1783)   228
Song ['Memory, hither come'] (before 1783)   229
Mad Song (before 1783)   230
Song ['Fresh from the dewy hill'] (before 1783)   230
Song ['When early morn'] (before 1783)   231
Song First by a Shepherd (1784)   232
Song Third by an Old Shepherd (1784)   232
[The Cynic's First Song] (*c*.1784)   232
[Miss Gittipin's First Song] (*c*.1784)   234
[The Cynic's Second Song] (*c*.1784)   234
[Obtuse Angle's Song] (*c*.1784)   235
[The Lawgiver's Song] (*c*.1784)   236
[Miss Gittipin's Second Song] (*c*.1784)   236

SONGS OF INNOCENCE (1789)
    Introduction   238
    A Dream   238
    The Little Girl Lost   239
    The Little Girl Found   241
    The Blossom   242
    The Lamb   243
    The Shepherd   243
    Infant Joy   244
    On Another's Sorrow   244
    Spring   245
    The Schoolboy   246

Laughing Song   247
The Little Black Boy   248
The Voice of the Ancient Bard   249
The Echoing Green   249
Nurse's Song   250
Holy Thursday   250
The Divine Image   251
The Chimney-Sweeper   252
A Cradle Song   252
The Little Boy Lost   254
The Little Boy Found   254
Night   254

MANUSCRIPT LYRICS BETWEEN *INNOCENCE* AND
EXPERIENCE

'I told my love'   256
'I laid me down'   256
'I saw a chapel'   256
'I asked a thief'   257
'I heard an angel'   257
A Cradle Song   258
'I feared the fury'   259
'Silent, silent night'   259
'Why should I care'   259
'O lapwing'   260
'Thou hast a lap full of seed'   260
In a Myrtle Shade   260
'As I wandered'   261
'Are not the joys'   261
To Nobodaddy   261
How to Know Love from Deceit   262
Soft Snow   262
An Ancient Proverb   262
To my Myrtle   263
Merlin's Prophecy   263
Day   263
The Fairy   263
'The sword sung on the barren heath'   264
'Abstinence sows sand'   264
'In a wife I would desire'   264
'If you trap the moment'   264

Eternity   265
The Question Answered   265
Lacedaemonian Instruction   265
Riches   265
An Answer to the Parson   265
'The look of love alarms'   265
'Her whole life is an epigram'   266
'An old maid early'   266
Motto to the Songs of Innocence and Experience   266

SONGS OF EXPERIENCE (1793)

Introduction   267
Earth's Answer   267
The Clod and the Pebble   268
Holy Thursday   269
The Chimney-Sweeper   269
Nurse's Song   270
The Sick Rose   270
The Fly   270
The Angel   271
The Tiger   271
My Pretty Rose Tree   272
Ah! Sunflower   273
The Lily   273
The Garden of Love   273
The Little Vagabond   274
London   274
The Human Abstract   275
Infant Sorrow   276
A Poison Tree   276
A Little Boy Lost   276
A Little Girl Lost   277
To Tirzah   278
A Divine Image   279

MANUSCRIPT LYRICS OF THE FELPHAM YEARS

'When a man has married a wife'   280
'A woman scaly'   280
'A fairy stepped upon my knee'   280
On the Virginity of the Virgin Mary and Joanna Southcott   281
The Golden Net   281

The Birds   282
The Grey Monk   282
Morning   284
'Terror in the house'   284
'Mock on'   284
'My Spectre around me'   285
'O'er my sins'   286
The Smile   287
The Mental Traveller   288
The Land of Dreams   291
Mary   292
The Crystal Cabinet   293
Long John Brown and Little Mary Bell   294
William Bond   295

LYRICS FROM THE EPIC POEMS

From the Preface to *Milton*   297
From Jerusalem (plate 27)   297
From Jerusalem (plate 41)   300
From Jerusalem (plate 52)   300
From Jerusalem (plate 77)   301

LATE LYRICS

'Grown old in love'   302
'Madman I have been called'   302
'He's a blockhead'   302
'I am no Homer's hero'   302
'The angel that presided o'er my birth'   302
'Some men created for destruction'   303
Imitation of Pope: A Compliment to the Ladies   303
'If I e'er grow'   303
'You don't believe'   303
'Great things are done'   304
'If you play a game of chance'   304
'I rose up'   304
'Why was Cupid a boy'   305
'Great men and fools'   306
To God   307
'Since all the riches'   307
'To Chloe's breast'   307
'Anger and wrath'   307

CONTENTS   xi

THE LOS POEMS

The Book of Urizen (1794)   309
The Book of Los (1795)   325
Milton: A Poem in Two Books (1804)   331
Jerusalem: The Emanation of the Giant Albion (1804–18)   381

Notes   515
Further Reading   587
Index of Titles and First Lines   589
Index of Names and Motifs   597

INTRODUCTION

GENERAL

THIS collection of William Blake's writings includes almost all his poetry and prose works, and a few of his letters. The only works of any substance omitted are 'An Island in the Moon', *The Book of Ahania*, and *The Four Zoas*. It differs, deliberately, from all recent editions of Blake. To start with, his very diverse output has been arranged under kinds of writing—described more fully below—rather than chronologically. A greater variety of his work is included here than in most editions, and it seemed inappropriate to make chronology the sole organizing principle of this disparate material, and thus thwart the reader wanting a concentrated experience of Blake as a lyricist, or insight into his views on a single important topic, such as art. There is also a strong negative justification for playing down chronology in Blake's case; the dates of composition of his texts are on the whole extremely uncertain, not least because 'composition' with Blake often means something very protracted: writing, etching, printing, hand-colouring, and finally arranging plates in sequence.

The second unusual feature of this anthology is that it offers a single, and entirely verbal, version of Blake's writings. Deleted or alternative readings are almost completely ignored, and there is no attempt to describe or reproduce Blake's illustrations to his poems. The first of these omissions needs little defence. Editors have fallen into the habit of transcribing what Blake crossed out, or improved on, for various reasons; interesting though the material can sometimes be, readers are entitled to an edition which makes a single choice of text (with Blake's endorsement) and is not cluttered with indications of what he rejected.

In recent years the doctrine has gained ground that Blake's text, where it was illustrated, can only be read adequately in conjunction with the illustration. This poses great problems for a modern editor—who has to strike a balance between the unsatisfactory extremes of direct facsimile and mere description of the illustration—but, more troublingly, the enhancement of our reading of Blake which was expected to flow from attention to his illustrations has simply not occurred. And the whole principle of the exercise may be questioned. Blake himself seems to have been less tender of his text than some of his editors. There are many formulations—ranging from phrases of a few words in length to whole lyrics and passages of blank verse—which he was happy to issue in more than one visual context. He evidently felt that they were robust and versatile, and might have regarded the anthologizers of the 'Jerusalem' lyric, and surely the manufacturers of posters carrying the 'Ancient of Days' image, more indulgently than the protective Blake experts do. After all, concentrating on a basic iconic element in his art, and adapting it to one's own imaginative needs, is precisely what he urges:

> If the spectator could enter into these images in his imagination, approaching them on the fiery chariot of his contemplative thought, if he could . . . make a friend and companion of one of these images of wonder . . . then would he arise from his grave, then would he meet the Lord in the air, and then he would be happy. ('Vision of the Last Judgement')

Finally, no attempt is made in this edition to summarize the content or message of individual prophetic works, nor is there any explication of Blake's mythology. I feel that both enterprises are mistaken. The drab paraphrases of the prophetic writing that have been provided by some editors would, if adequate, imply that Blake was a rather straightforward writer who inexplicably obscured his meaning. In practice, however, such paraphrases are not adequate: invariably they feel wrong measured against the actual experience of Blake's text. They are Procrustean devices that tend to fit badly at all points, and not just at some crux or isolated difficulty of interpretation which, once clarified, would make all the rest intelligible.

Students of Blake have been too ready to assume that his mythology is a well-formed system, with a systematic relationship to the real world, both capable of being reconstructed by a thorough and attentive study of his work. Actually, the more systematic Blake's prophecies become the less mythological they are: in the sense that in *Milton* and *Jerusalem* there is a good deal of pattern-forming with names of persons and places drawn from the Bible and elsewhere, but rather little myth-making with Blake's personal, invented machinery. The invented machinery is, after all, just that: something created by Blake out of a minimum of antecedent materials, and then added to (and sometimes reshaped) with almost every reappearance. Blake does not know everything about this mythology in advance; generally he is, at most, exploring it, not describing it. Equally, he does not have a fixed idea of what this personal mythology symbolizes in the non-mythological world.

A personal mythology was almost, it may be said, Blake's way of escaping from system, and not just because there was no inherited

framework to constrain him. Blake's cast of mind is often misunder-
stood; he is perceived as a man with passionately held beliefs on
religion, philosophy, politics, morality, and art whose propagation was
the purpose of his writings. The fact is that Blake was very ambivalent
in his beliefs (in practice there is no agreement among the critics as to
what his convictions were) and actually embraced intellectual and
emotional ambivalence. He is the great anti-simplifier, always probing
for contradiction, especially self-contradiction. To judge by some of his
remarks about the need for a perpetual splitting of one's nature, he
regarded the recognition of self-contradiction, of many personalities
within one, as a kind of moral imperative. Whatever the truth of this,
there is no doubt that Blake's personal mythology is complex (and
hence galling to the tidy-minded reader) because he used it precisely to
achieve complexity: as a device for virtually unlimited multiplication
and diversification. If Blake's mythology could be satisfactorily
summarized in a table it would have failed in its purpose.

There is, of course, a large and valuable body of commentary on
Blake's text aside from these topics, to which my own notes are heavily
indebted. This is the first new edition of Blake for several years, and the
most useful recent discoveries and suggestions, finding their way into
this kind of annotation for the first time, are cited more fully than more
familiar information. I have also allowed myself more commentary on
the prose texts than on the verse, since Blake's prose writings have been
almost ignored in modern annotated editions; my notes contain much
new material, some original and some derived from the work of other
scholars.

But, as far as the mythology is concerned, there is no substitute for
the text. The one function an editor of Blake can usefully perform is to
indicate where in that text particular mythological figures appear, and
another novel feature of this edition, the index of names and motifs, has
this as one of its aims. The index also cites all mythological, and some
legendary or quasi-historical, beings mentioned by Blake or referred to
in the footnotes. Blake believed his 'giant forms' to be drawn from the
common stock of the universal human imagination; hence he feels free
to plunge into their doings with so little explanation. Where the
relationship of other mythologies to his own is concerned his tendency
is no longer to subdivide and multiply, but to synthesize, just as
Chaucer, according to Blake, had recreated Silenus in his Host, and
Apollo in his Squire. The index also includes those recurrent motifs
(such as covering cherub, abomination of desolation, Daughters of
Memory) which come close to being mythological agents; explanations

of them, where called for, will be found in the footnotes to their first appearances as listed in the index. Unfortunately there are many other repeated motifs and terms which likewise could only be footnoted once, but which do not qualify for the index; in these cases readers will have to hunt down the relevant first footnote for themselves.

Finally, I hope that the general design of this anthology will be welcome to readers who feel intimidated by Blake's mythological writings. Blake is too commonly presented as a writer who offers just two kinds of text, between which the choice is stark: the lyrics of *Innocence* and *Experience*, and the gloomy, extensive labyrinth which has 'Prophetic Books' over its entrance. The choice and arrangement of material in this anthology may remind readers that the Prophetic Books are by no means all of a piece, that the *Innocence* and *Experience* poems are only part of a remarkable body of work in this vein, and that Blake was an exciting prose writer and an inventive adapter of traditional genres.

BLAKE ON RELIGION AND KNOWLEDGE

The supernatural, in some sense, is Blake's great theme, but his writings very clearly reflect—in a manner that has not lost its relevance—the difficulties of religious belief for an intellectually advanced person in his era. Commentators from the earliest years have been divided on the content of Blake's metaphysical views, and on the nature of his religious experience. In his own lifetime Blake's claims to eidetic supernatural experiences, or 'visions', were taken literally by some, and regarded by others as designed to challenge and provoke. In this section, and in the anthology as a whole, the reader will find numerous accounts, or implied accounts, of such eidetic experiences, but also much material to suggest that Blake would have wished, on ideological grounds, to lay claim to these even if they had not been granted to him. In the literary domain, for example, he was evidently moved and impressed by the ancient notion of a muse, or equivalent agent of inspiration. A writer as recent as Milton had written of his contact with such an agency with great freshness and authenticity. Blake could see little evidence that the tradition was alive in his own day, and he used Miltonic terms rather assertively about his own literary creativity. But we cannot tell if he actually, or only wishfully, experienced the illusion of supernatural dictation.

There is also much ambivalence and complexity in Blake's treatment of the underpinnings to religious experience: that is, in his treatment of the material world, of religious authority, and of Man's perceptual powers. While acutely conscious that the European philosophical and scientific climate since the mid-seventeenth century was inimical to religion, and more than eager to denounce this climate, Blake inherited its way of thinking, and his reassertions of a supernatural level of existence, and of the possibility of human acquaintance with this, concede much, in the way they are couched, to a sceptical and materialist vision of things. Blake believed that Man was God; by the same token, God is no more than Man. He denied that the natural world could be the grounds of religion; but 'everything that lives is holy'. For Blake Christianity is true; but the Old Testament is worse than unenlightened, and the New Testament must be read in an unorthodox, rather commonsensical spirit (so that Christ's miracles are natural events, for example). Blake thought Locke wrong to deny innate or non-experiential ideas; but for him the supernatural seems to be perceived by a sensory mechanism that Locke would have recognized. Blake, indeed, uses the terms 'perceive' and 'perception' and their variants with great frequency (surely more than any mystical writer before him), and to that extent is thoroughly a post-Lockean.

This complicated background is one reason why many of the texts in this section are Blake's reactions, and by no means uniformly hostile reactions, to other texts. For Blake religion was to a considerable extent a matter of denying certain other positions, and even this denial was seldom a root-and-branch affair.

BLAKE ON ART AND LITERATURE

Blake's major utterances about the visual and verbal arts date from the decade 1800–10, and they are not only reflexes of his work in that period but also responses to, and actions within, his professional predicament. Blake's period of most secure artistic employment—under the patronage of William Hayley at Felpham in the years 1800–3 —was also that in which he seems to have resolved more fully than before to try and establish himself as an independent artist, true to his own vision. This resolve, and Blake's utter failure to achieve recognition when he returned to London, form the context of most of the material in this section.

In literary terms, Blake now gave all his energies to mythical narratives in verse on a much larger scale than he had previously attempted. Pictorially (and here recognition mattered much more to him) he had to lay claim to being a 'painter' (hence his ill-fated exhibition of 1809) but did so in terms in which, by necessity and temperament, his background as a craftsman-engraver was strongly expressed. His championing of an art of outline rather than tonality flows from inexorable practical conditions (his years as apprentice and qualified engraver, and his lack of opportunity to encounter European art except in engraved reproductions), but also from a true enthusiasm for the neoclassical inspired by such stirring and colourful artistic personalities as Mortimer, Barry, and Fuseli (and by friends such as the latter and Flaxman), and from a commitment to an art of active, expressive human figures (which he thinks of as belonging to a major if controversial genre, history painting).

Most of the texts here have explicit relevance to Blake's vicissitudes as a working pictorial artist at a particular period, sometimes in terms that are historically local, but they cannot be neglected by the student of Blake whose main interest is, say, the verbal art of the 1790s. Blake wrote most fully about his art when circumstances goaded him, but less elaborated utterances on other occasions (in *The Marriage of Heaven and Hell*, for example) suggest that there was a consistency to his aesthetic aims and assumptions across all periods and all mediums (but see below, p. xxi). He simply decided around 1800 to become more completely the artist he had always believed in. And, however absorbed by professional difficulties, Blake always kept the larger picture in view: these texts are full of remarks that are intensely suggestive both for his own art, and for art in general.

EARLY VISIONARY AND NARRATIVE WRITINGS

The texts themselves are the only hard evidence we have about Blake's early activity as a writer. Those included here almost all come from a printed but unpublished collection of 1783 entitled *Poetical Sketches* (and the rest seem to belong with items printed there). There are statements of the time to the effect that the *Poetical Sketches* were youthful pieces, some very youthful; these have been widely doubted, but critics have nevertheless tended, for whatever reasons, to treat *Poetical Sketches* as immature. In particular, Blake's indebtedness to previous writers has been cited, but this is to misunderstand Blake's approach to his predecessors, which is not at all that of an insecure, unformed writer. To start with there is often considerable enterprise in Blake's choice of models, notably the seventeenth-century lyric writers—little appreciated in Blake's day—who seem to have influenced the lyrics in *Poetical Sketches* (see the items on pp. 227–31 below). If there is anything wrong with Blake's use of his models it is that they have too little power over him: in sonnets that follow the form in layout only, in an imitation of Spenser that bears no real resemblance, in personification pieces in the eighteenth-century manner that turn into something like nightmares. More positively, Blake generally takes some aspect of his source, ignoring the rest, and develops it with an extraordinary creativity for his own characteristic purposes. His adaptation, or hijacking, of the Tudor chronicle play in *Edward III* is the richest example of this. Only in the case of the 'Fair Elenor' ballad does Blake's vigorous taking-up of a genre fail to yield something unexpected and exciting.

SEPTENARY VERSE OF THE FRENCH REVOLUTIONARY PERIOD

A distinct phase of Blake's activity as a writer consists of the short blank-verse narrative or quasi-narrative texts which he wrote in the years when he was also producing the celebrated *Innocence* and *Experience* sequences. Though not united in medium (*Tiriel* was never etched, though the manuscript has accompanying illustrations, and with *The French Revolution* Blake tried for orthodox printed reproduction), and though unmistakably expressing several changes of approach in mode and materials, they cohere within the context of Blake's output. Their main formal link (the seven-stress line) is emphasized in the heading to this section. One text ('Asia') is not even intermittently in this metre, but it is included because Blake grouped it with three previous septenary poems, even though he had become more interested in the three/four-stress line used in *Urizen*. *Urizen* (with *Los* and *Ahania*) is itself a distinct phase in Blake's narrative verse; and the texts of Felpham and after, *Milton* and *Jerusalem*, are in their longer and more informal verse lines very different from the poems in this section.

These represent an episode (of no more than six years) of impressive creativity on Blake's part: a series of texts of extraordinary variety and novelty, mostly illustrated by the artist and reproduced by manual procedures, while he was writing and issuing in the same fashion a body of remarkable lyrics. The changes of direction—the gap that separates *Europe* from *Tiriel*, for example—may not have been as unwelcome to Blake, as much a token of wrong choices, as might appear. His restlessness and uncertainty, his taste for, and respect for, contradiction, may have been gratified in the extraordinary leaps from history to dream, from song to manifesto, from sex to epistemology—indeed some of these transitions occur within the poems themselves. On the other hand, the artist makes an equivocal appearance in these texts. At l. 37 of the Preludium to *America* Blake originally inserted four strangely disruptive lines according to which the 'bard ceased, ashamed of his own song' and 'enraged' dashed his harp to pieces (and, to complicate matters, Blake then deleted them); the fascinating lines about the writing of *Europe* reproduced here as the opening of the poem were not included in most of the copies that Blake issued. Despite the inventive, uninhibited quality of the series Blake ultimately seems to have felt thwarted by his texts, and these devices for surmounting alienation (which are essentially along the lines of the personal interpolations in the prose pieces of the 1780s) were apparently no solution to the problem.

THE LYRICS

In this section are reprinted all Blake's lyrics, other than unfinished, fragmentary, or very casual pieces. It is by far the most important body of short poems by any English writer since the first half of the seventeenth century, and is rich in problems of interpretation. With the exception of *Poetical Sketches* Blake never published an orthodox slim volume of short poems. The lyrics outside *Poetical Sketches* (amounting to all but the first six items in this section) are often intrinsically enigmatic poems, and the difficulty they offer is considerably increased by the contexts in which they have been transmitted to us. One group is scattered, often untitled and in fairly rough form, through Blake's manuscript remains. By contrast, the other group consists for the most part of etched and illustrated texts that belong to larger schemes; but these schemes—*Innocence*, *Experience*, and the two epics *Milton* and *Jerusalem*—have proved just as fertile of controversy over the meaning of individual lyrics as any intrinsic difficulty in the latter. Matters are not made easier by Blake's readiness to move his lyrics to different contexts: for example, 'An Island in the Moon' lyrics reappear in *Innocence*, and others are transferred from that collection to *Experience*. Blake seems to have been able to look at his lyrics at arm's length, with a detachment that permitted him to see different possibilities in them at different times. For any given poem a cluster or sequence of attitudes remains stable, but the kind of speaker that might have uttered it, and our judgement on that speaker, is not a fixed matter.

THE LOS POEMS

It is not clear by what stages Blake became committed to the aesthetic ideal of the large-scale. It seems likely that from early on he admired

Milton (the great English exponent of the major large form in literature, epic), and perhaps also Barry (probably the best known contemporary exponent of the pictorial equivalent, history painting), but there is no firm indication of a wish to emulate them in scale until *The Four Zoas* (*c*.1800). Nevertheless, the poems in this section all subscribe to the ideal of an art of large dimensions: either in themselves or (with *Urizen* and *Los*) in the scheme into which they were to be fitted.

This development is linked with the emergence of Los as the most important single figure in Blake's mythology, though one can only speculate about the nature of the connection. Whatever else Los's entry means, it is in part Blake's way of including himself in his narrative writing, a goal which he had apparently not been able to achieve to his own satisfaction by earlier more direct interventions. The figure of Los himself indicates something about Blake's commitment to large-scale art: he stands for effort, for exertion, for heroic, taxing persistence in emotionally and physically difficult tasks. Milton represented the writing of *Paradise Lost*, to some extent, in such a light; Barry had apparently talked to Blake in these terms about his Society of Arts project (see p. 74 below); but Blake, in producing the work of his later years, perhaps suffered more, and laboured more, than either. For him the traditional ideals of epic writing and history painting became bound up with a personal determination to surmount his extraordinarily adverse artistic circumstances after 1803. The Los figure, above all in *Jerusalem*, is also a haunting representation of one person's vital, fertile consciousness in a society—in Blake's case the consciousness of an energetic and adroit maker of verbal and visual imagery. The interest of the question whether Blake was 'mad' starts to drain away with the recognition of what is authentic and moving about what he made out of his 'madness'. In particular, it is telling to remind ourselves of how much this corresponds to how we, as observers in later generations, perceive artists in past societies, perhaps especially in the Romantic period. Accounts of the artistic culture of the years 1780 to 1840 have always tended to give an unusual importance to the inner lives of the individuals called Beethoven or Delacroix or Shelley or Keats or Berlioz (and so on) independently of their physical actions and achievements. There is more contemporary warrant for this in Blake's myth of Los than in any other artefact of the period.

In the editing of *Milton* and *Jerusalem*, which are by far the longest pieces in the anthology, there has been no departure from the rule of not offering decodings of mythological figures or accounts of the general meaning of a work, but I have provided substantial headnotes. Here a kind of geography of each work is described—both internally, and in terms of its relation as an artistic act to Blake's world— and its major preoccupations pointed out.

CHRONOLOGY

1757   Blake born (28 November) in Broad Street, Soho, where his father ran a hosiery business.

1759   Death of the poet William Collins.

1760   The first 'Ossian' poems of James Macpherson.

1762   Catherine Boucher, Blake's future wife, born (25 April). Her father was a financially hard-pressed market gardener in Battersea.

1763   Christopher Smart's *Song to David*.

1765   Thomas Percy's *Reliques of Ancient English Poetry*.

1767   Robert Blake, Blake's favourite brother, born (4 August).

1770   Death of George Whitefield.

1772   Death of Emanuel Swedenborg. Blake is apprenticed (4 August) for seven years to James Basire, the most distinguished topographical and antiquarian engraver in London.

1774   Jacob Bryant's *New System* starts to be published.

1776   Declaration of Independence in America. Death of the philosopher David Hume.

1777   Thomas Chatterton's *Poems . . . by Thomas Rowley*. Thomas Warton's *Poems*.

1778   Death of Voltaire.

1779   Death of the poet William Cowper. Blake admitted (8 October) as a student of the Royal Academy (founded in 1768), after submitting work to prove a basic proficiency in academic drawing. He is entitled to attend classes and use the Academy's resources for six years. For many further years Blake showed drawings in the Academy's annual exhibitions.

c.1780 Blake makes the acquaintance of Flaxman, Stothard, and Cumberland.

1782   Blake marries Catherine Boucher (18 August), and the couple take rooms near the modern Leicester Square.

1783   *Poetical Sketches* printed.

1784   Blake's father dies (4 July). Soon afterwards Blake tries unsuccessfully to start a print-selling business next door to his father's old establishment.

1785   The Blakes move to Poland Street, Soho.

1787   Thomas Taylor's *The . . . Hymns of Orpheus*. In February Blake's beloved brother Robert dies. At about this time Blake becomes intimate with Fuseli, and through him probably meets political radicals of the day.

1789   *The Book of Thel, Songs of Innocence,* and, probably, *Tiriel.* Blake and
        Catherine attend the first session of the general conference of the
        Swedenborgian New Jerusalem Church (13 April), and are among the
        sixty-odd signatories of a resolution to establish the church. Fall of the
        Bastille (14 July).

1790   *The Marriage of Heaven and Hell.* In the autumn the Blakes move to
        Hercules Buildings, Lambeth, and Blake enters a period of relative
        prosperity as a commercial engraver.

1791   *The French Revolution.* Death of John Wesley.

1792   Death of Sir Joshua Reynolds. Blake's mother dies (7 September).

1793   *Songs of Experience, Visions of the Daughters of Albion, America.*
        Execution of Louis XVI (21 January). In February war is declared
        between Britain and France.

1794   *Europe, The Book of Urizen.* Trial and acquittal of the radicals Horne
        Tooke, Holcroft, and Thelwall. Execution of Robespierre (28 July).

1795   *Africa, Asia, The Book of Los.* Blake is commissioned to do an
        elaborate series of illustrations for a new edition of Young's *Night
        Thoughts,* which were partly published, without success, about two
        years later. Passage of the Seditious Meetings and Treasonable
        Practices Acts.

1799   Outlawing of Corresponding Societies and other radical groups. In
        January Flaxman arranges for Blake to engrave two illustrations for
        his friend William Hayley's *Essay on Sculpture*: the beginning of
        Blake's association with the connoisseur and poet.

1800   Probable date of composition of *The Four Zoas.* Around June Blake
        visits Hayley at his house in Felpham, Sussex and becomes involved in
        painting a series of portraits of great authors to adorn the library, and
        in other projects of Hayley's. Blake moves into a cottage in Felpham
        (18 September) with his wife and sister-in-law to work under Hayley's
        patronage. By January 1802 he is expressing himself extremely
        discontented with his physical and artistic situation. By July 1803 he
        has resolved to leave.

1802   Britain signs the Peace of Amiens with France (March).

1803   War with France is renewed (May). On 12 August Blake is involved in
        an altercation outside his cottage with a private soldier, John Scolfield,
        who is quartered in the Felpham inn. On 15 August Scolfield deposes
        before a magistrate that Blake had assaulted him and uttered seditious
        opinions in the course of their altercation, 'damning' the King, and
        calling the army and the people 'slaves' whom Napoleon could easily
        conquer. Blake denies the charges, and is bailed by Hayley and a local
        printer. In September the Blakes return to London, and by October
        are living in South Molton Street.

1804   Probable date of composition of *Milton,* and of first work on *Jerusalem.*
        Blake stands trial at Chichester Quarter Sessions (11 January) and is
        acquitted. Hayley pays for his defence. Napoleon becomes Emperor (May).

1805/6 At the turn of the year, after two fairly prosperous years in London during which he charged more for his work than hitherto, Blake is let down or shabbily treated over several illustration projects. He now breaks with most old patrons and associates, including Flaxman and Hayley, and abandons commerical engraving for about ten years. Thomas Butts, who paid unhesitatingly for a large quantity of paintings, prints, and illuminated writing, was for a considerable period Blake's only known source of livelihood. Even this provision seems to have ceased around 1810, and thereafter the Blakes lived in poverty and obscurity for four or five years.

1806 Death of James Barry.

1808 Blake's illustrations to Blair's *The Grave*, one of the projects he had been partly ousted from, reviewed in a hostile and even abusive fashion in *The Examiner* and *The Antijacobin Review*.

1809 In May Blake mounts an exhibition of his work (for which he composes the *Descriptive Catalogue*) at his brother James's house, the old family home and hosiery establishment in Broad Street that James had inherited from his father. *The Examiner* runs the only review, which calls Blake 'an unfortunate lunatic'.

1814 Abdication of Napoleon.

1815 Final defeat of Napoleon. Blake starts to undertake a limited amount of commercial engraving again. From roughly this period also knowledge of at least his lyric poetry spreads in literary circles.

1818 Probable date of composition of *The Everlasting Gospel*. The painter John Linnell is introduced to Blake (June). He quickly obtains engraving work for Blake, and becomes a close friend, though some 35 years his junior. From now on there gathers round Blake a group of younger artists who genuinely admire his work and thought and do much on his behalf, though the Blakes remain relatively poor. The core of this group calls itself 'The Ancients', and includes Linnell, Samuel Palmer, and Edward Calvert. Several substantial (and superbly executed) illustration projects flow from the encouragement and assistance of Linnell: for Thornton's *The Pastorals of Virgil* (1819), for *The Book of Job* (1821), and for *The Divine Comedy* (1824).

1820 Death of Hayley.

1821 The Blakes move to a pair of rooms in Fountain Court, between the Strand and the Thames. Here Blake lived and worked until his death.

1825 Death of Fuseli.

1826 Death of Flaxman.

1827 Blake dies serenely on 12 August. Just before his death, according to one account, he 'burst out in singing of the things he saw in Heaven'.

NOTE ON THE TEXT

THE text throughout this edition may be thought of as a modernized version of the transcriptions in G. E. Bentley Jr.'s *William Blake's Writings* (1978). It is this text which I have consistently relied on, generally only checking it against facsimiles of Blake's originals where it seemed problematical, and only departing from it where it was obviously doubtful, or where I was obliged to for the sake of the readability aimed at in this anthology. The modernizing has been thorough, but all original spellings in the verse (and in some of the prose) which have metrical consequences are retained, and, if necessary, an apostrophe added to indicate the elision of a syllable.

The degree sign (°) indicates a note at the end of the book. General headnotes, giving details of publication, etc., are not cued.

BLAKE ON RELIGION
AND KNOWLEDGE

From the annotations to John Caspar Lavater's
Aphorisms on Man (*1788*)

I DO not believe there is such a thing literally, but Hell is the being shut up in the possession of corporeal desires which shortly weary the man, for all life is holy.

I do not allow that there is such a thing as superstition taken in the strict sense of the word.

A man must first deceive himself before he is thus superstitious and so he is a hypocrite.

True superstition is ignorant honesty and this is beloved of God and Man.

Hypocrisy is as distant from superstition, as the wolf from the lamb.

No man was ever truly superstitious who was not truly religious as far as he knew.

Deduct from a rose its redness, from a lily its whiteness, from a diamond its hardness, from a sponge its softness, from an oak its height, from a daisy its lowness and rectify everything in Nature as the philosophers do, and then we shall return to chaos and God will be compelled to be eccentric if he creates. O happy philosopher.

Variety does not necessarily suppose deformity, for a rose and a lily are various and both beautiful.

Beauty is exuberant,° but not of ugliness but of beauty, and if ugliness is adjoined to beauty it is not the exuberance of beauty; so if Rafael is hard and dry it is not his genius but an accident acquired, for how can substance and accident be predicated of the same essence! I cannot conceive.

But the substance gives tincture to the accident and makes it physiognomic.

It is the God in *all* that is our companion and friend, for our God himself says: 'You are my brother, my sister and my mother', and St John: 'Whoso dwelleth in love dwelleth in God and God in him',° and such an one cannot judge of any but in love and his feelings will be attractions or repulses.

God is in the lowest effects as well as in the highest causes for he is become a worm that he may nourish the weak.

For let it be remembered that creation is God descending according to the weakness of Man, for our Lord is the word of God and everything on earth is the word of God and in its essence is God.

Those who are offended with anything in this book would be offended with the innocence of a child and for the same reason: because it reproaches him with the errors of acquired folly.

I hope no one will call what I have written cavilling because he may think my remarks of small consequence. For I write from the warmth of my heart, and cannot resist the impulse I feel to rectify what I think false in a book I love so much, and approve so generally.

Man is bad or good as he unites himself with bad or good spirits. Tell me with whom you go and I'll tell you what you do.

As we cannot experience pleasure but by means of others who experience either pleasure or pain through us, and as all of us on earth are united in thought (for it is impossible to think without images of somewhat on earth) so it is impossible to know God or heavenly things without conjunction with those who know God and heavenly things. Therefore all who converse in the spirit, converse with spirits. For these reasons I say that this book is written by consultation with good spirits because it is good, and that the name Lavater is the amulet° of those who purify the heart of Man.

There is a strong objection to Lavater's principles (as I understand them) and that is he makes everything originate in its accident: he makes the vicious propensity not only a leading feature of the man but the stamina° on which all his virtues grow. But as I understand vice it is a negative. It does not signify what the laws of kings and priests have called 'vice'. We who are philosophers ought not to call the staminal virtues of humanity by the same name that we call the omissions of intellect springing from poverty.

Every man's leading propensity ought to be called his leading virtue and his Good Angel, but the philosophy of causes and consequences° misled Lavater as it has all his contemporaries. Each thing is its own cause and its own effect. Accident is the omission of act in self and the hindering of act in another. This is vice, but all act is virtue. To hinder another is not an act; it is the contrary. It is a restraint on action both in ourselves and in the person hindered, for he who hinders another omits his own duty at the time.

Murder is hindering another.

Theft is hindering another.

Backbiting, undermining, circumventing and whatever is negative is vice.

But the origin of this mistake in Lavater and his contemporaries is, they suppose that woman's love is sin. In consequence all the loves and graces with them are sin.

From the annotations to Emanuel Swedenborg's
Wisdom of Angels Concerning Divine Love
and Divine Wisdom (c.*1789*)

Understanding or thought is not natural to Man. It is acquired by means of suffering and distress i.e. experience. Will, desire, love, rage, envy, and all other affections are natural, but understanding is acquired.

Man can have no idea of any thing greater than Man as a cup cannot contain more than its capaciousness.

But God is a man not because he is so perceived by Man but because he is the creator of Man.

Think of a white cloud as being holy—you cannot love it. But think of a holy man within the cloud—love springs up in your thought. For to think of holiness distinct from Man is impossible to the affections. Thought alone can make monsters, but the affections cannot.

All Religions are One (c.*1788*)

The voice of one crying in the wilderness°

THE ARGUMENT

As the true method of knowledge is experiment the true faculty of knowing must be the faculty which experiences. This faculty I treat of.

PRINCIPLE I

That the Poetic Genius is the true Man, and that the body or outward form of Man is derived from the Poetic Genius. Likewise that the forms of all things are derived from their genius, which by the ancients was called an Angel and Spirit and Demon.

PRINCIPLE 2

As all men are alike in outward form, so (and with the same infinite variety) all are alike in the Poetic Genius.

PRINCIPLE 3

No man can think, write or speak from his heart, but he must intend truth. Thus all sects of philosophy are from the Poetic Genius adapted to the weaknesses of every individual.

PRINCIPLE 4

As none by travelling over known lands can find out the unknown, so from already acquired knowledge Man could not acquire more. Therefore an universal Poetic Genius exists.

PRINCIPLE 5

The religions of all nations are derived from each nation's different reception of the Poetic Genius, which is everywhere called the Spirit of Prophecy.°

PRINCIPLE 6

The Jewish and Christian testaments are an original derivation from the Poetic Genius. This is necessary from the confined nature of bodily sensation.

PRINCIPLE 7

As all men are alike (though infinitely various) so all religions—and, as all similars, have one source.

The true Man is the source, he being the Poetic Genius.

There is No Natural Religion (c.1788) (1)

THE ARGUMENT

Man has no notion of moral fitness but from education. Naturally he is only a natural organ subject to sense.

I

Man cannot naturally perceive but through his natural or bodily organs.

II

Man by his reasoning power can only compare and judge of what he has already perceived

III

From a perception of only 3 senses or 3 elements none could deduce a fourth or fifth.

IV

None could have other than natural or organic thoughts if he had none but organic perceptions.

V

Man's desires are limited by his perceptions; none can desire what he has not perceived.

VI

The desires and perceptions of Man untaught by anything but organs of sense must be limited to objects of sense.

There is No Natural Religion (c.*1788*) (2)

I

Man's perceptions are not bounded by organs of perception; he perceives more than sense (though ever so acute) can discover.

II

Reason or the ratio° of all we have already known is not the same that it shall be when we know more.

[*The third proposition in this sequence seems not to have survived*]

IV

The bounded is loathed by its possessor. The same dull round even of a universe would soon become a mill with complicated wheels.

V

If the many become the same as the few, when possessed, 'More! More!' is the cry of a mistaken soul. Less than all cannot satisfy Man.

VI

If any could desire what he is incapable of possessing, despair must be his eternal lot.

VII

The desire of Man being infinite the possession is infinite and himself infinite.

APPLICATION

He who sees the infinite in all things sees God. He who sees the ratio only sees himself only.

CONCLUSION

If it were not for the poetic or prophetic character the philosophic and experimental would soon be at the ratio of all things, and stand still, unable to do other than repeat the same dull round over again.

The Marriage of Heaven and Hell (1790)

THE ARGUMENT°

Rintrah roars and shakes his fires in the burdened air;
Hungry clouds swag on the deep.°

Once meek, and in a perilous path,
The just man kept his course along
The vale of death.°
Roses are planted where thorns grow,
And on the barren heath
Sing the honey bees.

Then the perilous path was planted,
And a river and a spring   10
On every cliff and tomb,
And on the bleached bones°
Red clay brought forth.

Till the villain left the paths of ease
To walk in perilous paths, and drive
The just man into barren climes.

Now the sneaking serpent walks°
In mild humility,
And the just man rages in the wilds
Where lions roam.   20

Rintrah roars and shakes his fires in the burdened air;
Hungry clouds swag on the deep.

As a new Heaven is begun, and it is now thirty-three years° since its advent, the eternal Hell revives. And lo! Swedenborg is the angel sitting at the tomb; his writings are the linen clothes folded up.° Now is the dominion of Edom,° and the return of Adam into Paradise; see Isaiah xxxiv and xxxv.°

Without contraries is no progression. Attraction and repulsion, reason and energy, love and hate, are necessary to human existence.°

From these contraries spring what the religious call Good and Evil. Good is the passive that obeys reason, Evil is the active springing from energy.

Good is Heaven. Evil is Hell.

THE VOICE OF THE DEVIL

All bibles or sacred codes have been the causes of the following errors:

1. That Man has two real existing principles, viz, a body and a soul.

2. That energy, called Evil, is alone from the body, and that reason, called Good, is alone from the soul.

3. That God will torment Man in eternity for following his energies.

But the following contraries to these are true:

1. Man has no body distinct from his soul, for that called body is a portion of soul discerned by the five senses, the chief inlets of soul in this age.

2. Energy is the only life and is from the body, and reason is the bound or outward circumference of energy.

3. Energy is eternal delight.

Those who restrain desire do so because theirs is weak enough to be restrained; and the restrainer, or reason, usurps its place and governs the unwilling.

And being restrained it by degrees becomes passive, till it is only the shadow of desire.

The history of this is written in *Paradise Lost*, and the governor, or reason, is called Messiah.

And the original archangel, or possessor of the command of the heavenly host, is called the Devil or Satan, and his children are called Sin and Death.

But in the *Book of Job* Milton's Messiah is called Satan.°

For this history has been adopted by both parties.

It indeed appeared to reason as if desire was cast out, but the Devil's account is that the Messiah fell, and formed a heaven of what he stole from the abyss.

This is shown in the Gospel, where he prays to the Father to send the comforter,° or desire, that reason may have ideas to build on, the Jehovah of the Bible being no other than he who dwells in flaming fire. Know that after Christ's death he became Jehovah.

But in Milton the Father is destiny, the Son a ratio of the five senses, and the Holy Ghost vacuum!

Note. The reason Milton wrote in fetters when he wrote of angels and God, and at liberty when of devils and Hell, is because he was a true poet, and of the Devil's party without knowing it.

A MEMORABLE FANCY

As I was walking among the fires of Hell, delighted with the enjoyments of genius, which to angels look like torment and insanity, I collected some of their proverbs, thinking that, as the sayings used in a nation mark its character, so the proverbs of Hell show the nature of infernal wisdom better than any description of buildings or garments.

When I came home, on the abyss of the five senses, where a flat-sided steep frowns over the present world, I saw a mighty devil folded in black clouds, hovering on the sides of the rock. With corroding fires he wrote the following sentence now perceived by the minds of men, and read by them on earth:

> How do you know but ev'ry bird that cuts the airy way°
> Is an immense world of delight, closed by your senses five?

PROVERBS OF HELL

In seed time learn, in harvest teach, in winter enjoy.
Drive your cart and your plough over the bones of the dead.
The road of excess leads to the palace of wisdom.
Prudence is a rich ugly old maid courted by incapacity.
He who desires but acts not breeds pestilence.

The cut worm forgives the plough.

Dip him in the river who loves water.

A fool sees not the same tree that a wise man sees.

He whose face gives no light shall never become a star.

Eternity is in love with the productions of time.

The busy bee has no time for sorrow.

The hours of folly are measured by the clock, but of wisdom no clock can measure.

All wholesome food is caught without a net or a trap.

Bring out number, weight and measure in a year of dearth.

No bird soars too high, if he soars with his own wings.

A dead body revenges not injuries.

The most sublime act is to set another before you.

If the fool would persist in his folly he would become wise.

Folly is the cloak of knavery.

Shame is pride's cloak.

Prisons are built with stones of law, brothels with bricks of religion.

The pride of the peacock is the glory of God.

The lust of the goat is the bounty of God.

The wrath of the lion is the wisdom of God.

The nakedness of Woman is the work of God.

Excess of sorrow laughs. Excess of joy° weeps.

The roaring of lions, the howling of wolves, the raging of the stormy sea, and the destructive sword are portions of eternity too great for the eye of Man.

The fox condemns the trap, not himself.

Joys impregnate. Sorrows bring forth.

Let Man wear the fell of the lion, Woman the fleece of the sheep.

The bird a nest, the spider a web, Man friendship.

The selfish smiling fool and the sullen frowning fool shall be both thought wise, that they may be a rod.

What is now proved was once only imagined.

The rat, the mouse, the fox, the rabbit, watch the roots. The lion, the tiger, the horse, the elephant, watch the fruits.

The cistern contains; the fountain overflows.

One thought fills immensity.

Always be ready to speak your mind, and a base man will avoid you.

Everything possible to be believed is an image of truth.

The eagle never lost so much time as when he submitted to learn of the crow.

The fox provides for himself, but God provides for the lion.

Think in the morning; act in the noon; eat in the evening; sleep in the night.

He who has suffered you to impose on him, knows you.

As the plough follows words, so God rewards prayers.

The tigers of wrath are wiser than the horses of instruction.

Expect poison from the standing water.

You never know what is enough unless you know what is more than enough.

Listen to the fool's reproach! It is a kingly title!

The eyes of fire, the nostrils of air, the mouth of water, the beard of earth.

The weak in courage is strong in cunning.

The apple tree never asks the beech how he shall grow, nor the lion the horse, how he shall take his prey.

The thankful receiver bears a plentiful harvest.

If others had not been foolish, we should be so.

The soul of sweet delight can never be defiled.

When thou seest an eagle, thou seest a portion of genius; lift up thy head!

As the caterpillar chooses the fairest leaves to lay her eggs on, so the priest lays his curse on the fairest joys.

To create a little flower is the labour of ages.

Damn braces. Bless relaxes.

The best wine is the oldest. The best water the newest.

Prayers plough not! Praises reap not!

Joys laugh not! Sorrows weep not!

The head Sublime,° the heart Pathos, the genitals Beauty, the hands and feet Proportion.

As the air to a bird or the sea to a fish, so is contempt to the contemptible.

The crow wished everything was black; the owl, that everything was white.

Exuberance is beauty.

If the lion was advised by the fox, he would be cunning.

Improvement makes straight roads, but the crooked roads without improvement are roads of genius.

Sooner murder an infant in its cradle than nurse unacted desires.

Where Man is not, Nature is barren.

Truth can never be told so as to be understood, and not be believed.

                Enough! Or too much!

The ancient poets animated all sensible objects with gods or geniuses, calling them by the names, and adorning them with the properties of woods, rivers, mountains, lakes, cities, nations, and whatever their enlarged and numerous senses could perceive.

And particularly they studied the genius of each city and country, placing it under its mental deity.

Till a system was formed, which some took advantage of and enslaved the vulgar by attempting to realise or abstract the mental deities from their objects. Thus began priesthood: choosing forms of worship from poetic tales.

And at length they pronounced that the gods had ordered such things.

Thus men forgot that all deities reside in the human breast.

A MEMORABLE FANCY

The prophets Isaiah and Ezekiel dined with me, and I asked them how they dared so roundly to assert that God spake to them; and whether they did not think at the time that they would be misunderstood, and so be the cause of imposition.

Isaiah answered: 'I saw no God, nor heard any, in a finite organical perception; but my senses discovered the infinite in everything, and as I was then persuaded, and remain confirmed, that the voice of honest indignation is the voice of God, I cared not for consequences but wrote.'

Then I asked: 'Does a firm persuasion that a thing is so, make it so?'

He replied: 'All poets believe that it does, and in ages of imagination the firm persuasion removed mountains; but many are not capable of a firm persuasion of anything.'

Then Ezekiel said: 'The philosophy of the east taught the first principles of human perception. Some nations held one principle for the origin and some another. We of Israel taught that the Poetic Genius (as you now call it) was the first principle, and all the others merely derivative—which was the cause of our despising the priests and philosophers of other countries, and prophesying that all gods would at last be proved to originate in ours° and to be the tributaries of the Poetic Genius. It was this that our great poet King David° desired so fervently and invoked so pathetic'ly, saying by this he conquers enemies and governs kingdoms. And we so loved our god, that we cursed in his name all the deities of surrounding nations, and asserted that they had rebelled. From these opinions the vulgar came to think that all nations would at last be subject to the Jews.

'This,' said he, 'like all firm persuasions, is come to pass, for all nations believe the Jews' code and worship the Jews' god, and what greater subjection can be?'

I heard this with some wonder, and must confess my own conviction. After dinner I asked Isaiah to favour the world with his lost works; he said none of equal value was lost. Ezekiel said the same of his.

I also asked Isaiah what made him go naked and barefoot for three years.° He answered: 'The same that made our friend Diogenes the Grecian.'°

I then asked Ezekiel why he ate dung, and lay so long on his right and left side. He answered, 'The desire of raising other men into a perception of the infinite.° This the North American tribes practise, and is he honest who resists his genius or conscience only for the sake of present ease or gratification?'

The ancient tradition that the world will be consumed in fire at the end of six thousand years° is true, as I have heard from Hell.

For the cherub with his flaming sword is hereby commanded to leave his guard at the Tree of Life,° and when he does the whole creation will be consumed, and appear infinite and holy, whereas it now appears finite and corrupt.

This will come to pass by an improvement of sensual enjoyment.

But first the notion that Man has a body distinct from his soul is to be expunged. This I shall do by printing in the infernal method by corrosives,° which in Hell are salutary and medicinal, melting apparent surfaces away, and displaying the infinite which was hid.

If the doors of perception were cleansed everything would appear to Man as it is: infinite.

For Man has closed himself up, till he sees all things through narrow chinks of his cavern.

A MEMORABLE FANCY

I was in a printing-house in Hell and saw the method in which knowledge is transmitted from generation to generation.

In the first chamber was a dragon-man, clearing away the rubbish from a cave's mouth; within, a number of dragons were hollowing the cave.

In the second chamber was a viper folding round the rock and the cave, and others adorning it with gold, silver and precious stones.

In the third chamber was an eagle with wings and feathers of air; he caused the inside of the cave to be infinite. Around were numbers of eagle-like men, who built palaces in the immense cliffs.

In the fourth chamber were lions of flaming fire, raging around and melting the metals into living fluids.

In the fifth chamber were unnamed forms, which cast the metals into the expanse.

There they were received by men who occupied the sixth chamber, and took the forms of books and were arranged in libraries.

The giants° who formed this world into its sensual existence, and now seem to live in it in chains, are in truth the causes of its life and the sources of all activity; but the chains are the cunning of weak and tame minds, which have power to resist energy, according to the proverb: 'The weak in courage is strong in cunning'.

Thus one portion of being is the Prolific, the other, the Devouring. To the devourer it seems as if the producer was in his chains, but it is not so; he only takes portions of existence and fancies that the whole.

But the Prolific would cease to be prolific unless the devourer as a sea received the excess of his delights.

Some will say: 'Is not God alone the Prolific?' I answer: 'God only acts and is in existing beings or men.'

These two classes of men are always upon earth, and they should be enemies; whoever tries to reconcile them seeks to destroy existence.

Religion is an endeavour to reconcile the two.

Note. Jesus Christ did not wish to unite but to separate them, as in the parable of sheep and goats! And he says, 'I came not to send peace, but a sword.'°

Messiah or Satan or Tempter° was formerly thought to be one of the antediluvians who are our energies.

A MEMORABLE FANCY

An angel came to me and said: 'O pitiable foolish young man! Oh, horrible! Oh, dreadful state! Consider the hot burning dungeon thou art preparing for thyself to all eternity, to which thou art going in such career.'

I said: 'Perhaps you will be willing to show me my eternal lot, and we will contemplate together upon it, and see whether your lot or mine is most desirable.'

So he took me° through a stable and through a church and down into the church vault, at the end of which was a mill. Through the mill we went, and came to a cave. Down the winding cavern we groped our tedious way, till a void boundless as a nether sky appeared beneath us, and we held by the roots of trees and hung over this immensity. But I said: 'If you please we will commit ourselves to this void, and see whether providence is here also. If you will not, I will!' But he answered: 'Do not presume, O young man; but as we here remain, behold thy lot which will soon appear when the darkness passes away.'

So I remained with him sitting in the twisted root of an oak. He was suspended in a fungus which hung with the head downward into the deep.

By degrees we beheld the infinite abyss, fiery as the smoke of a burning city. Beneath us at an immense distance was the sun, black but shining. Round it were fiery tracks on which revolved vast spiders, crawling after their prey, which flew or rather swum in the infinite deep, in the most terrific shapes of animals sprung from corruption; and the air was full of them, and seemed composed of them. These are devils, and are called Powers of the Air. I now asked my companion which was my eternal lot. He said: 'Between the black and white spiders'.

But now, from between the black and white spiders, a cloud and fire burst and rolled through the deep, black'ning all beneath, so that the nether deep grew black as a sea, and rolled with a terrible noise. Beneath us was nothing now to be seen but a black tempest, till looking east between the clouds and the waves, we saw a cataract of blood mixed with fire, and not many stones' throw from us appeared and sunk again the scaly fold of a monstrous serpent. At last to the east, distant about three degrees, appeared a fiery crest above the waves. Slowly it reared like a ridge of golden rocks, till we discovered two globes of crimson fire, from which the sea fled away in clouds of smoke; and now we saw it was the head of Leviathan. His forehead was divided into streaks of green and purple like those on a tiger's forehead. Soon we saw his mouth and red gills hang just above the raging foam, tingeing the black deep with beams of blood, advancing toward us with all the fury of a spiritual existence.

My friend the angel climbed up from his station into the mill. I remained alone, and then this appearance was no more, but I found myself sitting on a pleasant bank beside the river by moonlight, hearing a harper who sung to the harp; and his theme was: 'The man who never alters his opinion is like standing water, and breeds reptiles of the mind'.

But I arose, and sought for the mill, and there I found my angel, who, surprised, asked me how I escaped.

I answered: 'All that we saw was owing to your metaphysics; for when you ran away, I found myself on a bank by moonlight hearing a harper. But now we have seen my eternal lot, shall I show you yours?' He laughed at my proposal; but I by force suddenly caught him in my arms, and flew westerly through the night, till we were elevated above the earth's shadow. Then I flung myself with him directly into the body of the sun. Here I clothed myself in white, and taking in my hand Swedenborg's volumes sunk from the glorious clime, and passed all the planets till we came to Saturn. Here I stayed to rest and then leaped into the void between Saturn and the fixed stars.°

'Here', said I, 'is your lot, in this space, if space it may be called.' Soon we saw the stable and the church, and I took him to the altar and opened the Bible, and lo! it was a deep pit, into which I descended, driving the angel before me. Soon we saw seven houses of brick. One we entered; in it were a number of monkeys, baboons, and all of that species, chained by the middle, grinning and snatching at one another, but withheld by the shortness of their chains. However I saw that they sometimes grew numerous, and then the weak were caught by the strong and, with a grinning aspect, first coupled with them and then devoured, by plucking off first one limb and then another till the body was left a helpless trunk. This after grinning and kissing it with seeming fondness they devoured too; and here and there I saw one savourily picking the flesh off of his own tail. As the stench terribly annoyed us both we went into the mill, and I in my hand brought the skeleton of a body, which in the mill was Aristotle's *Analytics*.°

So the angel said: 'Thy fantasy has imposed upon me and thou oughtest to be ashamed.'

I answered: 'We impose on one another, and it is but lost time to converse with you whose works are only Analytics.'

OPPOSITION IS TRUE FRIENDSHIP

I have always found that angels have the vanity to speak of themselves as the only wise; this they do with a confident insolence sprouting from systematic reasoning.

Thus Swedenborg boasts that what he writes is new, though it is only the contents or index of already-published books.

A man carried a monkey about for a show, and because he was a little wiser than the monkey, grew vain, and conceived himself as much wiser than seven men. It is so with Swedenborg; he shows the folly of churches and exposes hypocrites, till he imagines that all are religious, and himself the single one on earth that ever broke a net.

Now hear a plain fact: Swedenborg has not written one new truth. Now hear another: he has written all the old falsehoods.

And now hear the reason: he conversed with angels, who are all religious, and conversed not with devils who all hate religion, for he was incapable through his conceited notions.

Thus Swedenborg's writings are a recapitulation of all superficial opinions and an analysis of the more sublime, but no further.

Have now another plain fact: any man of mechanical talents may from the writings of Paracelsus or Jacob Behmen° produce ten thousand volumes of equal value with Swedenborg's, and from those of Dante or Shakespeare an infinite number.

But when he has done this, let him not say that he knows better than his master, for he only holds a candle in sunshine.

A MEMORABLE FANCY

Once I saw a devil in a flame of fire, who arose before an angel that sat on a cloud, and the devil uttered these words:

'The worship of God is honouring his gifts in other men, each according to his genius, and loving the greatest men best. Those who envy or calumniate great men hate God, for there is no other God.'

The angel, hearing this, became almost blue, but mastering himself he grew yellow, and at last white pink° and smiling, and then replied:

'Thou idolater, is not God one, and is not he visible in Jesus Christ? And has not Jesus Christ given his sanction to the law of ten commandments; and are not all other men fools, sinners, and nothings?'

The devil answered: 'Bray a fool in a mortar with wheat, yet shall not his folly be beaten out of him.° If Jesus Christ is the greatest man, you ought to love him in the greatest degree. Now hear how he has given his sanction to the law of ten commandments:° did he not mock at the sabbath, and so mock the sabbath's god? murder those who were murdered because of him? turn away the law from the woman taken in adultery? steal the labour of others to support him? bear false witness when he omitted making a defence before Pilate? covet when he prayed for his disciples, and when he bid them shake off the dust of their feet against such as refused to lodge them? I tell you, no virtue can exist without breaking these ten commandments. Jesus was all virtue, and acted from impulse, not from rules.'

When he had so spoken I beheld the angel, who stretched out his arms, embracing the flame of fire, and he was consumed and arose, as Elijah.°

Note. This angel, who is now become a devil, is my particular friend. We often read the Bible together in its infernal or diabolical sense, which the world shall have if they behave well.

I have also the Bible of Hell, which the world shall have whether they will or no.

One law for the lion and ox is oppression.

A SONG OF LIBERTY°

1. The Eternal Female groaned!° It was heard over all the earth.°

2. Albion's coast is sick, silent; the American meadows faint.

3. Shadows of prophecy shiver along by the lakes and the rivers, and mutter across the ocean. France, rend down thy dungeon;°

4. Golden Spain, burst the barriers of old Rome;

5. Cast thy keys, O Rome,° into the deep, down falling, even to eternity down falling—

6. And weep!

7. In her trembling hands she took the new-born terror, howling.

8. On those infinite mountains of light, now barred out by the Atlantic sea, the new-born fire stood before the starry king!

9. Flagged° with grey-browed snows and thunderous visages, the jealous wings waved over the deep.

10. The speary hand burned aloft, unbuckled was the shield, forth went the hand of jealousy among the flaming hair, and hurled the new-born wonder through the starry night.

11. The fire, the fire, is falling!

12. Look up! Look up! O citizen of London, enlarge thy countenance! O Jew, leave counting gold! Return to thy oil and wine, O African, black African! (Go, winged thought, widen his forehead.)

13. The fiery limbs, the flaming hair shot like the sinking sun into the western sea.

14. Waked from his eternal sleep, the hoary element roaring fled away.

15. Down rushed, beating his wings in vain, the jealous king: his grey browed counsellors, thunderous warriors, curled veterans among helms, and shields, and chariots, horses, elephants, banners, castles, slings and rocks,

16. Falling, rushing, ruining! Buried in the ruins, on Urthona's dens.

17. All night beneath the ruins; then, their sullen flames faded, emerge round the gloomy king.

18. With thunder and fire, leading his starry hosts through the waste

wilderness, he promulgates his ten commands, glancing his beamy eyelids over the deep in dark dismay,

19. Where the son of fire in his eastern cloud, while the morning plumes her golden breast,

20. Spurning the clouds written with curses, stamps the stony law° to dust, loosing the eternal horses from the dens of night,° crying: Empire° is no more! And now the lion and wolf shall cease.

CHORUS

Let the priests of the raven of dawn no longer, in deadly black, with hoarse note curse the sons of joy. Nor his accepted brethren whom, tyrant, he calls free, lay the bound or build the roof. Nor pale religious lechery call that virginity that wishes but acts not.

For everything that lives is holy.

From the annotations to Richard Watson's
An Apology for the Bible (*1798*)

To defend the Bible in this year 1798 would cost a man his life.

It is an easy matter for a bishop to triumph over Paine's attack, but it is not so easy for one who loves the Bible.

The perversions of Christ's words and acts are attacked by Paine, and also the perversions of the Bible. Who dare defend either the acts of Christ or the Bible unperverted? But to him who sees this mortal pilgrimage in the light that I see it, duty to his country is the first consideration and safety the last.

Read patiently; take not up this book in an idle hour. The consideration of these things is the whole duty of Man,° and the affairs of life and death trifles, sports of time. These considerations [are the] business of eternity.

I have been commanded from Hell not to print this as it is what our enemies wish.

Read Chapter XXIII of Matthew° and then condemn Paine's hatred of priests if you dare.

Paine is either a devil or an inspired man. Men who give themselves to their energetic genius in the manner that Paine does are no examiners.° If they are not determinately wrong they must be right, or the Bible is false. As to examiners, in these points they will be spewed out. The man who pretends to be a modest enquirer into the truth of a self-evident thing is a knave. The truth and certainty of virtue and honesty (i.e. inspiration) needs no one to prove it; it is evident as the sun and moon. He who stands doubting of what he intends, whether it is virtuous or vicious, knows not what virtue means. No man can do a vicious action and think it to be virtuous. No man can take darkness for light. He may pretend to do so, and may pretend to be a modest enquirer, but he is a knave.

To me who believe the Bible and profess myself a Christian, a defence of the wickedness of the Israelites in murdering so many thousands° under pretence of a command from God is altogether abominable and blasphemous. Wherefore did Christ come? Was it not to abolish the Jewish imposture? Was not Christ martyred because he taught that God loved all men and was their father, and forbade all contention for wordly prosperity—in opposition to the Jewish scriptures, which are only an example of the wickedness and deceit of the Jews, and were written as an example of the possibility of human beastliness in all its branches? Christ died as an unbeliever, and if the bishops had their will so would Paine. But he who speaks a word against the Son of Man shall be forgiven; let the bishop prove that he has not spoken against the Holy Ghost,° who in Paine strives with Christendom as in Christ he strove with the Jews.

That mankind are in a less distinguished situation with regard to mind than they were in the time of Homer, Socrates, Phidias, Glycon,° Aristotle let all their works witness. Paine says that Christianity put a stop to improvement, and the bishop has not shewn the contrary.

Read the Edda of Iceland, the Songs of Fingal,° the accounts of North American savages (as they are called). Likewise read Homer's *Iliad*; he was certainly a savage in the bishop's sense. He knew nothing of God in the bishop's sense of the word and yet he was no fool.

The Bible, or peculiar word of God, exclusive of conscience or the word of God universal, is that abomination which, like the Jewish ceremonies, is for ever removed, and henceforth every man may converse with God and be a king and priest in his own house.

The trifles which the bishop has combated in the following letters are such as do nothing against Paine's arguments, none of which the bishop has dared to consider. One for instance, which is that the books of the Bible were never believed willingly by any nation, and that none but designing villains ever pretended to believe: that the Bible is all a state trick, through which though the people at all times could see, they never had the power to throw off. Another argument is that all the commentators on the Bible are dishonest designing knaves, who in hopes of a good living adopt the state religion. This he has shown with great force, which calls upon his opponent loudly for an answer. I could name an hundred such.

Jesus could not do miracles where unbelief hindered;° hence we must conclude that the man who holds miracles to be ceased puts it out of his own power to ever witness one. The manner of a miracle being performed is in modern times considered as an arbitrary command of the agent upon the patient, but this is an impossibility° not a miracle. Neither did Jesus ever do such a miracle. Is it a greater miracle to feed five thousand men with five loaves than to overthrow all the armies of Europe with a small pamphlet?° Look over the events of your own life, and if you do not find that you have both done such miracles and lived by such you do not see as I do. True, I cannot do a miracle through experiment, and to domineer over and prove to others my superior power—as neither could Christ. But I can and do work such as both astonish and comfort me and mine. How can Paine the worker of miracles ever doubt Christ's in the above sense of the word miracle? But how can Watson ever believe the above sense of a miracle, who considers it as an arbitrary act of the agent upon an unbelieving patient—whereas the Gospel says that Christ could not do a miracle because of unbelief? If Christ could not do miracles because of unbelief the reason alleged by priests for miracles is false; for those who believe want not to be confounded° by miracles. Christ and his prophets and apostles were not ambitious miracle-mongers.

Prophets in the modern sense of the word have never existed. Jonah was no prophet in the modern sense, for his prophecy of Nineveh failed. Every honest man is a prophet. He utters his opinion both of private and public matters thus: 'If you go on so, the result is so'. He never says: 'Such a thing shall happen let you do what you will.' A prophet is a seer not an arbitrary dictator. It is Man's fault if God is not able to do him good, for he gives to the just and to the unjust but the unjust reject his gift.

All penal laws court transgression, and therefore are cruelty and murder.

The laws of the Jews were (both ceremonial and real) the basest and most oppressive of human codes, and being, like all other codes, given under pretence of divine command were what Christ pronounced them: 'the abomination that maketh desolate',° i.e. state religion, which is the source of all cruelty.

The bishop never saw the everlasting gospel° any more than Tom Paine.

It appears to me now that Tom Paine is a better Christian than the bishop.

I have read this book with attention, and find that the bishop has only hurt Paine's heel while Paine has broken his head. The bishop has not answered one of Paine's grand objections.

'To my friend Butts' (1800)

> To my friend Butts I write
> My first vision of light.
> On the yellow sands sitting
> The sun was emitting
> His glorious beams
> From heaven's high streams.
> Over sea, over land,
> My eyes did expand
> Into regions of air,
> Away from all care,   10
> Into regions of fire,
> Remote from desire.
> The light of the morning
> Heaven's mountains adorning,
> In particles bright°
> The jewels of light
> Distinct shone and clear.
> Amazed and in fear
> I each particle gazed—
> Astonished, amazed,   20

For each was a man
Human-formed. Swift I ran,
For they beckoned to me,
Remote by the sea,
Saying: 'Each grain of sand,
Every stone on the land,
Each rock and each hill,
Each fountain and rill,
Each herb and each tree,
Mountain, hill, earth and sea,   30
Cloud, meteor and star
Are men seen afar'.
I stood in the streams
Of heaven's bright beams
And saw Felpham sweet
Beneath my bright feet,
In soft female charms.
And in her fair arms
My shadow I knew,
And my wife's shadow too,   40
And my sister and friend.°
(We like infants descend
In our shadows on earth,
Like a weak mortal birth.)
My eyes more and more
Like a sea without shore
Continue expanding,
The heavens commanding,
Till the jewels of light,
Heavenly men beaming bright,   50
Appeared as one man;
Who complacent began
My limbs to enfold
In his beams of bright gold,
Like dross purged away
All my mire and my clay.
Soft-consumed in delight
In his bosom sun-bright
I remained. Soft he smiled,
And I heard his voice mild,   60
Saying: 'This is my fold,
O thou ram horned with gold,

Who awakest from sleep
On the sides of the deep.
On the mountains around
The roarings resound
Of the lion and wolf,
The loud sea and deep gulf.
These are guards of my fold,
O thou ram horned with gold.'   70
And the voice faded mild.
I remained as a child.
All I ever had known
Before me bright shone.
I saw you and your wife
By the fountains of life.
Such the vision to me
Appeared on the sea.

'With happiness stretched across the hills' (1802)

With happiness stretched across the hills
In a cloud that dewy sweetness distils,
With a blue sky spread over with wings
And a mild sun that mounts and sings,
With trees and fields full of fairy elves
And little devils who fight for themselves
(Rememb'ring the verses that Hayley sung°
When my heart knocked against the root of my tongue),
With angels planted in hawthorn bowers
And God himself in the passing hours,   10
With silver angels across my way
And golden demons that none can stay,
With my father hovering upon the wind
And my brother Robert just behind
And my brother John, the evil one,
In a black cloud making his moan
(Though dead, they appear upon my path
Notwithstanding my terrible wrath;
They beg, they entreat, they drop their tears,
Filled full of hopes, filled full of fears),   20

With a thousand angels upon the wind
Pouring disconsolate from behind
To drive them off. And before my way
A frowning thistle implores my stay.
What to others a trifle appears
Fills me full of smiles or tears.
For double the vision my eyes do see,
And a double vision is always with me.
With my inward eye 'tis an old man grey,
With my outward a thistle across my way.   30
'If thou goest back,' the thistle said,
'Thou art to endless woe betrayed.
For here does Theotormon lower,
And here is Enitharmon's bower,
And Los the terrible thus hath sworn,
Because thou backward dost return,
Poverty, envy, old age and fear
Shall bring thy wife upon a bier,
And Butts shall give what Fuseli gave,°
A dark black rock and a gloomy cave.'   40

I struck the thistle with my foot
And broke him up from his delving root.
'Must the duties of life each other cross?
Must every joy be dung and dross?
Must my dear Butts feel cold neglect,
Because I give Hayley his due respect?
Must Flaxman look upon me as wild°
And all my friends be with doubts beguiled?
Must my wife live in my sister's bane,
Or my sister survive on my love's pain?   50
The curses of Los, the terrible shade,
And his dismal terrors make me afraid.'

So I spoke and struck in my wrath
The old man weltering upon my path.°
Then Los appeared in all his power;
In the sun he appeared, descending before
My face in fierce flames. In my double sight
'Twas outward a sun, inward Los in his might.°

'My hands are laboured day and night,
And ease comes never in my sight.   60
My wife has no indulgence given,
Except what comes to her from Heaven.
We eat little, we drink less.
This earth breeds not our happiness.
Another sun feeds our life's streams.
We are not warmed with thy beams.
Thou measurest not the time to me
Nor yet the space that I do see.
My mind is not with thy light arrayed.
Thy terrors shall not make me afraid.'   70

When I had my defiance given,
The sun stood trembling in heaven.
The moon that glowed remote below
Became leprous and white as snow,
And every soul of men on the earth
Felt affliction and sorrow and sickness and dearth.
Los flamed in my path and the sun was hot.
With the bows of my mind and the arrows of thought
My bowstring fierce with ardour breathes;
My arrows glow in their golden sheaves.   80
My brothers and father march before.
The heavens drop with human gore.

Now I a fourfold vision see,
And a fourfold vision is given to me.
'Tis fourfold in my supreme delight
And threefold in soft Beulah's night°
And twofold always. May God us keep
From single vision and Newton's sleep.

Auguries of Innocence (c.1804)

To see a world in a grain of sand,
And a heaven in a wild flower—
Hold infinity in the palm of your hand,
And eternity in an hour.

A robin redbreast in a cage
Puts all Heaven in a rage;
A dove-house filled with doves and pigeons
Shudders Hell through all its regions.
A dog starved at his master's gate
Predicts the ruin of the state.   10
A horse misused upon the road
Calls to Heaven for human blood.
Each outcry of the hunted hare
A fibre from the brain does tear.
A skylark wounded in the wing,
A cherubim does cease to sing.
The gamecock clipped and armed for fight°
Does the rising sun affright.
Every wolf's and lion's howl
Raises from Hell a human soul.   20
The wild deer wand'ring here and there
Keeps the human soul from care.
The lamb misused breeds public strife,
And yet forgives the butcher's knife.
The bat that flits at close of eve
Has left the brain that won't believe.
The owl that calls upon the night
Speaks the unbeliever's fright.
He who shall hurt the little wren
Shall never be beloved by men.   30
He who the ox to wrath has moved
Shall never be by woman loved.
The wanton boy that kills the fly°
Shall feel the spider's enmity.
He who torments the chafer's sprite
Weaves a bower in endless night.
The caterpillar on the leaf
Repeats to thee thy mother's grief.
Kill not the moth nor butterfly,
For the Last Judgement draweth nigh.   40
He who shall train the horse to war
Shall never pass the polar bar.°
The beggar's dog and widow's cat:
Feed them and thou wilt grow fat.
The gnat that sings his summer's song
Poison gets from slander's tongue.

The poison of the snake and newt
Is the sweat of envy's foot;
The poison of the honey bee
Is the artist's jealousy.   50
The prince's robes and beggar's rags
Are toadstools on the miser's bags.

A truth that's told with bad intent
Beats all the lies you can invent;
It is right it should be so.
Man was made for joy and woe,
And when this we rightly know
Through the world we safely go.
Joy and woe are woven fine,
A clothing for the soul divine.   60
Under every grief and pine
Runs a joy with silken twine.

The babe is more than swaddling bands.
Throughout all these human lands,
Tools were made and born were hands°
(Every farmer understands).
Every tear from every eye
Becomes a babe in eternity;
This is caught by females bright
And returned to its own delight.   70
The bleat, the bark, bellow and roar
Are waves that beat on Heaven's shore.
The babe that weeps the rod beneath
Writes 'Revenge' in realms of death.
The beggar's rags fluttering in air
Does to rags the heavens tear.
The soldier armed with sword and gun
Palsied strikes the summer's sun.
The poor man's farthing is worth more
Than all the gold on Afric's shore.   80
One mite wrung from the lab'rer's hands
Shall buy and sell the miser's lands;
Or, if protected from on high,
Does that whole nation sell and buy.

He who mocks the infant's faith
Shall be mocked in age and death;
He who shall teach the child to doubt
The rotting grave shall ne'er get out;
He who respects the infant's faith
Triumphs over hell and death.   90
The child's toys and the old man's reasons
Are the fruits of the two seasons.
The questioner who sits so sly
Shall never know how to reply.
He who replies to words of doubt
Doth put the light of knowledge out.
The strongest poison ever known°
Came from Caesar's laurel crown.
Nought can deform the human race
Like to the armour's iron brace.   100
When gold and gems adorn the plough
To peaceful arts shall envy bow.
A riddle, or the cricket's cry,
Is to doubt a fit reply.
The emmet's inch and eagle's mile°
Make lame philosophy to smile.
He who doubts from what he sees
Will ne'er believe, do what you please.
If the sun and moon should doubt
They'd immediately go out.   110

To be in a passion you good may do,
But no good if a passion is in you.
The whore and gambler, by the state
Licenced, build that nation's fate.
The harlot's cry from street to street
Shall weave old England's winding sheet;
The winner's shout, the loser's curse,
Dance before dead England's hearse.
Every night and every morn
Some to misery are born;   120
Every morn and every night
Some are born to sweet delight.
Some are born to sweet delight,
Some are born to endless night.

We are led to believe a lie
When we see not through the eye—
Which was born in a night to perish in a night,°
When the soul slept in beams of light.
God appears, and God is light
To those poor souls who dwell in night;   130
But does a human form display
To those who dwell in realms of day.

Prefaces to Chapters II, III and IV of Jerusalem (1804-18)

[Preface to Chapter II]

TO THE JEWS.

Jerusalem the Emanation of the giant Albion! Can it be? Is it a truth that the learned have explored? Was Britain the primitive seat of the patriarchal religion?° If it is true, my title-page is also true: that Jerusalem was and is the Emanation of the giant Albion. It is true, and cannot be controverted. Ye are united, O ye inhabitants of earth, in one religion, the religion of Jesus, the most ancient, the eternal, and the everlasting gospel. The wicked will turn it to wickedness, the righteous to righteousness. Amen! Huzza!° Selah!°

'All things begin and end in Albion's ancient Druid° rocky shore.'

Your ancestors derived their origin from Abraham, Heber, Shem, and Noah,° who were Druids,° as the Druid temples (which are the patriarchal pillars and oak groves) over the whole earth° witness to this day.

You have a tradition,° that Man anciently contained in his mighty limbs all things in heaven and earth; this you received from the Druids.

'But now the starry heavens are fled from the mighty limbs of Albion.'

Albion was the parent of the Druids, and in his chaotic state of sleep Satan and Adam and the whole world was created by the Elohim.°

The fields from Islington to Marylebone [*see below, p. 297*]

If humility is Christianity, you O Jews, are the true Christians—if your tradition that Man contained in his limbs all animals is true and they were separated from him by cruel sacrifices, and when compulsory cruel sacrifices had brought humanity into a feminine tabernacle, in the loins of Abraham and David, the Lamb of God, the Saviour, became apparent on earth as the prophets had foretold. The return of Israel is a return to mental sacrifice and war. Take up the cross, O Israel, and follow Jesus.

[*Preface to Chapter III*]

Rahab°is an }
eternal state }   TO THE DEISTS.°   { The spiritual states of the soul are all eternal. Distinguish between the man and his present state.

He never can be a friend to the human race who is the preacher of natural morality or natural religion. He is a flatterer who means to betray, to perpetuate tyrant pride and the laws of that Babylon which he foresees shall shortly be destroyed, with the spiritual and not the natural sword. He is in the state named Rahab, which state must be put off before he can be the friend of Man.

You, O deists, profess yourselves the enemies of Christianity; and you are so. You are also the enemies of the human race and of universal Nature. Man is born a Spectre or Satan, and is altogether an evil, and requires a new selfhood continually, and must continually be changed into his direct contrary. But your Greek philosophy (which is a remnant of Druidism) teaches that Man is righteous in his vegetated Spectre: an opinion of fatal and accursed consequence to Man, as the ancients saw plainly by revelation, to the entire abrogation of experimental theory (and many believed what they saw, and prophesied of Jesus).

Man must and will have some religion. If he has not the religion of Jesus he will have the religion of Satan, and will erect the synagogue of Satan,° calling the prince of this world God, and destroying all who do not worship Satan under the name of God. Will any one say, 'Where are those who worship Satan under the name of God?' Where are they? Listen! Every religion that preaches vengeance for sin is the religion of the enemy and avenger, and not of the forgiver of sin; and their God is Satan, named by the divine name. Your religion, O deists, deism, is the worship of the god of this world° by the means of what you call natural religion and natural philosophy,° and of natural morality or self-righteousness, the selfish virtues of the natural heart. This was the religion of the Pharisees, who murdered Jesus. Deism is the same and ends in the same.

Voltaire, Rousseau, Gibbon, Hume charge the spiritually religious with hypocrisy. But how a monk, or a Methodist° either, can be a hypocrite, I cannot conceive. We are men of like passions with others, and pretend not to be holier than others. Therefore, when a religious man falls into sin he ought not to be called a hypocrite; this title is more properly to be given to a player° who falls into sin, whose profession is virtue and morality, and the making men self-righteous. Foote° in calling Whitefield 'hypocrite' was himself one! For Whitefield pretended not to be holier than others, but confessed his sins before all the world. Voltaire! Rousseau! you cannot escape my charge that you are Pharisees and hypocrites, for you are constantly talking of the virtues of the human heart, and particularly of your own, that you may accuse others and especially the religious, whose errors you, by this display of pretended virtue, chiefly design to expose. Rousseau thought men good by nature; he found them evil and found no friend. Friendship cannot exist without forgiveness of sins continually. The book written by Rousseau called his *Confessions* is an apology and cloak for his sin and not a confession.

But you also charge the poor monks and religious with being the causes of war, while you acquit and flatter the Alexanders and Caesars, the Louises and Fredericks,° who alone are its causes and its actors. But the religion of Jesus, forgiveness of sin, can never be the cause of a war nor of a single martyrdom.

Those who martyr others or who cause war are deists, but never can be forgivers of sin. The glory of Christianity is to conquer by forgiveness. All the destruction therefore in Christian Europe has arisen from deism, which is natural religion.

I saw a monk of Charlemagne [*see below*, p. *300*]

[*Preface to Chapter IV*]

TO THE CHRISTIANS.

Devils are	I give you the end of a golden string.
false religions.	Only wind it into a ball:
'Saul, Saul,	It will lead you in at Heaven's gate,
Why persecutest thou me?'°	Built in Jerusalem's wall.

We are told to abstain from fleshly desires that we may lose no time from the work of the Lord. Every moment lost is a moment that cannot be redeemed: every pleasure that intermingles with the duty of our station is a folly unredeemable, and is planted like the seed of a wild flower among our wheat. All the tortures of repentance are tortures of self-reproach on account of our leaving the divine harvest to the enemy, the struggles of entanglement with incoherent roots. I know of no other Christianity, and of no other Gospel, than the liberty both of body and mind to exercise the divine arts of imagination: imagination, the real and eternal world of which this vegetable universe is but a faint shadow, and in which we shall live in our eternal or imaginative bodies when these vegetable mortal bodies are no more. The apostles know of no other Gospel. What were all their spiritual gifts? What is the divine spirit? Is the Holy Ghost any other than an intellectual fountain? What is the harvest of the Gospel and its labours? What is that talent which it is a curse to hide?° What are the treasures of Heaven which we are to lay up for ourselves?° Are they any other than mental studies and performances? What are all the gifts of the Gospel? Are they not all mental gifts? Is God a spirit who must be worshipped in spirit and in truth, and are not the gifts of the spirit everything to Man? O ye religious, discountenance everyone among you who shall pretend to despise art and science!° I call upon you in the name of Jesus! What is the life of Man but art and science? Is it meat and drink? Is not the body more than raiment?° What is mortality but the things relating to the body, which dies? What is immortality but the things relating to the spirit, which lives eternally? What is the joy of Heaven but improvement in the things of the spirit? What are the pains of Hell but ignorance, bodily lust, idleness and devastation of the things of the spirit? Answer this to yourselves, and expel from among you those who pretend to despise the labours of art and science, which alone are the labours of the Gospel. Is not this plain and manifest to the thought? Can you think at all and not pronounce heartily: that to labour in knowledge is to build up Jerusalem and to despise knowledge is to despise Jerusalem and her builders? And remember! He who despises and mocks a mental gift in another, calling it pride and selfishness and sin, mocks Jesus, the giver of every mental gift (which always appear to the ignorance-loving hypocrite as sins). But that which is a sin in the sight of cruel Man is not so in the sight of our kind God. Let every Christian, as much as in him lies, engage himself openly and publicly before all the world in some mental pursuit for the building up of Jerusalem.

I stood among my valleys of the south
And saw a flame of fire, even as a wheel

Of fire surrounding all the heavens. It went
From west to east against the current of
Creation, and devoured all things in its loud
Fury and thundering course round heaven and earth.
By it the sun was rolled into an orb;
By it the moon faded into a globe
Travelling through the night. For from its dire
And restless fury Man himself shrunk up   10
Into a little root a fathom long.
And I asked a watcher and a holy one°
Its name? He answered, 'It is the wheel of religion.'
I wept and said: 'Is this the law of Jesus,
This terrible devouring sword turning every way?'
He answered: 'Jesus died because he strove
Against the current of this wheel. Its name
Is Caiaphas, the dark preacher of death,
Of sin, of sorrow, and of punishment,
Opposing Nature! It is natural religion;   20
But Jesus is the bright preacher of life,
Creating Nature from this fiery law
By self-denial and forgiveness of sin.
Go therefore, cast out devils in Christ's name;
Heal thou the sick of spiritual disease;
Pity the evil, for thou art not sent
To smite with terror and with punishments
Those that are sick, like to the Pharisees
Crucifying and encompassing sea and land
For proselytes to tyranny and wrath.   30
But to the publicans and harlots go!
Teach them true happiness; but let no curse
Go forth out of thy mouth to blight their peace.
For Hell is opened to Heaven: thine eyes beheld
The dungeons burst and the prisoners set free.'

England! awake, awake, awake [*see below, p. 301*]

From 'Vision of the Last Judgement' (c.*1810*)

The Last Judgement: when all those are cast away who trouble religion
with questions concerning good and evil, or eating of the tree of those

knowledges, or reasonings which hinder the vision of God, turning all into a consuming fire, when imaginative art and science, and all intellectual gifts—all the gifts of the Holy Ghost—are looked upon as of no use, and only contention remains to Man, then the Last Judgement begins, and its vision is seen by everyone according to the situation he holds.

The Last Judgement is not fable or allegory, but vision. Fable or allegory are a totally distinct and inferior kind of poetry. Vision or imagination is a representation of what eternally exists, really and unchangeably. Fable or allegory is formed by the Daughters of Memory.° Imagination is surrounded by the Daughters of Inspiration who in the aggregate are called Jerusalem.

Fable is allegory, but what critics call the fable° is vision itself.

The Hebrew Bible and the Gospel of Jesus are not allegory but eternal vision, or imagination of all that exists.

Note here that fable or allegory is seldom without some vision. *Pilgrim's Progress* is full of it; the Greek poets the same. But allegory and vision ought to be known as two distinct things and so called, for the sake of eternal life.

Plato has made Socrates say° that poets and prophets do not know or understand what they write or utter. This is a most pernicious falsehood. If they do not, pray is an inferior kind to be called knowing? Plato confutes himself.

The Last Judgement is one of these stupendous visions. I have represented it as I saw it. To different people it appears differently, as everything else does. For though on earth things seem permanent they are less permanent than a shadow, as we all know too well.

The nature of visionary fancy or imagination is very little known, and the eternal nature and permanence of its ever existent images is considered as less permanent than the things of vegetable and generative nature. Yet the oak dies as well as the lettuce; but its eternal image and individuality never dies, but renews by its seed. Just so the imaginative image returns by the seed of contemplative thought.

Let it here be noted that the Greek fables originated in spiritual mystery and real vision, which are lost and clouded in fable and allegory; while the Hebrew Bible and the Greek Gospel are genuine, preserved by the Saviour's mercy. The nature of my work is visionary or imaginative. It is an endeavour to restore what the ancients called the Golden Age.

This world of imagination is the world of Eternity. It is the divine bosom into which we shall all go after the death of the vegetated body. This world of imagination is infinite and eternal, whereas the world of generation or vegetation is finite and temporal. There exist in that eternal world the permanent realities of every thing, which we see reflected in this vegetable glass of Nature. All things are comprehended in their eternal forms in the divine body of the Saviour, the true vine° of Eternity, the human imagination, who appeared to me as coming to judgement among his saints and throwing off the temporal that the eternal might be established. Around him were seen the images of existences according to a certain order suited to my imaginative eye.

Abel kneels on a bloody cloud descriptive of those churches before the flood, that they were filled with blood and fire and vapour of smoke; even till Abraham's time the vapour and heat was not extinguished.° These states exist now. Man passes on but states remain for ever. He passes through them like a traveller: who may as well suppose that the places he has passed through exist no more as a man may suppose that the states he has passed through exist no more. Everything is eternal.

In Eternity one thing never changes into another thing. Each identity is eternal. Consequently Apuleius's *Golden Ass*, and Ovid's *Metamorphoses*,° and others of the like kind are fable, yet they contain vision in a sublime degree—being derived from real vision in more ancient writings. Lot's wife° being changed into a pillar of salt alludes to the mortal body being rendered a permanent statue, but not changed or transformed into another identity while it retains its own individuality. A man can never become ass nor horse. Some are born with shapes of men who may be both, but eternal identity is one thing and corporeal vegetation is another thing. Changing water into wine by Jesus, and into blood by Moses,° relates to vegetable nature also.

It ought to be understood that the persons Moses and Abraham are not here meant, but the states signified by those names, the individuals being representatives or visions of those states as they were revealed to mortal man in the series of divine revelations, as they are written in the

Bible. These various states I have seen in my imagination. When distant they appear as one man, but as you approach they appear multitudes of nations. Abraham hovers above his posterity which appear as multitudes of children ascending from the Earth surrounded by stars, as it was said: 'as the stars of Heaven for multitude'.° Jacob and his twelve sons hover beneath the feet of Abraham and receive their children from the Earth. I have seen when at a distance multitudes of men in harmony appear like a single infant, sometimes in the arms of a female; this represented the Church. . . .

The ladies will be pleased to see that I have represented the Furies by three men and not by three women. It is not because I think the ancients wrong, but they will be pleased to remember that mine is vision and not fable. The spectator may suppose them clergymen in the pulpit scourging sin instead of forgiving it.

Between the figures° of Adam and Eve appears a fiery gulf descending from the sea of fire before the throne. In this cataract four angels descend headlong with four trumpets to awake the dead. Beneath these is the seat of the harlot named Mystery in the Revelations. She is seized by two beings each with three heads; they represent vegetative existence. As it is written in Revelations, they strip her naked and burn her with fire.

It represents the eternal consummation of vegetable life and death with its lusts. The wreathed torches in their hands represent eternal fire, which is the fire of generation or vegetation; it is an eternal consummation. Those who are blessed with imaginative vision see this Eternal Female and tremble at what others fear not while they despise and laugh at what others fear.

Her kings and councillors and warriors descend in flames, lamenting and looking upon her in astonishment and terror, and Hell is opened beneath her seat on the left hand.

Beneath her feet is a flaming cavern in which is seen the great red dragon with seven heads and ten horns.

He has Satan's book of accusations lying on the rock open before him. He is bound in chains by two strong demons; they are Gog and Magog,° who have been compelled to subdue their master (*Ezekiel* xxxviii.8)—with their hammers and tongs about to new create the seven-headed kingdoms. The graves beneath are opened and the dead awake and obey the call of the trumpet. Those on the right hand awaken in joy: those on the left in horror. Beneath the dragon's cavern a skeleton begins to animate, starting into life at the trumpet's sound while the wicked contend with each other on the brink of perdition. On the right a youthful couple are awaked by their children. An aged patriarch is awaked by his aged wife. He is Albion our ancestor, patriarch of the Atlantic continent, whose history preceded that of the Hebrews, and in whose sleep, or 'chaos', creation began. At their head the aged woman is Britannica,° the wife of Albion; Jerusalem is their daughter. Little infants creep out of the flowery mould° into the green fields of the blessed, who in various joyful companies embrace and ascend to meet Eternity.

Also on the right hand of Noah a female descends to meet her lover or husband, representative of that love, called friendship, which looks for no other heaven than their beloved—and in him sees all reflected as in a glass of eternal diamond.

On the right hand of these rise the diffident and humble, and on their left a solitary woman with her infant. These are caught up by three aged men, who appear as suddenly emerging from the blue sky for their help. These three aged men represent divine providence as opposed to and distinct from divine vengeance—represented by three aged men on the side of the picture, among the wicked, with scourges of fire.

If the spectator could enter into these images in his imagination, approaching them on the fiery chariot of his contemplative thought; if he could enter into Noah's rainbow or into his bosom, or could make a friend and companion of one of these images of wonder, which always entreats him to leave mortal things (as he must know)—then would he arise from his grave, then would he meet the Lord in the air,° and then he would be happy. General knowledge is remote knowledge; it is in particulars that wisdom consists and happiness too, both in art and in life. General masses are as much art as a pasteboard man is human. Every man has eyes, nose and mouth. This every idiot knows, but he who enters into and discriminates most minutely the manners and intentions, the characters in all their branches, is the alone wise or sensible man, and on this discrimination all art is founded. I entreat then that the spectator will attend to the hands and feet, to the lineaments of the countenances. They are all descriptive of character, and not a line is drawn without intention, and that most discriminate and particular. As poetry admits not a letter that is insignificant, so painting admits not a grain of sand or a blade of grass insignificant—much less an insignificant blur or mark.

A Last Judgement is necessary because fools flourish. Nations flourish

under wise rulers and are depressed under foolish rulers; it is the same
with individuals as nations. Works of art can only be produced in
perfection where the man is either in affluence, or is above the care of it.
Poverty is the fool's rod, which at last is turned upon his own back. This
is a Last Judgement when men of real art govern and pretenders fall.
Some people, and not a few artists, have asserted that the painter of this
picture would not have done so well if he had been properly
encouraged. Let those who think so reflect on the state of nations under
poverty, and their incapability of art. Though art is above either, the
argument is better for affluence than poverty, and though he would not
have been a greater artist, yet he would have produced greater works of
art in proportion to his means. A Last Judgement is not for the purpose
of making bad men better, but for the purpose of hindering them from
oppressing the good with poverty and pain by means of such vile
arguments and insinuations.

Around the throne Heaven is opened, and the nature of eternal things
displayed, all springing from the divine humanity. All beams from him.
He is the bread and the wine; he is the water of life. Accordingly on each
side of the opening heaven appears an apostle. That on the right
represents Baptism; that on the left represents the Lord's Supper.° All
life consists of these two: throwing off error and knaves from our
company continually, and receiving truth or wise men into our
company continually. He who is out of the Church and opposes it is no
less an agent of religion than he who is in it. To be an error and to be
cast out is a part of God's design. No man can embrace true art till he
has explored and cast out false art—such is the nature of mortal
things—or he will be himself cast out by those who have already
embraced true art. Thus my picture is a history of art and science, the
foundation of society, which is humanity itself. What are all the gifts of
the spirit but mental gifts? Whenever any individual rejects error and
embraces truth a last judgement passes upon that individual.

Over the head of the Saviour and Redeemer, the Holy Spirit like a dove
is surrounded by a blue heaven, in which are the two cherubim that
bowed over the ark;° for here the temple is opened in Heaven, and the
ark of the covenant is as a dove of peace. The curtains are drawn apart,
Christ having rent the veil. The candlestick and the table of shewbread
appear. On each side a glorification of angels with harps surround the
dove.
    The temple stands on the mount of God. From it flows on each side

the River of Life, on whose banks grows the Tree of Life,° among whose branches temples and pinnacles, tents and pavilions, gardens and groves display Paradise, with its inhabitants walking up and down in conversations concerning mental delights.

Here they are no longer talking of what is good and evil, or of what is right or wrong, and puzzling themselves in Satan's labyrinth,° but are conversing with eternal realities as they exist in the human imagination. We are in a world of generation and death, and this world we must cast off if we would be painters such as Rafael, Michelangelo and the ancient sculptors. If we do not cast off this world we shall be only Venetian painters, who will be cast off and lost from art.

Jesus is surrounded by beams of glory in which are seen all around him infants emanating from him. These represent the eternal births of intellect from the Divine Humanity. A rainbow surrounds the throne and the glory,° in which youthful nuptials receive the infants in their hands. In Eternity Woman is the Emanation of Man; she has no will of her own. There is no such thing in Eternity as a female will.

On the side next Baptism are seen those called in the Bible nursing fathers and nursing mothers.° They represent education. On the side next the Lord's Supper the Holy Family, consisting of Mary, Joseph, John the Baptist, Zacharias and Elizabeth, receiving the bread and wine among other spirits of the just made perfect. Beneath these a cloud of women and children are taken up, fleeing from the rolling cloud which separates the wicked from the seats of bliss. These represent those who, though willing, were too weak to reject error without the assistance and countenance of those already in the truth. For a man can only reject error by the advice of a friend, or by the immediate inspiration of God. It is for this reason among many others that I have put the Lord's Supper on the left hand of the throne, for it appears so at the Last Judgement for a protection.

The combats of good and evil is eating of the Tree of Knowledge. The combats of truth and error is eating of the Tree of Life. These are not only universal but particular. Each are personified. There is not an error but it has a man for its agent: that is, it is a man. There is not a truth but it has also a man. Good and evil are qualities in every man, whether a good or evil man. These are enemies and destroy one another by every means in their power, both of deceit and of open violence. The deist and the Christian are but the results of these opposing natures. Many are deists who would in certain circumstances have been

Christians in outward appearance. Voltaire was one of this number; he was as intolerant as an inquisitor. Manners make the man, not habits. It is the same in art: by their works ye shall know them.° The knave who is converted to deism, and the knave who is converted to Christianity, is still a knave, but he himself will not know it though everybody else does. Christ comes as he came at first to deliver those who were bound under the knave, not to deliver the knave; he comes to deliver Man the accused, and not Satan the accuser.° We do not find anywhere that Satan is accused of sin; he is only accused of unbelief, and thereby drawing Man into sin, that he may accuse him. Such is the Last Judgement: a deliverance from Satan's accusation. Satan thinks that sin is displeasing to God. He ought to know that nothing is displeasing to God but unbelief and eating of the Tree of Knowledge of Good and Evil.

Men are admitted into Heaven not because they have curbed and governed their passions, or have no passions, but because they have cultivated their understandings. The treasures of Heaven are not negations of passion, but realities of intellect from which all the passions emanate uncurbed in their eternal glory. The fool shall not enter into Heaven, let him be ever so holy. Holiness is not the price of entrance into Heaven. Those who are cast out are all those who, having no passions of their own because no intellect, have spent their lives in curbing and governing other peoples' by the various arts of poverty and cruelty of all kinds. Woe, woe, woe to you hypocrites!° Even murder the courts of justice (more merciful than the Church) are compelled to allow is not done in passion but in cool-blooded design and intention.

The modern Church crucifies Christ with the head downwards.

Many persons such as Paine and Voltaire, with some of the ancient Greeks, say, 'we will not converse concerning good and evil; we will live in Paradise and liberty'. You may do so in spirit, but not in the mortal body as you pretend, till after the Last Judgement. For in Paradise they have no corporeal and mortal body; that originated with the fall and was called Death, and cannot be removed but by a Last Judgement. While we are in the world of mortality we must suffer. The whole creation groans to be delivered.° There will always be as many hypocrites born as honest men, and they will always have superior power in mortal things. You cannot have liberty in this world without what you call

Moral Virtue, and you cannot have Moral Virtue without the slavery of that half of the human race who hate what you call Moral Virtue.

The nature of hatred and envy, and of all the mischiefs in the world are here depicted. No one envies or hates one of his own party. Even the devils love one another in their way; they torment one another for other reasons than hate or envy; these are only employed against the just. Neither can Seth° envy Noah, or Elijah envy Abraham, but they may both of them envy the success of Satan, or of Og,° or Molech.° The horse never envies the peacock nor the sheep the goat, but they envy a rival in life and existence whose ways and means exceed their own, let him be of what class of animals he will. A dog will envy a cat who is pampered at the expense of his comfort, as I have often seen. The Bible never tells us that devils torment one another through envy. It is through this that they torment the just, but for what do they torment one another? I answer, for the coercive laws of Hell, moral hypocrisy. They torment a hypocrite when he is discovered; they punish a failure in the tormentor who has suffered the subject of his torture to escape. In Hell all is self-righteousness; there is no such thing there as forgiveness of sin. He who does forgive sin is crucified as an abettor of criminals, and he who performs works of mercy in any shape whatever is punished and if possible destroyed—not through envy or hatred or malice, but through self-righteousness that thinks it does God service, which God is Satan. They do not envy one another. They contemn and despise one another. Forgiveness of sin is only at the judgement seat of Jesus the Saviour, where the accuser is cast out: not because he sins but because he torments the just and makes them do what he condemns as sin, and what he knows is opposite to their own identity.

It is not because angels are holier than men or devils that makes them angels, but because they do not expect holiness from one another, but from God only.

The player is a liar when he says, 'angels are happier than men because they are better'. Angels are happier than men and devils because they are not always prying after good and evil in one another, and eating the Tree of Knowledge for Satan's gratification.

Thinking as I do that the creator of this world is a very cruel being, and being a worshipper of Christ, I cannot help saying: 'The Son, Oh how unlike the Father! First God Almighty comes with a thump on the head; then Jesus Christ comes with a balm to heal it'.

The Last Judgement is an overwhelming of bad art and science. Mental things are alone real. What is called corporeal nobody knows of its dwelling-place; it is in fallacy, and its existence an imposture. Where is the existence out of mind or thought? Where is it but in the mind of a fool? Some people flatter themselves that there will be no Last Judgement, and that bad art will be adopted and mixed with good art. That error or experiment will make a part of truth, and they boast that it is its foundation. These people flatter themselves. I will not flatter them. Error is created. Truth is eternal. Error or creation will be burned up, and then and not till then truth or Eternity will appear. It is burnt up the moment men cease to behold it. I assert for myself that I do not behold the outward creation, and that to me it is hindrance and not action. It is as the dirt upon my feet: no part of me. 'What', it will be questioned, 'when the sun rises do you not see a round disk of fire somewhat like a guinea?' Oh no, no! I see an innumerable company of the heavenly host crying 'Holy, holy, holy° is the Lord God Almighty!' I question not my corporeal or vegetative eye any more than I would question a window concerning a sight. I look through it and not with it.

Many suppose that before the creation all was solitude and chaos. This is the most pernicious idea that can enter the mind, as it takes away all sublimity from the Bible and limits all existence to creation and to chaos—to the time and space fixed by the corporeal vegetative eye—and leaves the man who entertains such an idea the habitation of unbelieving demons. Eternity exists, and all things in Eternity, independent of creation, which was an act of mercy. I have represented those who are in Eternity by some in a cloud, within the rainbow that surrounds the throne. They merely appear as in a cloud when anything of creation, redemption, or judgement are the subjects of contemplation (though their whole contemplation is concerning these things). The reason they so appear is the humiliation of the reason and doubting selfhood, and the giving all up to inspiration. By this it will be seen that I do not consider either the just or the wicked to be in a supreme state, but to be every one of them states of the sleep which the soul may fall into in its deadly dreams of good and evil, when it leaves Paradise following the serpent.

The Greeks represent Chronos or Time as a very aged man. This is fable, but the real vision of Time is in eternal youth. I have, however, somewhat accommodated my figure of Time to the common opinion, as I myself am also infected with it and my visions also infected, and I see time aged—alas, too much so.

Allegories are things that relate to moral virtues. Moral virtues do

not exist; they are allegories and dissimulations. But Time and Space
are real beings, a male and a female. Time is a man; Space is a woman
and her masculine portion is death.

Annotation to *J. G. Spurzheim's*
Observations on . . . Insanity (c.*1818*)

Cowper° came to me and said: 'Oh, that I were insane, always! I will
never rest. Cannot you make me truly insane? I will never rest till I am
so. Oh, that in the bosom of God I was hid! You retain health and yet
are mad as any of us all—over us all—mad as a refuge from unbelief,
from Bacon, Newton, and Locke.'

The Everlasting Gospel (c.*1818*)

If moral virtue was Christianity
Christ's pretensions were all vanity,
And Caiaphas and Pilate men°
Praiseworthy, and the lion's den,
And not the sheepfold, allegories
Of God and Heaven and their glories.
The moral Christian is the cause
Of the unbeliever and his laws.
The Roman virtues, warlike fame,
Take Jesus' and Jehovah's name.   10
For what is Antichrist, but those°
Who against sinners Heaven close
With iron bars in virtuous state,
And Rhadamanthus at the gate?°

What can this Gospel of Jesus be?
What life and immortality?
What was it that he brought to light
That Plato and Cicero did not write?°
The heathen deities wrote them all,
These moral virtues, great and small.   20

What is the accusation of sin
But moral virtues' deadly gin?
The moral virtues in their pride
Did o'er the world triumphant ride
In wars and sacrifice for sin,
And souls to Hell ran trooping in—
The accuser, holy God of all
This pharisaic worldly ball,
Amidst them in his glory beams
Upon the rivers and the streams.   30
Then Jesus rose and said to me,
'Thy sins are all forgiven thee.'
Loud Pilate howled, loud Caiaphas yelled
When they the Gospel light beheld.
It was when Jesus said to me,
'Thy sins are all forgiven thee'.
The Christian trumpets loud proclaim
Through all the world in Jesus' name
Mutual forgiveness of each vice,
And oped the gates of Paradise.   40
The moral virtues in great fear
Formed the cross and nails and spear,
And the accuser standing by
Cried out 'Crucify, crucify.
Our moral virtues ne'er can be,
Nor warlike pomp and majesty,
For moral virtues all begin
In the accusations of sin,
And all the heroic virtues end
In destroying the sinners' friend.   50
Am I not Lucifer the Great,
And you my daughters in great state,
The fruit of my mysterious Tree
Of Good and Evil, and misery,
And death, and Hell, which now begin
On everyone who forgives sin?'

Was Jesus born of a virgin pure
With narrow soul and looks demure?
If he intended to take on sin
The mother should an harlot been,   60

Just such a one as Magdalen
With seven devils in her pen.°
Or were Jew virgins still more cursed,°
And more sucking devils nursed?
Or what was it which he took on
That he might bring salvation?
A body subject to be tempted,
From neither pain nor grief exempted,
Or such a body as might not feel
The passions that with sinners deal?   70
Yes, but they say he never fell;
Ask Caiaphas, for he can tell:
'He mocked the Sabbath, and he mocked°
The Sabbath's God, and he unlocked°
The evil spirits from their shrines,
And turned fishermen to divines:
O'erturned the tent of secret sins,
And its golden cords and pins.
'Tis the bloody shrine of war
Pinned around from star to star,   80
Halls of justice, hating vice,
Where the devil combs his lice.
He turned the devils into swine°
That he might tempt the Jews to dine
(Since which a pig has got a look
That for a Jew may be mistook).
"Obey your parents." What says he?°
"Woman, what have I to do with thee?"°
He scorned earth's parents, scorned earth's God,
And mocked the one and the other's rod;   90
His seventy disciples sent°
Against religion and government.
They by the sword of justice fell,
And him their cruel murderer tell.
He left his father's trade to roam,
A wand'ring vagrant without home,
And thus he others' labour stole,
That he might live above control.
The publicans and harlots he
Selected for his company,   100
And from the adulteress turned away
God's righteous law, that lost its prey.'°

Was Jesus gentle, or did he
Give any marks of gentility?
When twelve years old he ran away,
And left his parents in dismay.
When after three days' sorrow found,
Loud as Sinai's trumpet sound:°
'No earthly parents I confess—
My heavenly Father's business.°   110
Ye understand not what I say,
And angry, force me to obey.
Obedience is a duty then,
And favour gains with God and men.'
John from the wilderness loud cried;
Satan gloried in his pride.
'Come,' said Satan, 'come away;
I'll soon see if you'll obey.°
John for disobedience bled,°
But you can turn the stones to bread.°   120
God's high king and God's high priest
Shall plant their glories in your breast,
If Caiaphas you will obey,
If Herod you, with bloody prey
Feed with the sacrifice, and be
Obedient, fall down, worship me.'
Thunders and lightnings broke around,
And Jesus' voice in thunder's sound:°
'Thus I seize the spiritual prey.
Ye smiters with disease, make way.   130
I come your king and god to seize;
Is God a smiter with disease?'
The god of this world raged in vain.
He bound old Satan in his chain,°
And, bursting forth, his furious ire
Became a chariot of fire.
Throughout the land he took his course,
And traced diseases to their source.
He cursed the scribe and Pharisee,°
Trampling down hypocrisy.   140
Where'er his chariot took its way
There gates of death let in the day,°
Broke down from every chain and bar,
And Satan in his spiritual war

Dragged at his chariot wheels. Loud howled
The god of this world; louder rolled
The chariot wheels, and louder still
His voice was heard from Zion's hill,
And in his hand the scourge shone bright.
He scourged the merchant Canaanite   150
From out the temple of his mind,°
And in his body tight does bind
Satan and all his hellish crew.
And thus with wrath he did subdue
The serpent bulk of Nature's dross,
Till he had nailed it to the cross.
He took on sin in the virgin's womb
And put it off on the cross and tomb,
To be worshipped by the Church of Rome.°

Did Jesus teach doubt, or did he   160
Give any lessons of philosophy,
Charge visionaries with deceiving
Or call men wise for not believing?

Was Jesus humble, or did he
Give any proofs of humility,
Boast of high things with humble tone
And give with charity a stone?°
When the rich learned Pharisee°
Came to consult him secretly,
Upon his heart with iron pen   170
He wrote: 'Ye must be born again.'
He was too proud to take a bribe.
He spoke with authority, not like a scribe.°
He says with most consummate art:
'Follow me; I am meek and lowly of heart'°—
As that is the only way to escape
The miser's net and the glutton's trap.
What can be done with such desperate fools
Who follow after the heathen schools
(I was standing by when Jesus died;   180
What I called 'humility' they called 'pride')?
He who loves his enemies betrays his friends;
This surely is not what Jesus intends,

But the sneaking pride of heroic schools
And the scribes' and Pharisees' virtuous rules.
For he acts with honest triumphant pride,
And this is the cause that Jesus died.
He did not die with Christian ease,
Asking pardon of his enemies.
If he had, Caiaphas would forgive;   190
Sneaking submission can always live.
He had only to say that God was the Devil,
And the Devil was God, like a Christian civil,
Mild Christian regrets to the Devil confess
For affronting him thrice in the wilderness—
He had soon been bloody Caesar's elf,
And at the last he would have been Caesar himself—
Like Dr Priestley and Bacon and Newton.°
Poor spiritual knowledge is not worth a button,
For thus the Gospel Sir Isaac confutes: 200
'God can only be known by his attributes,
And as for the indwelling of the Holy Ghost
Or of Christ and his Father—it's all a boast,
And pride and vanity of the imagination,
That disdains to follow this world's fashion.'
To teach doubt and experiment
Certainly was not what Christ meant.
What was he doing all that time
From twelve years old to manly prime?
Was he then idle, or the less   210
About his father's business?
Or was his wisdom held in scorn
Before his wrath began to burn
In miracles throughout the land
That quite unnerved Caiaphas' hand?
If he had been Antichrist, creeping Jesus,
He'd have done any thing to please us—
Gone sneaking into synagogues,
And not used the elders and priests like dogs,
But humble as a lamb or ass   220
Obeyed himself to Caiaphas.
God wants not Man to humble himself;
This is the trick of the ancient elf.
This is the race that Jesus ran:
Humble to God, haughty to Man,

Cursing the rulers before the people
Even to the Temple's highest steeple.
And when he humbled himself to God,
Then descended the cruel rod:
'If thou humblest thyself thou humblest me;   230
Thou also dwell'st in eternity.°
Thou art a man. God is no more.
Thy own humanity learn to adore,
For that is my spirit of life.
Awake! Arise to spiritual strife,
And thy revenge abroad display
In terrors at the Last Judgement day.
God's mercy and long-suffering
Is but the sinner to judgement to bring.
Thou on the cross for them shalt pray,°   240
And take revenge at the last day.'
Jesus replied, and thunders hurled:
'I never will pray for the world.°
Once I did so when I prayed in the garden;°
I wished to take with me a bodily pardon.'
Can that?—which was of woman born°
In the absence of the morn,
When the soul fell into sleep
And archangels round it weep,
Shooting out against the light   250
Fibres of a deadly night,
Reasoning upon its own dark fiction°
In doubt, which is self-contradiction.
Humility is only doubt,
And does the sun and moon blot out,
Rooting over with thorns and stems
The buried soul and all its gems
(This life's dim windows of the soul),
Distorts the heavens from pole to pole,
And leads you to believe a lie   260
When you see with, not through, the eye—
That was born in a night, to perish in a night,
When the soul slept in the beams of light.
Was Jesus chaste, or did he
Give any lessons of chastity?
The morning blushed fiery red;
Mary was found in adulterous bed.°

Earth groaned beneath, and Heaven above°
Trembled at discovery of love.
Jesus was sitting in Moses' chair;   270
They brought the trembling woman there.
'Moses commands she be stoned to death';
What was the sound of Jesus breath?
He laid his hand on Moses' law;
The ancient heavens in silent awe,°
Writ with curses from pole to pole,
All away began to roll.
The earth trembling and naked lay
In secret bed of mortal clay,
On Sinai felt the hand divine   280
Putting back the bloody shrine.
And she heard the breath of God
As she heard by Eden's flood.
'Good and evil are no more.°
Sinai's trumpets, cease to roar!
Cease, finger of God, to write!°
The heavens are not clean in thy sight;°
Thou art good and thou alone,
Nor may the sinner cast one stone.
To be good only is to be   290
A devil, or else a Pharisee.°
Thou angel of the presence divine,°
That didst create this body of mine,
Wherefore hast thou writ these laws
And created Hell's dark jaws?
My presence I will take from thee:
A cold leper thou shalt be.
Though thou wast so pure and bright
That Heaven was impure in thy sight;
Though thy oath turned Heaven pale;   300
Though thy covenant built Hell's jail;°
Though thou didst all to chaos roll
With the serpent for its soul—
Still the breath divine does move,°
And the breath divine is love.
Mary, fear not; let me see
The seven devils that torment thee.
Hide not from my sight thy sin,
That forgiveness thou may'st win.

Has no man condemned thee?'   310
'No man, Lord!' 'Then what is he
Who shall accuse thee? Come ye forth,
Fallen fiends of heav'nly birth,
That have forgot your ancient love,
And driven away my trembling dove.
You shall bow before her feet,
You shall lick the dust for meat,°
And though you cannot love, but hate,
Shall be beggars at love's gate.°
What was thy love? Let me see it.   320
Was it love, or dark deceit?'
'Love too long from me has fled.
'Twas dark deceit to earn my bread.
'Twas covet, or 'twas custom, or
Some trifle not worth caring for:
That they may call a shame and sin
Love's temple that God dwelleth in,
And hide in secret hidden shrine
The naked human form divine,
And render that a lawless thing   330
On which the soul expands its wing.
But this, O Lord, this was my sin,
When first I let these devils in
In dark pretence to chastity:
Blaspheming love, blaspheming thee.
Thence rose secret adulteries,
And thence did covet also rise.
My sin thou hast forgiven me;
Canst thou forgive my blasphemy?
Canst thou return to this dark hell,   340
And in my burning bosom dwell,
And canst thou die, that I may live,
And canst thou pity, and forgive?'
Then rolled the shadowy man away
From the limbs of Jesus to make them his prey,
An ever-devouring appetite
Glittering with festering venoms bright,
Crying: 'Crucify this cause of distress,
Who don't keep the secrets of holiness!
All mental powers by diseases we bind,   350
But he heals the deaf and the dumb and the blind.

Whom God has afflicted for secret ends
He comforts and heals, and calls them friends.'
But when Jesus was crucified,
Then was perfected his glitt'ring pride;
In three nights he devoured his prey,
And still he devours the body of clay.
For dust and clay is the serpent's meat,
Which never was made for Man to eat.

I am sure this Jesus will not do,   360
Either for Englishman or Jew.
The vision of Christ that thou dost see
Is my vision's greatest enemy.
Thine has a great hook nose like thine;
Mine has a snub nose like to mine.
Thine is the friend of all mankind;
Mine speaks in parables to the blind.
Thine loves the same world that mine hates.
Thy Heaven-doors are my Hell-gates.
Socrates taught what Melitus°   370
Loathed as a nation's bitterest curse,
And Caiaphas was, in his own mind,
A benefactor to mankind.
Both read the Bible day and night,
But thou read'st black where I read white.

From the annotations to Bishop Berkeley's Siris (c.*1820*)

Jesus considered imagination to be the real man and says, I will not
leave you orphans and I will manifest myself to you.° He says also the
spiritual body, or angel, as little children always behold the face of the
heavenly Father.°

Harmony and proportion are qualities, and not things. The harmony
and proportion of a horse are not the same with those of a bull.
Everything has its own harmony and proportion, two inferior qualities
in it. For its reality is its imaginative form.°

Knowledge is not by deduction, but immediate by perception or sense°
at once. Christ addresses himself to the man, not to his reason. Plato did
not bring life and immortality to light; Jesus only did this.

Jesus supposes everything to be evident to the child and to the poor and
unlearned. Such is the Gospel.

The whole Bible is filled with imagination and visions from end to end,
and not with moral virtues. That is the baseness of Plato, and the Greek
and all warriors. The moral virtues are continual accusers of sin, and
promote eternal wars and dominancy over others.

God is not a mathematical diagram.

The natural body is an obstruction to the soul or spiritual body.

Man is all imagination. God is Man and exists in us and we in him.

What Jesus came to remove was the heathen or platonic philosophy
which blinds the eye of imagination, the real man.

From the annotations to R. J. Thornton's
The Lord's Prayer, Newly Translated (1827)

I look upon this as a most malignant and artful attack upon the kingdom
of Jesus by the classical-learned through the instrumentality of Dr
Thornton. The Greek and Roman classics is the Antichrist. I say 'is'
and not 'are' as most expressive and correct too.

Christ and his apostles were illiterate men. Caiaphas, Pilate and Herod
were learned.

The only thing for Newtonian and Baconian philosophers to consider is
this: whether Jesus did not suffer himself to be mocked by Caesar's
soldiers willingly, and to consider this to all eternity will be comment
enough.

This is saying the Lord's Prayer backwards, which they say raises the Devil.

Doctor Thornton's Tory translation translated out of its disguise in the classical and Scotch° language into the vulgar English:

Our Father Augustus Caesar who art in this thy substantial astronomical telescopic heavens, holiness to thy name or title, and reverence to thy shadow. Thy kingship come upon Earth first and then in Heaven. Give us day by day our real, taxed, substantial money-bought bread. Deliver from the Holy Ghost whatever cannot be taxed, for all is debts and taxes between Caesar and us and one another. Lead us not to read the Bible but let our Bible be Virgil and Shakspeare, and deliver us from poverty in Jesus, that evil one. For thine is the kingship (or allegoric godship) and the power (or war) and the glory (or law) ages after ages in thy descendants; for God is only an allegory of kings° and nothing else.

Amen

I swear that Basileia (Βασιλεία) is not kingdom but kingship.° I, Nature, hermaphroditic priest and king, live in real, substantial, natural-born Man, and that spirit is the ghost of matter or Nature; and God is the ghost of the priest and king—who exist, whereas God exists not except from their effluvia.

BLAKE ON ART AND LITERATURE

Letter of 16 August 1799

To Dr Trusler

Reverend Sir,

I find more and more that my style of designing is a species by itself, and in this which I send you have been compelled by my genius or angel to follow where he led. If I were to act otherwise it would not fulfill the purpose for which alone I live: which is, in conjunction with such men as my friend Cumberland,° to renew the lost art of the Greeks.°

I attempted every morning for a fortnight together to follow your dictate,° but when I found my attempts were in vain resolved to show an independence which I know will please an author better than slavishly following the track of another, however admirable that track may be. At any rate my excuse must be: I could not do otherwise; it was out of my power!

I know I begged of you to give me your ideas, and promised to build on them. Here I counted without my host. I now find my mistake.

The design I have sent is:

A father taking leave of his wife and child is watched by two fiends incarnate, with intention that when his back is turned they will murder the mother and her infant. If this is not Malevolence with a vengeance I have never seen it on earth, and if you approve of this I have no doubt of giving you Benevolence with equal vigour, as also Pride and Humility, but cannot previously describe in words what I mean to design for fear I should evaporate the spirit of my invention. But I hope that none of my designs will be destitute of infinite particulars which will present themselves to the contemplator. And though I call them mine I know that they are not mine, being of the same opinion with Milton when he says° that the Muse visits his slumbers and awakes and governs his song when morn purples the east; and being also in the predicament of that prophet who says, I cannot go beyond the command of the Lord to speak good or bad.°

If you approve of my manner, and it is agreeable to you, I would rather paint pictures in oil of the same dimensions than make drawings, and on the same terms. By this means you will have a number of cabinet pictures° which I flatter myself will not be unworthy of a scholar of Rembrandt and Teniers—whom I have studied no less than Rafael and Michelangelo. Please to send me your orders respecting this, and in my next effort I promise more expedition.

I am reverend Sir
Your very humble servant,
WILLIAM BLAKE

Letter of 23 August 1799

To Dr Trusler

Reverend Sir,

I really am sorry that you are fallen out with the spiritual world, especially if I should have to answer for it. I feel very sorry that your ideas and mine on moral painting differ so much° as to have made you angry with my method of study. If I am wrong I am wrong in good company. I had hoped your plan comprehended all species of this art, and especially that you would not reject that species which gives existence to every other, namely, visions of eternity. You say that I want somebody to elucidate my ideas, but you ought to know that what is grand is necessarily obscure to weak men. That which can be made explicit to the idiot is not worth my care. The wisest of the ancients considered what is not too explicit as the fittest for instruction because it rouses the faculties to act. I name Moses, Solomon, Aesop, Homer, Plato.

But as you have favoured me with your remarks on my design permit me in return to defend it against a mistaken one, which is that I have supposed malevolence without a cause. Is not merit in one a cause of envy in another, and serenity and happiness and beauty a cause of malevolence? But want of money and the distress of a thief can never be alleged as the cause of his thieving, for many honest people endure greater hardships with fortitude. We must therefore seek the cause elsewhere than in want of money, for that is the miser's passion, not the thief's.

I have therefore proved your reasonings ill-proportioned, which you can never prove my figures to be. They are those of Michelangelo, Rafael and the antique, and of the best living models. I perceive that your eye is perverted by caricature prints, which ought not to abound so much as they do. Fun I love, but too much fun is of all things the most loathsome. Mirth is better than fun, and happiness is better than mirth. I feel that a man may be happy in this world. And I know that this world is a world of imagination and vision. I see everything I paint in this world, but everybody does not see alike. To the eyes of a miser a guinea is more beautiful than the sun, and a bag worn with the use of money has more beautiful proportions than a vine filled with grapes. The tree which moves some to tears of joy is in the eyes of others only a green thing that stands in the way. Some see Nature all ridicule and deformity (and by these I shall not regulate my proportion), and some scarce see Nature at all. But to the eyes of the man of imagination

Nature is imagination itself. As a man is so he sees. As the eye is formed such are its powers. You certainly mistake when you say that the visions of fancy are not to be found in this world. To me this world is all one continued vision of fancy or imagination, and I feel flattered when I am told so. What is it sets Homer, Virgil and Milton in so high a rank of art? Why is the Bible more entertaining and instructive than any other book? Is it not because they are addressed to the imagination, which is spiritual sensation, and but mediately° to the understanding or reason? Such is true painting, and such was alone valued by the Greeks and the best modern artists. Consider what Lord Bacon says: 'Sense sends over to imagination before reason have judged, and reason sends over to imagination before the decree can be acted'. (See *Advancement of Learning*, Part 2, page 47 of first edition.)

But I am happy to find a great majority of fellow mortals who can elucidate my visions; and particularly they have been elucidated by children, who have taken a greater delight in contemplating my pictures than I even hoped. Neither youth nor childhood is folly or incapacity. Some children are fools and so are some old men. But there is a vast majority on the side of imagination or spiritual sensation.

To engrave after another painter is infinitely more laborious than to engrave one's own inventions. And of the size you require my price has been thirty guineas and I cannot afford to do it for less. I had twelve for the head I sent you as a specimen, but after my own designs I could do at least six times the quantity of labour in the same time—which will account for the difference of price, as also that chalk engraving is at least six times as laborious as aquatint. I have no objection to engraving after another artist. Engraving is the profession I was apprenticed to and should never have attempted to live by anything else if orders had not come in for my designs and paintings—which I have the pleasure to tell you are increasing every day. Thus if I am a painter it is not to be attributed to seeking after. But I am contented whether I live by painting or engraving.

I am reverend Sir your obedient servant,
WILLIAM BLAKE

Letter of 26 August 1799

To George Cumberland

Dear Cumberland,
I ought long ago to have written to you to thank you for your kind

recommendation to Dr Trusler which, though it has failed of success, is not the less to be remembered by me with gratitude.

I have made him a drawing in my best manner. He sent it back with a letter full of criticisms, in which he says it accords not with his intentions—which are to reject all fancy from his work. How far he expects to please I cannot tell. But, as I cannot paint dirty rags and old shoes where I ought to place naked beauty or simple ornament, I despair of ever pleasing one class of men. Unfortunately our authors of books are among this class. How soon we shall have a change for the better I cannot prophesy. Dr Trusler says, '*Your Fancy* from what I have seen of it (and I have seen variety at Mr Cumberland's) seems to be in the other world, or the world of spirit, which accords not with my intentions which, whilst living in this world, wish to follow *the Nature of it*.' I could not help smiling at the difference between the doctrines of Dr Trusler and those of Christ.° But, however, for his own sake I am sorry that a man should be so enamoured of Rowlandson's caricatures as to call them copies from life and manners, or fit things for a clergyman to write upon.

Pray let me entreat you to persevere in your designing; it is the only source of pleasure. All your other pleasures depend upon it. It is the tree; your pleasures are the fruit. Your inventions of intellectual visions are the stamina of everything you value. Go on; if not for your own sake yet for ours who love and admire your work, but above all for the sake of the arts. Do not throw aside for any long time the honour intended you by Nature, to revive the Greek workmanship. I study your outlines° as usual just as if they were antiques.

As to myself, about whom you are so kindly interested, I live by miracle. I am painting small pictures from the Bible. For as to engraving, in which art I cannot reproach myself with any neglect, yet I am laid by in a corner as if I did not exist, and since my Young's *Night Thoughts*° have been published even Johnson° and Fuseli have discarded my graver. But, as I know that he who works and has his health cannot starve, I laugh at fortune and go on and on. I think I foresee better things than I have ever seen. My work pleases my employer,° and I have an order for fifty small pictures at one guinea each, which is something better than mere copying after another artist. But above all I feel myself happy and contented, let what will come. Having passed now near twenty years in ups and downs I am used to them, and perhaps a little practice in them may turn out to benefit. It is now exactly twenty years since I was upon the ocean of business° and, though I laugh at Fortune, I am persuaded that she alone is the governor of worldly riches, and when it is fit she will call on me. Till then I wait with patience, in hopes that she is busied among my friends.

With mine and my wife's best compliments to Mrs Cumberland, I remain,

Yours sincerely,
WILLIAM BLAKE

Letter of 2 July 1800

To George Cumberland

Dear Cumberland,

I have to congratulate you on your plan for a National Gallery being put into execution.° All your wishes shall in due time be fulfilled. The immense flood of Grecian light and glory which is coming on Europe will more than realize our warmest wishes. Your honours will be unbounded when your plan shall be carried into execution, as it must be if England continues a nation. I hear that it is now in the hands of ministers, that the King shows it great countenance and encourage-ment, that it will soon be before Parliament, and that it *must* be extended and enlarged to take in originals° both of painting and sculpture—by considering every valuable original that is brought into England or can be purchased abroad as its objects of acquisition. Such is the plan as I am told, and such must be the plan if England wishes to continue at all worth notice, as you have yourself observed—only now we must possess originals as well as France,° or be nothing.

Excuse I entreat you my not returning thanks at the proper moment for your kind present. No persuasion could make my stupid head believe that it was proper for me to trouble you with a letter of mere compliment and expression of thanks. I begin to emerge from a deep pit of melancholy, melancholy without any real reason for it—a disease which God keep you from and all good men. Our artists of all ranks praise your outlines and wish for more. Flaxman is very warm in your commendation, and more and more of a Grecian. Mr Hayley has lately mentioned your work on outline in 'Notes' to *An Essay on Sculpture in Six Epistles to John Flaxman.*° I have been too little among friends, which I fear they will not excuse and I know not how to apologize for. Poor Fuseli, sore from the lash of envious tongues, praises you and dispraises with the same breath. He is not naturally good-natured, but he is artificially very ill-natured; yet even from him I learn the estimation you are held in among artists and connoisseurs.

I am still employed in making designs and little pictures, with now and then an engraving, and find that in future to live will not be so difficult as it has been. It is very extraordinary that London in so few years, from a city of mere necessaries, or at least a commerce of the lowest order of luxuries, should have become a city of elegance in some degree, and that its once stupid inhabitants should enter into an emulation of Grecian manners. There are now I believe as many booksellers as there are butchers, and as many printshops as of any other trade. We remember when a printshop was a rare bird in London, and I myself remember when I thought my pursuits of art a kind of criminal dissipation and neglect of the main chance, which I hid my face for not being able to abandon—as a passion which is forbidden by law and religion. But now it appears to be law and gospel too; at least I hear so from the few friends I have dared to visit in my stupid melancholy. Excuse this communication of sentiments, which I felt necessary to my repose at this time. I feel very strongly that I neglect my duty to my friends, but it is not want of gratitude or friendship, but perhaps an excess of both. Let me hear of your welfare. Remember my and my wife's respectful compliments to Mrs Cumberland and family,

> And believe me to be for ever,
> Yours,
> WILLIAM BLAKE

To my Dearest Friend, John Flaxman (1800)

I bless thee, O Father of Heaven and Earth, that ever I saw Flaxman's face.

Angels stand round my spirit in Heaven; the blessed of Heaven are my friends upon earth.

When Flaxman was taken to Italy, Fuseli was given to me for a season,°

And now Flaxman hath given me Hayley, his friend to be mine. Such my lot upon Earth.

Now my lot in the Heavens is this: Milton loved me in childhood and showed me his face;

Ezra came with Isaiah the prophet, but Shakespeare in riper years gave me his hand;

Paracelsus and Behmen appeared to me; terrors appeared in the Heavens above

And in Hell beneath, and a mighty and awful change threatened the earth.

The American War began. All its dark horrors passed before my face Across the Atlantic to France. Then the French Revolution commenced in thick clouds.

And my angels have told me that seeing such visions I could not subsist on the earth

But by my conjunction with Flaxman, who knows to forgive nervous fear.

From the annotations to Henry Boyd's
A Translation of the Inferno of Dante Alighieri
(c.1800)

The grandest poetry is immoral, the grandest characters wicked, very Satan: Capaneus,° Othello (a murderer), Prometheus, Jupiter, Jehovah, Jesus (a wine-bibber).° Cunning and morality are not poetry, but philosophy. The poet is independent and wicked; the philosopher is dependent and good.

Poetry is to excuse vice, and show its reason and necessary purgation.

Letter of 11 September 1801

To Thomas Butts

My dear Sir,

I hope you will continue to excuse my want of steady perseverance (by which want I am still so much your debtor and you so much my creditor), but such as I can be I will. I can be grateful; and I can soon send you some of your designs which I have nearly completed. In the meantime by my sister's hands I transmit to Mrs Butts an attempt at your likeness, which I hope she who is the best judge will think like. Time flies faster (as seems to me) here than in London.° I labour incessantly, and accomplish not one half of what I intend because my abstract folly hurries me often away while I am at work, carrying me over mountains and valleys which are not real, in a land of abstraction where spectres of the dead wander. This I endeavour to prevent and with my whole might chain my feet to the world of duty and reality, but in vain! The faster I bind the better is the ballast, for I, so far from being bound down, take the world with me in my flights and often it seems lighter than a ball of wool rolled by the wind. Bacon and Newton would prescribe ways of making the world heavier to me, and Pitt would prescribe distress° for a medicinal potion, but as none on earth can give me mental distress,—and I know that all distress inflicted by Heaven is a mercy—a fig for all corporeal! Such distress is my mock and scorn. Alas, wretched, happy, ineffectual labourer of time's moments that I am! Who shall deliver me from this spirit of abstraction and improvidence? Such my dear Sir is the truth of my state, and I tell it you in palliation of my seeming neglect of your most pleasant orders. But I have not neglected them, and yet a year is rolled over and only now I approach the prospect of sending you some, which you may expect soon. I should have sent them by my sister, but as the coach goes three times a week to London and they will arrive as safe as with her, I shall have an opportunity of enclosing several together which are not yet completed. I thank you again and again for your generous forbearance of which I have need. And now I must express my wishes to see you at Felpham, and to show you Mr Hayley's library, which is still unfinished, but is in a finishing way and looks well. I ought also to mention my extreme disappointment at Mr Johnson's° forgetfulness, who appointed to call on you but did not. He is also a happy abstract, known by all his friends as the most innocent forgetter of his own interests. He is nephew to the late Mr Cowper the poet. You would like him much. I continue painting miniatures, and improve more and more as all my friends tell me, but my principal labour at this time is engraving plates for Cowper's *Life*: a work of magnitude which Mr Hayley is now labouring with all his matchless industry, and which will be a most valuable acquisition to literature—not only on account of Mr Hayley's composition but also as it will contain letters of Cowper to his friends, perhaps, or rather certainly, the very best letters that ever were published.

My wife joins with me in love to you and Mrs Butts, hoping that her joy is now increased, and yours also, in an increase of family and of health and happiness.

I remain dear Sir,
Ever yours sincerely,
WILLIAM BLAKE

Felpham Cottage,
of cottages the prettiest,
September 11 1801.

Next time I have the happiness to see you I am determined to paint another portrait of you from life in my best manner, for memory will not do in such minute operation. For I have now discovered that without Nature before the painter's eye he can never produce anything in the walks of natural painting. Historical designing° is one thing and portrait painting another, and they are as distinct as any two arts can be. Happy would that man be who could unite them.

From a letter of 10 January 1802

To Thomas Butts

... I find on all hands great objections to my doing anything but the mere drudgery of business, and intimations that if I do not confine myself to this I shall not live. This has always pursued me. ...

... that I cannot live without doing my duty to lay up treasures in Heaven is certain and determined, and to this I have long made up my mind; and why this should be made an objection to me, while drunkenness, lewdness, gluttony and even idleness itself does not hurt other men, let Satan himself explain. The thing I have most at heart— more than life or all that seems to make life comfortable without—is the interest of true religion and science, and whenever anything appears to affect that interest (especially if I myself omit any duty to my station as a soldier of Christ)° it gives me the greatest of torments. I am not ashamed, afraid, or averse to tell you what ought to be told: that I am under the direction of messengers from Heaven daily and nightly. But the nature of such things is not, as some suppose, without trouble or care. Temptations are on the right hand and left. Behind the sea of time and space roars and follows swiftly. He who keeps not right onward is lost, and if our footsteps slide in clay how can we do otherwise than fear and tremble? But I should not have troubled you with this account of my spiritual state unless it had been necessary in explaining the actual cause of my uneasiness, into which you are so kind as to enquire; for I never obtrude such things on others unless questioned, and then I never disguise the truth. But if we fear to do the dictates of our angels and tremble at the tasks set before us; if we refuse to do spiritual acts because of natural fears or natural desires! Who can describe the dismal torments of such a state? I too well remember the threats I heard! 'If you who are organized by Divine Providence for spiritual communion refuse, and bury your talent in the earth even though you should want natural bread, sorrow and desperation pursues you through life, and after death shame and confusion of face to eternity! Everyone in eternity will leave you, aghast at the man who was crowned with glory and honour° by his brethren, and betrayed their cause to their enemies. You will be called the base Judas who betrayed his friend!' Such words would make any stout man tremble, and how then could I be at ease? But I am now no longer in that state, and now go on again with my task fearless and though my path is difficult. I have no fear of stumbling while I keep it.

Letter of 22 November 1802

To Thomas Butts

Dear Sir

My brother° tells me that he fears you are offended with me. I fear so too because there appears some reason why you might be so. But when you have heard me out you will not be so.

I have now given two years to the intense study of those parts of the art which relate to light and shade and colour, and am convinced that either my understanding is incapable of comprehending the beauties of colouring, or the pictures which I painted for you are equal in every part of the art, and superior in one, to anything that has been done since the age of Rafael. All Sir J. Reynolds's discourses to the Royal Academy will show, that the Venetian finesse in art can never be united with the majesty of colouring necessary to historical beauty, and in a letter° to the Rev. Mr Gilpin, author of a work on picturesque scenery, he says thus: 'It may be worth consideration whether the epithet picturesque is not applicable to the excellencies of the inferior schools rather than to the higher. The works of Michelangelo, Rafael, etc. appear to me to have nothing of it; whereas Rubens and the Venetian painters may almost be said to have nothing else. Perhaps "picturesque" is somewhat synonymous to the word "taste", which we should think improperly applied to Homer or Milton, but very well to Prior or Pope. I suspect that the application of these words are to excellencies of an inferior order, and which are incompatible with the grand style. You are certainly right in saying that variety of tints and forms is picturesque, but it must be remembered on the other hand that the reverse of this ('*uniformity of colour*' and a '*long continuation of lines*') produces grandeur'. So says Sir Joshua, and so say I, for I have now proved that the parts of the art which I neglected to display in those little pictures and drawings which I had the pleasure and profit to do for you, are incompatible with the designs. There is nothing in the art which our painters do that I can confess myself ignorant of. I also know and understand, and can assuredly affirm, that the works I have done for you are equal to Carracci° or Rafael (and I am now seven years older than Rafael was when he died). I say they are equal to Carracci or Rafael, or else I am blind, stupid, ignorant, and incapable in two years' study to understand those things which a boarding school miss can comprehend in a fortnight. Be assured, my dear friend, that there is not one touch in those drawings and pictures but what came from my head and my heart in unison, that I am proud of being their author and grateful to you my employer, and that I look upon you as the chief of my friends—whom I would endeavour to please because you among all men have enabled me to produce these things. I would not send you a drawing or a picture till I had again reconsidered my notions of art, and had put myself back as if I was a learner. I have proved that I am right, and shall now go on with the vigour I was in my childhood famous for.

But I do not pretend to be perfect—but if my works have faults Carracci, Correggio, and Rafael's have faults also. Let me observe that the yellow leather flesh of old men, the ill-drawn and ugly young women, and above all the daubed black and yellow shadows that are found in most fine, aye and the finest, pictures I altogether reject as ruinous to effect, though connoisseurs may think otherwise.

Let me also notice that Carracci's pictures are not like Correggio's, nor Correggio's like Rafael's, and if neither of them was to be encouraged till he did like any of the others he must die without encouragement. My pictures are unlike any of these painters, and I would have them to be so. I think the manner I adopt more perfect than any other; no doubt they thought the same of theirs.

You will be tempted to think that as I improve the pictures etc. that I did for you are not what I would now wish them to be. On this I beg to say that they are what I intended them, and that I know I never shall do better; for if I was to do them over again they would lose as much as they gained, because they were done in the heat of my spirits.

But you will justly enquire: why I have not written all this time to you? I answer I have been very unhappy, and could not think of troubling you about it or any of my real friends (I have written many letters to you which I burned and did not send)—and why I have not before now finished the miniature I promised to Mrs Butts? I answer I have not till now in any degree pleased myself, and now I must entreat you to excuse faults—for portrait painting is the direct contrary to designing and historical painting in every respect. If you have not Nature before you for every touch you cannot paint portraits, and if you have Nature before you at all you cannot paint history. It was Michel-angelo's opinion° and is mine. Pray give my wife's love, with mine, to Mrs Butts. Assure her that it cannot be long before I have the pleasure of painting from you in person, and then that she may expect a likeness, but now I have done all I could and know she will forgive any failure in consideration of the endeavour.

And now let me finish with assuring you that though I have been very unhappy I am so no longer. I am again emerged into the light of day. I still and shall to eternity embrace Christianity, and adore him who is the express image° of God, but I have travelled through perils and darkness not unlike a champion.° I have conquered and shall still go on conquering. Nothing can withstand the fury of my course among the stars of God,° and in the abysses of the accuser. My enthusiasm° is still what it was, only enlarged and confirmed.

I now send two pictures, and hope you will approve of them. I have enclosed the account of money received and work done, which I ought long ago to have sent you. Pray forgive errors and omissions of this kind. I am incapable of many attentions which it is my duty to observe towards you through multitude of employment, and through hope of soon seeing you again. I often omit to inquire of you, but pray let me now hear how you do and of the welfare of your family.

Accept my sincere love and respect.

I remain, yours sincerely,
WILLIAM BLAKE

A piece of seaweed serves for a barometer; it gets wet and dry as the weather gets so.

From a letter of 25 April 1803

To Thomas Butts

. . . Now I may say to you what perhaps I should not dare to say to any one else: that I can alone carry on my visionary studies in London unannoyed, and that I may converse with my friends in eternity, see visions, dream dreams, and prophesy and speak parables, unobserved and at liberty from the doubts of other mortals—perhaps doubts proceeding from kindness, but doubts are always pernicious, especially when we doubt our friends. Christ is very decided on this point: 'He who is not with me is against me.'° There is no medium or middle state, and if a man is the enemy of my spiritual life while he pretends to be the friend of my corporeal he is a real enemy. But the man may be the friend of my spiritual life while he seems the enemy of my corporeal— but not *vice versa*.

What is very pleasant, everyone who hears of my going to London again applauds it as the only course for the interest of all concerned in my works, observing that I ought not to be away from the opportunities London affords of seeing fine pictures, and the various improvements in works of art going on in London.

But none can know the spiritual acts of my three years slumber on the banks of the ocean° unless he has seen them in the spirit—or unless he should read my long poem° descriptive of those acts. For I have in these three years composed an immense number of verses on one grand theme, similar to Homer's *Iliad* or *Paradise Lost*, the persons and machinery° entirely new to the inhabitants of Earth (some of the persons excepted). I have written this poem from immediate dictation, twelve or sometimes twenty or thirty lines at a time, without premeditation° and even against my will. The time it has taken in writing was thus rendered non-existent, and an immense poem exists which seems to be the labour of a long life—all produced without labour or study. I mention this to show you what I think the grand reason of my being brought down here. . . .

Preface *to* Milton (*1804*)

The stolen and perverted° writings of Homer and Ovid, of Plato and Cicero, which all men ought to contemn, are set up by artifice against the sublime of the Bible. But when the new age is at leisure to pronounce all will be set right, and these grand works of the more ancient, and consciously and professedly inspired men will hold their proper rank, and the Daughters of Memory shall become the Daughters of Inspiration. Shakespeare and Milton were both curbed by the general malady and infection from the silly Greek and Latin slaves of the sword. Rouse up, O young men of the new age! Set your foreheads against the ignorant hirelings! For we have hirelings in the camp, the court and the university who would, if they could, for ever depress mental and prolong corporeal war. Painters, on you I call! Sculptors! Architects! Suffer not the fashionable fools to depress your powers by the prices they pretend to give for contemptible works, or the expensive advertising boasts that they make of such works. Believe Christ and his apostles that there is a class of men whose whole delight is in destroying.° We do not want either Greek or Roman models, if we are but just and true to our own imaginations, those worlds of eternity in which we shall live for ever, in Jesus our Lord.

And did those feet in ancient time [*see below, p. 297*]

'Would to God that all the Lord's people were prophets!' *Numbers* xi 29.

Preface to Chapter I of Jerusalem (*1804–18*)

TO THE PUBLIC

After my three years' slumber on the banks of the ocean, I again display my giant forms to the public. My former giants and fairies having received the highest reward possible, the love and friendship of those with whom to be connected is to be blessed, I cannot doubt that this more consolidated° and extended work will be as kindly received.

The enthusiasm of the following poem the author hopes no reader will think presumptuousness or arrogance, when he is reminded that the ancients entrusted their love to their writing to the full as enthusiastically as I have who acknowledge mine for my Saviour and Lord, for they were wholly absorbed in their gods. I also hope the reader will be with me, wholly one in Jesus our Lord, who is the God of Fire° and Lord of Love to whom the ancients looked, and saw his day afar off with trembling and amazement.

The spirit of Jesus is continual forgiveness of sin. He who waits to be righteous before he enters into the Saviour's kingdom, the divine body, will never enter there. I am perhaps the most sinful of men; I pretend not to holiness. Yet I pretend to love, to see, to converse with daily, as man with man and, the more, to have an interest in the Friend of Sinners. Therefore dear reader, forgive what you do not approve, and love me for this energetic exertion of my talent.

Reader, lover of books, lover of Heaven,
And of that God from whom all books are given,
Who in mysterious Sinai's awful cave
To Man the wondrous art of writing gave.
Again he speaks in thunder and in fire:
Thunder of thought, and flames of fierce desire.
Even from the depths of Hell his voice I hear,
Within the unfathomed caverns of my ear.
Therefore I print, nor vain my types° shall be;
Heaven, Earth and Hell henceforth shall live in harmony.

OF THE MEASURE IN WHICH THE FOLLOWING POEM IS WRITTEN

We who dwell on earth can do nothing of ourselves; everything is
conducted by spirits, no less than digestion or sleep. To note the last
words of Jesus, Ἐδόθη μοι πᾶσα ἐξουσία ἐν οὐρανῷ καὶ ἐπὶ γῆς.°

When this verse was first dictated to me° I considered a monotonous
cadence like that used by Milton and Shakespeare and all writers of
English blank verse, derived from the modern bondage of rhyming,° to
be a necessary and indispensable part of verse. But I soon found that in
the mouth of a true orator such monotony was not only awkward, but as
much a bondage as rhyme itself. I therefore have produced a variety in
every line, both of cadences and number of syllables. Every word and
every letter is studied and put into its fit place: the terrific° numbers°
are reserved for the terrific parts, the mild and gentle for the mild and
gentle parts, and the prosaic for inferior parts. All are necessary to each
other. Poetry fettered, fetters the human race. Nations are destroyed,
or flourish, in proportion as their poetry, painting and music are
destroyed or flourish. The primeval state of Man was wisdom, art and
science.

From the annotations to Volume 1 of Sir Joshua Reynolds's Works (mainly c.*1808*)

This man was hired to depress art.

This is the opinion of Will. Blake. My proofs of this opinion are
given in the following notes.

Advice of the popes who succeeded the age of Rafael:

> Degrade first the arts if you'd mankind degrade;
> Hire idiots to paint with cold light and hot shade;
> Give high price for the worst, leave the best in disgrace;
> And with labours of ignorance fill every place.

Having spent the vigour of my youth and genius under the oppression of Sir Joshua and his gang of cunning hired knaves—without employment and as much as could possibly be without bread—the reader° must expect to read in all my remarks on these books nothing but indignation and resentment. While Sir Joshua was rolling in riches Barry° was poor and unemployed (except by his own energy), Mortimer° was called a madman, and only portrait painting applauded and rewarded by the rich and great. Reynolds and Gainsborough blotted and blurred one against the other, and divided all the English world between them. Fuseli, indignant, almost hid himself. I am hid.

The arts and sciences are the destruction of tyrannies or bad governments. Why should good government endeavour to repress what is its chief and only support?

The foundation of empire° is art and science. Remove them or degrade them and the empire is no more. Empire follows art and not *vice versa* as Englishmen suppose.

'On peut dire° que le Pape Léon Xme° en encourageant les études donna des armes contre lui-même. J'ai oui dire à un Seigneur Anglais, qu'il avait vû une lettre du Seigneur Polus, ou de la Pole, depuis Cardinal,° à ce Pape, dans laquelle, en le félicitant sur ce qu'il etendait le progrès des Sciences en Europe, il l'avertissait *qu'il était dangereux de rendre les hommes trop savans.*'

Voltaire, Moeurs des Nations, Tome 4.

O Englishmen! why are you still of this foolish cardinal's opinion?

Who will dare to say that polite art is encouraged, or either wished or tolerated, in a nation where The Society for the Encouragement of Art suffered Barry to give them his labour for nothing°—a society composed of the flower of the English nobility and gentry—a society suffering an artist to starve while he supported really what they, under pretence of encouraging, were endeavouring to depress. Barry told me that while he did that work he lived on bread and apples.

O Society for Encouragement of Art! O King and nobility of England! Where have you hid Fuseli's Milton?° Is Satan troubled at his exposure?

Let not that nation where less than nobility is the reward pretend that art is encouraged by that nation. Art is first in intellectuals° and ought to be first in nations.

Invention depends altogether upon execution or organization. As that is right or wrong so is the invention perfect or imperfect. Whoever is set to undermine the execution of art is set to destroy art. Michelangelo's art depends on Michelangelo's execution altogether.

I am happy I cannot say that Rafael ever was from my earliest childhood hidden from me. I saw and I knew immediately the difference between Rafael and Rubens.

> Some look to see the sweet outlines
> And beauteous forms that love does wear.
> Some look to find out patches, paint,
> Bracelets, and stays, and powdered hair.

I was once looking over the prints from Rafael and Michelangelo in the library of the Royal Academy. Moser° came to me and said, 'You should not study these old, hard, stiff and dry unfinished works of art. Stay a little and I will show you what you should study.' He then went and took down Le Brun's° and Rubens' galleries.° How I did secretly rage! I also spoke my mind . . . I said to Moser: 'These things that you call finished are not even begun. How can they then be finished? The man who does not know the beginning, never can know the end of art.'

To generalize is to be an idiot. To particularize is the alone distinction of merit. General knowledges are those knowledges that idiots possess.

I consider Reynolds's discourses to the Royal Academy as the simulations of the hypocrite who smiles particularly where he means to betray. His praise of Rafael is like the hysteric smile of revenge—his softness and candour° the hidden trap and poisoned feast. He praises Michelangelo° for qualities which Michelangelo abhorred; and he blames Rafael° for the only qualities which Rafael valued. Whether Reynolds knew what he was doing, is nothing to me; the mischief is just the same whether a man does it ignorantly or knowingly. I always considered true art and true artists to be particularly insulted and degraded by the reputation of these discourses, as much as they were degraded by the reputation of Reynolds's paintings, and that such artists as Reynolds are at all times hired by the Satans, for the depression of art. A pretence of art, to destroy art.

The neglect of Fuseli's Milton in a country pretending to the encouragement of art is a sufficient apology for my vigorous indignation, if indeed the neglect of my own powers had not been. Ought not the patrons and employers of fools to be execrated in future ages? They will and shall. Foolish men! your own real greatness depends on your encouragement of the arts, and your fall will depend on their neglect and depression. What you fear is your true interest. Leo X was advised not to encourage the arts; he was too wise to take this advice.

The rich men of England form themselves into a society° to sell and not to buy pictures. The artist who does not throw his contempt on such trading exhibitions does not know either his own interest or his duty.

> When nations grow old the arts grow cold,
> And commerce settles on every tree,
> And the poor and the old can live upon gold—
> For all are born poor, aged sixty-three.

Reynolds's opinion was that genius may be taught, and that all pretence to inspiration is a lie and a deceit, to say the least of it. For if it is a deceit the whole Bible is madness. This opinion originates in the Greeks calling the muses Daughters of Memory.

The enquiry in England is not whether a man has talent and genius, but whether he is passive, and polite, and a virtuous ass, and obedient to noblemen's opinions in art and science. If he is he is a good man. If not he must be starved.

Minute discrimination° is not accidental. All sublimity is founded on minute discrimination.

I do not believe that Rafael taught Michelangelo, or that Michelangelo taught Rafael, any more than I believe that the rose teaches the lily how to grow, or the apple tree teaches the pear tree how to bear fruit. I do not believe the tales of anecdote writers when they militate against individual character.

The lives of painters say that Rafael died of dissipation. Idleness is one thing and dissipation another. He who has nothing to dissipate cannot dissipate. The weak man may be virtuous enough but will never be an artist.

Painters are noted for being dissipated and wild.

The difference between a bad artist and a good one is the bad artist seems to copy a great deal. The good one really does copy a great deal.

Servile copying is the great merit of copying.

Every eye sees differently. As the eye such the object.

Mere enthusiasm° is the all in all!—Bacon's philosophy has ruined England. Bacon is only Epicurus° over again.

The man who asserts that there is no such thing as softness in art, and that everything in art is definite and determinate, has not been told this by practice but by inspiration and vision—because vision is determinate and perfect and he copies that without fatigue, everything being definite and determinate. Softness is produced alone by comparative strength and weakness in the marking out of the forms. I say these principles could never be found out by the study of Nature with con- or innate science.°

The following discourse° is particularly interesting to blockheads, as it endeavours to prove that there is no such thing as inspiration, and that any man of a plain understanding may by thieving from others become a Michelangelo.

Without minute neatness of execution the sublime cannot exist! Grandeur of ideas is founded on precision of ideas.

It is evident that Reynolds wished none but fools to be in the arts, and in order to this he calls all others vague enthusiasts or madmen.

What has reasoning to do with the art of painting?

Knowledge of ideal beauty is not to be acquired. It is born with us. Innate ideas are in every man.° Born with him, they are truly himself. The man who says that we have no innate ideas must be a fool and knave, having no con-science or innate science.

All forms are perfect in the poet's mind, but these are not abstracted nor compounded from Nature, but are from imagination.

What is general nature? Is there such a thing? What is general knowledge? Is there such a thing? All knowledge is particular.

Distinct general form cannot exist. Distinctness is particular not general.°

All but names of persons and places is invention both in poetry and painting.

Why should Titian and the Venetians be named in a discourse on art? Such idiots are not artists.

> Venetian! all thy colouring is no more
> That bolstered° plasters on a crooked whore.

A history painter paints the hero, and not Man in general, but most minutely in particular.

Gainsborough told a gentleman of rank and fortune that the worst painters always chose the grandest subjects. I desired the gentleman to set Gainsborough about one of Rafael's grandest subjects, namely, Christ delivering the keys to St Peter, and he would find that in Gainsborough's hands it would be a vulgar subject of poor fishermen and a journeyman carpenter.

Passion and expression is beauty itself. The face that is incapable of passion and expression is deformity itself. Let it be painted, and patched, and praised, and advertised for ever, it will only be admired by fools.

The ancients were chiefly attentive to complicated and minute discrimination of character; it is the whole of art.

The great style is always novel or new in all its operations.

Original and characteristical are the two grand merits of the great style.

To my eye Rubens's colouring is most contemptible. His shadows are of a filthy brown somewhat of the colour of excrement; these are filled with tints and messes of yellow and red. His lights° are all the colours of the rainbow laid on indiscriminately and broken one into another. Altogether his colouring is contrary to the colouring of real art and science.

Opposed to Rubens's colouring Sir Joshua has placed Poussin, but he ought to put all men of genius who ever painted. Rubens and the Venetians are opposite in everything to true art and they meant to be so; they were hired for this purpose.

When a man talks of acquiring invention, and of learning how to produce original conception, he must expect to be called a fool by men of understanding. But such a hired knave cares not for the few. His eye is on the many, or rather the money.

How ridiculous it would be to see the sheep endeavouring to walk like the dog, or the ox striving to trot like the horse; just as ridiculous it is to see one man striving to imitate another. Man varies from man more than animal from animal of different species.

If art was progressive we should have had Michelangelos and Rafaels to succeed and to improve upon each other. But it is not so. Genius dies with its possessor and comes not again till another is born with it.

Reynolds thinks that Man learns all that he knows. I say on the contrary that Man brings all that he has or can have into the world with him. Man is born like a garden ready planted and sown. This world is too poor to produce one seed.

He who can be bound down is no genius. Genius cannot be bound. It may be rendered indignant and outrageous:° 'Oppression makes the wise man mad.'°

The purpose of the following discourse° is to prove that taste and genius are not of heavenly origin, and that all who have supposed that they are so are to be considered as weak-headed fanatics. The obligations Reynolds has laid on bad artists of all classes will at all times make them his admirers, but most especially for this discourse: in which it is proved that the stupid are born with faculties equal to other men—only they have not cultivated them because they thought it not worth the trouble.

Obscurity is neither the source of the sublime nor of anything else.

The ancients did not mean to impose when they affirmed their belief in vision and revelation. Plato was in earnest. Milton was in earnest. They believed that God did visit Man really and truly, and not as Reynolds pretends.°

How very anxious Reynolds is to disprove and contemn spiritual perception!

It is not in terms that Reynolds and I disagree. Two contrary opinions can never by any language be made alike. I say taste and genius are not teachable or acquirable but are born with us; Reynolds says the contrary.

God forbid that truth should be confined to mathematical demonstration.

He who does not know truth at sight is unworthy of her notice.

Burke's treatise on the Sublime and Beautiful° is founded on the opinions of Newton and Locke. On this treatise Reynolds has grounded many of his assertions in all his discourses. I read Burke's treatise when very young. At the same time I read Locke on Human Understanding and Bacon's *Advancement of Learning*. On every one of these books I wrote my opinions, and on looking them over find that my notes on Reynolds in this book are exactly similar. I felt the same contempt and abhorrence then that I do now. They mock inspiration and vision. Inspiration and vision was then, and now is, and I hope will always remain my element, my eternal dwelling place. How can I then hear it contemned without returning scorn for scorn?

Rembrandt was a generalizer; Poussin was a particularizer.

If you endeavour to please the worst you will never please the best. To please all is impossible.

From A Descriptive Catalogue of Pictures (1809)

PREFACE

The eye that can prefer the colouring of Titian and Rubens to that of Michelangelo and Rafael, ought to be modest and to doubt its own powers. Connoisseurs talk as if Rafael and Michelangelo had never seen the colouring of Titian or Correggio. They ought to know that Correggio was born two years before Michelangelo, and Titian but four years after.° Both Rafael and Michelangelo knew the Venetian,° and contemned and rejected all he did with the utmost disdain, as that which is fabricated for the purpose to destroy art.

Mr. B. appeals to the public, from the judgement of those narrow blinking eyes that have too long governed art in a dark corner. The eyes of stupid cunning never will be pleased with the work any more than with the look of self-devoting genius. The quarrel of the Florentine with the Venetian is not because he does not understand drawing, but because he does not understand colouring. How should he who does not know how to draw a hand or a foot, know how to colour it?

Colouring does not depend on where the colours are put, but on where the lights and darks are put, and all depends on form or outline, on where that is put. Where that is wrong, the colouring never can be right; and it is always wrong in Titian and Correggio, Rubens and Rembrandt. Till we get rid of Titian and Correggio, Rubens and Rembrandt, we never shall equal Rafael and Albert Durer, Michelangelo, and Julio Romano.

Clearness and precision have been the chief objects in painting these pictures: clear colours unmudded by oil,° and firm and determinate lineaments unbroken by shadows, which ought to display and not to hide form, as is the practice of the latter schools of Italy and Flanders.

This picture° also is a proof of the power of colours unsullied with oil or with any cloggy vehicle. Oil has falsely been supposed to give strength to colours, but a little consideration must show the fallacy of this opinion. Oil will not drink or absorb colour enough to stand the test of very little time and of the air. It deadens every colour it is mixed with, at its first mixture, and in a little time becomes a yellow mask over all that it touches. Let the works of modern artists since Rubens's time witness the villainy of someone at that time, who first brought oil painting into general opinion and practice°—since which we have never had a picture painted that could show itself by the side of an earlier production. Whether Rubens or Vandyke, or both, were guilty of this villainy is to be enquired in another work on painting,° and who first forged the silly story and known falsehood about John of Bruges° inventing oil colours. In the mean time let it be observed that before Vandyke's time, and in his time, all the genuine pictures are on plaster or whiting grounds° and none since.

The two pictures of Nelson and Pitt are compositions of a mythological cast, similar to those apotheoses° of Persian, Hindu and Egyptian antiquity, which are still preserved on rude monuments, being copies from some stupendous originals now lost or perhaps buried till some happier age. The artist having been taken in vision into the ancient republics, monarchies and patriarchates of Asia, has seen those wonderful originals called in the sacred scriptures the Cherubim,° which were sculptured and painted on walls of temples, towers, cities, palaces, and erected in the highly cultivated states of Egypt, Moab, Edom, Aram,° among the rivers of Paradise—being originals from which the Greeks and Etrurians copied Hercules Farnese, Venus of the Medicis, Apollo Belvedere, and all the grand works of ancient art. They were executed in a very superior style to those justly admired copies, being with their accompaniments terrific and grand in the highest degree. The artist has endeavoured to emulate the grandeur of those seen in his vision, and to apply it to modern heroes, on a smaller scale.

No man can believe that either Homer's mythology, or Ovid's, were the production of Greece, or of Latium. Neither will anyone believe that the Greek statues, as they are called, were the invention of Greek artists. Perhaps the Torso° is the only original work remaining; all the rest are evidently copies, though fine ones, from greater works of the Asiatic patriarchs. The Greek muses are daughters of Mnemosyne, or memory, and not of inspiration or imagination, therefore not authors of such sublime conceptions. Those wonderful originals seen in my visions were some of them one hundred feet in height; some were painted as pictures, and some carved as basso relievos, and some as groups of statues, all containing mythological and recondite meaning, where more is meant than meets the eye. The artist wishes it was now the fashion to make such monuments, and then he should not doubt of having a national commission to execute these two pictures on a scale that is suitable to the grandeur of the nation, who is the parent of his heroes, in high-finished fresco, where the colours would be as pure and as permanent as precious stones though the figures were one hundred feet in height.

All frescos are as high-finished as miniatures or enamels, and they are known to be unchangeable; but oil, being a body itself, will drink or absorb very little colour, and changing yellow, and at length brown, destroys every colour it is mixed with, especially every delicate colour. It turns every permanent white to a yellow and brown putty, and has compelled the use of that destroyer of colour, white lead, which, when its protecting oil is evaporated, will become lead again. This is an awful thing to say to oil painters; they may call it madness, but it is true. All the genuine old little pictures, called cabinet pictures, are in fresco and not in oil. Oil was not used, except by blundering ignorance, till after Vandyke's time; but the art of fresco painting being lost, oil became a fetter to genius, and a dungeon to art. But one convincing proof among many others that these assertions are true is that real gold and silver cannot be used with oil, as they are in all the old pictures and in Mr. B.'s frescos.

Sir Jeffery Chaucer and the nine and twenty pilgrims on their journey to Canterbury.°

The time chosen is early morning,° before sunrise, when the jolly company are just quitting the Tabarde Inn. The Knight and Squire with the Squire's Yeoman lead the procession. Next follow the youthful° Abbess, her nun and three priests; her greyhounds attend her:

> Of small hounds had she that she fed
> With roast flesh, milk and wastel bread.

Next follow the Friar and Monk; then the Tapiser, the Pardoner, and the Somner and Manciple. After these 'Our host', who occupies the centre of the cavalcade, directs them to the Knight as the person who would be likely to commence their task of each telling a tale in their order. After the Host follow the Shipman, the Haberdasher, the Dyer, the Franklin, the Physician, the Plowman, the Lawyer, the poor Parson, the Merchant, the Wife of Bath, the Miller, the Cook, the Oxford Scholar, Chaucer himself, and the Reeve comes as Chaucer has described:

> And ever he rode hinderest of the rout.

These last are issuing from the gateway of the inn; the Cook and the Wife of Bath are both taking their morning's draught of comfort. Spectators stand at the gateway of the inn, and are composed of an old man, a woman and children.

The landscape is an eastward view of the country, from the Tabarde Inn in Southwark, as it may be supposed to have appeared in Chaucer's time, interspersed with cottages and villages. The first beams of the sun are seen above the horizon. Some buildings and spires indicate the situation of the great city. The inn is a gothic building, which Thynne in his glossary° says was the lodging of the Abbot of Hyde by Winchester. On the inn is inscribed its title, and a proper advantage is taken of this circumstance to describe the subject of the picture. The words written over the gateway of the inn are as follows: 'The Tabarde Inn, by Henry Baillie, the lodgynge-house for Pilgrims, who journey to Saint Thomas's Shrine at Canterbury.'

The characters of Chaucer's pilgrims are the characters which compose all ages and nations. As one age falls another rises, different to mortal sight, but to immortals only the same; for we see the same characters repeated again and again, in animals, vegetables, minerals, and in men. Nothing new occurs in identical existence.° Accident ever varies; Substance can never suffer change nor decay.

Of Chaucer's characters, as described in his Canterbury Tales, some of the names or titles are altered by time, but the characters themselves for ever remain unaltered, and consequently they are the physiognomies or lineaments of universal human life, beyond which Nature never steps. Names alter; things never alter. I have known multitudes of those who would have been monks in the age of monkery, who in this deistical age are deists. As Newton numbered the stars, and as Linneus numbered the plants, so Chaucer numbered the classes of men.

The painter has consequently varied the heads and forms of his personages into all Nature's varieties. The horses he has also varied to accord to their riders. The costume is correct according to authentic monuments.

The Knight and Squire with the Squire's Yeoman lead the procession, as Chaucer has also placed them first in his prologue. The Knight is a true hero, a good, great and wise man; his whole length portrait on horseback, as written by Chaucer, cannot be surpassed. He has spent his life in the field, has ever been a conqueror, and is that species of character which in every age stands as the guardian of Man against the oppressor. His son is like him, with the germ of perhaps greater perfection still: as he blends literature and the arts with his warlike studies. Their dress and their horses are of the first rate, without ostentation, and with all the true grandeur that unaffected simplicity when in high rank always displays. The Squire's Yeoman is also a great character, a man perfectly knowing in his profession:

> And in his hand he bare a mighty bow.

Chaucer describes here a mighty man: one who in war is the worthy attendant on noble heroes.

The Prioress follows these with her female chaplain:

> Another Nonne also with her had she,
> That was her Chaplaine and Priests three.

This lady is described also as of the first rank, rich and honoured. She has certain peculiarities and little delicate affectations, not unbecoming in her, being accompanied with what is truly grand and really polite. Her person and face, Chaucer has described with minuteness; it is very elegant, as was the beauty of our ancestors, till after Elizabeth's time, when voluptuousness and folly began to be accounted beautiful.

Her companion and her three priests were no doubt all perfectly delineated in those parts of Chaucer's work which are now lost;° we ought to suppose them suitable attendants on rank and fashion.

The Monk follows these with the Friar. The painter has also grouped with these the Pardoner and the Sompnour and the Manciple, and has here also introduced one of the rich citizens of London°— characters likely to ride in company, all being above the common rank in life, or attendants on those who were so.

For the Monk is described by Chaucer as a man of the first rank in society: noble, rich, and expensively attended.° He is a leader of the age, with certain humorous accompaniments in his character that do not degrade, but render him an object of dignified mirth—but also with other accompaniments not so respectable.

The Friar is a character also of a mixed kind:

> A friar there was, a wanton and a merry.

But in his office he is said to be a 'full solemn man'. Eloquent, amorous, witty, and satirical, young, handsome, and rich, he is a complete rogue—with constitutional gaiety enough to make him a master of all the pleasures of the world:

> His neck was white as the flour de lis,
> Thereto strong he was as a champioun.

It is necessary here to speak of Chaucer's own character, that I may set certain mistaken critics right in their conception of the humour and fun that occurs on the journey. Chaucer is himself the great poetical observer of men, who in every age is born to record and eternize its acts. This he does as a master, as a father, and superior, who looks down on their little follies, from the emperor to the miller—sometimes with severity, oftener with joke and sport.

Accordingly Chaucer has made his Monk a great tragedian, one who studied poetical art. So much so, that the generous Knight is, in the compassionate dictates of his soul, compelled to cry out:

> Ho quoth the Knyght, good Sir, no more of this,
> That ye have said, is right ynough I wis;
> And mokell more, for little heaviness,
> Is right enough for much folk as I guesse.
> I say for me, it is a great disease,
> Whereas men have been in wealth and ease;
> To heare of their sudden fall alas,
> And the contrary is joy and solas.

The Monk's definition of tragedy in the proem to his tale is worth repeating:

> Tragedie is to tell a certain story,
> As old books us maken memory;
> Of hem that stood in great prosperity.
> And be fallen out of high degree,
> Into miserie and ended wretchedly.

Though a man of luxury, pride and pleasure, he is a master of art and learning, though affecting to despise it. Those who can think that the proud huntsman, and noble housekeeper,° Chaucer's Monk, is intended for a buffoon or burlesque character, know little of Chaucer.

For the Host who follows this group, and holds the centre of the cavalcade, is a first rate character, and his jokes are no trifles; they are always—though uttered with audacity, and equally free with the lord and the peasant—they are always substantially and weightily express-ive of knowledge and experience. Henry Baillie, the keeper of the greatest inn, of the greatest city (for such was the Tabarde Inn in Southwark, near London)—our Host was also a leader of the age.

By way of illustration, I instance Shakespeare's witches in *Macbeth*. Those who dress them for the stage consider them as wretched old women and not, as Shakespeare intended, the Goddesses of Destiny.° This shows how Chaucer has been misunderstood in his sublime work. Shakespeare's fairies also are the rulers of the vegetable world, and so are Chaucer's.° Let them be so considered, and then the poet will be understood, and not else.

But I have omitted to speak of a very prominent character, the Pardoner, the age's knave, who always commands and domineers over the high and low vulgar. This man is sent in every age for a rod and scourge, and for a blight, for a trial of men, to divide the classes of men; he is in the most holy sanctuary,° and he is suffered by Providence for wise ends, and has also his great use, and his grand leading destiny.

His companion, the Sompnour, is also a devil of the first magnitude, grand, terrific, rich and honoured in the rank of which he holds the destiny. The uses to society are perhaps equal of the devil and of the angel; their sublimity who can dispute?

> In daunger had he at his own gise,
> The young girls of his diocese,
> And he knew well their counsel, etc.

The principal figure in the next group is the Good Parson, an apostle, a real messenger of Heaven,° sent in every age for its light and its warmth. This man is beloved and venerated by all, and neglected by all. He serves all, and is served by none. He is, according to Christ's definition, the greatest of his age.° Yet he is a poor parson of a town. Read Chaucer's description of the Good Parson, and bow the head and the knee to him who in every age sends us such a burning and a shining light. Search, O ye rich and powerful! for these men and obey their counsel, then shall the golden age return. But alas! you will not easily distinguish him from the Friar or the Pardoner; they also are 'full solemn men' and their counsel you will continue to follow.

I have placed by his side the Sergeant at Lawe, who appears delighted to ride in his company, and between him and his brother, the Plowman; as I wish men of law would always ride with them, and take their counsel, especially in all difficult points. Chaucer's lawyer is a character of great venerableness, a judge, and a real master of the jurisprudence of his age.

The Doctor of Physic is in this group, and the Franklin, the voluptuous country gentleman, contrasted with the Physician and, on his other hand, with two citizens of London.° Chaucer's characters live age after age. Every age is a Canterbury Pilgrimage. We all pass on, each sustaining one or other of these characters; nor can a child be born, who is not one of these characters of Chaucer. The Doctor of Physic is described as the first of his profession, perfect, learned, completely master and doctor in his art. Thus the reader will observe that Chaucer makes every one of his characters perfect in his kind; every one is an antique statue, the image of a class, and not of an imperfect individual.

This group also would furnish substantial matter, on which volumes might be written. The Franklin is one who keeps open table, who is the genius of eating and drinking, the Bacchus; as the Doctor of Physic is the Esculapius, the Host is the Silenus, the Squire is the Apollo, the Miller is the Hercules etc. Chaucer's characters are a description of the eternal principles that exist in all ages. The Franklin is voluptuousness itself most nobly portrayed:

> It snewed in his house of meat and drink.

The Plowman is simplicity itself, with wisdom and strength for its stamina. Chaucer has divided the ancient character of Hercules between his Miller and his Plowman. Benevolence is the plowman's great characteristic. He is thin with excessive labour, and not with old age, as some have supposed:

> He would thresh and thereto dike and delve
> For Christe's sake, for every poore wight,
> Withouten hire, if it lay in his might.

Visions of these eternal principles or characters of human life appear to poets in all ages. The Grecian gods were the ancient Cherubim of Phoenicia; but the Greeks, and since them the Moderns, have neglected to subdue the gods of Priam.° These gods are visions of the eternal attributes, or divine names, which, when erected into gods, become destructive to humanity. They ought to be the servants and not the masters of Man, or of society. They ought to be made to sacrifice to Man, and not Man compelled to sacrifice to them; for when separated from Man or humanity, who is Jesus the Saviour, the vine of eternity, they are thieves and rebels, they are destroyers.

The Plowman of Chaucer is Hercules in his supreme eternal state, divested of his spectrous shadow—which is the Miller, a terrible fellow, such as exists in all times and places for the trial of men, to astonish every neighbourhood with brutal strength and courage, to get rich and powerful to curb the pride of Man.

The Reeve and the Manciple are two characters of the most consummate wordly wisdom. The Shipman, or Sailor, is a similar genius of Ulyssean art but with the highest courage superadded.

The citizens and their Cook are each leaders of a class. Chaucer has been somehow made to number four citizens,° which would make his whole company, himself included, thirty-one. But he says there was but nine and twenty in his company:

> Full nine and twenty in a company.

(The Webbe, or Weaver, and the Tapiser, or Tapestry Weaver, appear to me to be the same person; but this is only an opinion, for 'full nine and twenty' may signify one more or less. But I dare say that Chaucer wrote 'A Webbe Dyer', that is, a Cloth Dyer:

> A Webbe Dyer and a Tapiser.

The Merchant cannot be one of the three citizens, as his dress is different, and his character is more marked, whereas Chaucer says of his rich citizens:

> All were yclothed in o liverie.)

The characters of women Chaucer has divided into two classes, the Lady Prioress and the Wife of Bath. Are not these leaders of the ages of men? The Lady Prioress in some ages predominates, and in some the Wife of Bath—in whose character Chaucer has been equally minute and exact, because she is also a scourge and a blight. I shall say no more of her, nor expose what Chaucer has left hidden. Let the young reader study what he has said of her: it is useful as a scarecrow. There are of such characters born too many for the peace of the world.

I come at length to the Clerk of Oxenford. This character varies from that of Chaucer, as the contemplative philosopher varies from the poetical genius. There are always these two classes of learned sages, the poetical and the philosophical. The painter has put them side by side, as if the youthful clerk had put himself under the tuition of the mature poet. Let the philosopher always be the servant and scholar of inspiration and all will be happy.

Such are the characters that compose this picture, which was painted in self-defence against the insolent and envious imputation of unfitness for finished and scientific art, and this imputation most artfully and industriously endeavoured to be propagated among the public by ignorant hirelings. The painter courts comparison with his competitors° who, having received fourteen hundred guineas and more from the profits of his designs in that well-known work, Designs for Blair's *Grave*, have left him to shift for himself, while others, more obedient to an employer's opinions and directions, are employed at a great expense to produce works, in succession to his, by which they acquired public patronage. This has hitherto been his lot: to get patronage for others, and then to be left and neglected, and his work, which gained that patronage, cried down as eccentricity and madness—as unfinished and neglected by the artist's violent temper. He is sure the works now exhibited will give the lie to such aspersions.

Those who say that men are led by interest are knaves. A knavish character will often say, 'Of what interest is it to me to do so and so?' I answer, 'Of none at all, but the contrary, as you well know. It is of malice and envy that you have done this. Hence I am aware of you, because I know that you act not from interest but from malice, even to your own destruction.' It is therefore become a duty which Mr. B. owes to the public, who have always recognized him and patronized him, however hidden by artifices, that he should not suffer such things to be done, or be hindered from the public exhibition of his finished productions by any calumnies in future.

The character and expression in this picture could never have been produced with Rubens's light and shadow, or with Rembrandt's, or any thing Venetian or Flemish. The Venetian and Flemish practice is broken lines, broken masses, and broken colours. Mr. B.'s practice is unbroken lines, unbroken masses, and unbroken colours. Their art is to lose form; his art is to find form, and to keep it. His arts are opposite to theirs in all things.

As there is a class of men whose whole delight is in the destruction of men, so there is a class of artists whole whose art and science is fabricated for the purpose of destroying art. Who these are is soon known: 'by their works ye shall know them.'° All who endeavour to raise up a style against Rafael, Michelangelo, and the antique; those who separate painting from drawing; who look if a picture is well drawn and, if it is, immediately cry out that it cannot be well-coloured—those are the men.

But to show the stupidity of this class of men, nothing need be done but to examine my rival's prospectus.

The two first characters in Chaucer, the Knight and the Squire, he has put among his rabble; and indeed his prospectus calls the Squire the fop of Chaucer's age. Now hear Chaucer:

> Of his Stature, he was of even length,
> And wonderly deliver, and of great strength;
> And he had be sometime in Chivauchy,
> In Flanders, in Artois, and in Picardy,
> And borne him well as of so litele space.

Was this a fop?

> Well could he sit a horse, and faire ride,
> He could songs make, and eke well indite
> Just, and eke dance, pourtray, and well write.

Was this a fop?

> Curteis he was, and meek, and serviceable;
> And kerft before his fader at the table.

Was this a fop?

It is the same with all his characters: he has done all by chance, or perhaps his fortune—money, money. According to his prospectus he has three Monks; these he cannot find in Chaucer, who has only one Monk, and that no vulgar character, as he has endeavoured to make him. When men cannot read they should not pretend to paint. To be sure, Chaucer is a little difficult to him who has only blundered over novels and catchpenny trifles of book-sellers. Yet a little pains ought to be taken even by the ignorant and weak. He has put the Reeve, a vulgar fellow, between his Knight and Squire, as if he was resolved to go contrary in everything to Chaucer, who says of the Reeve:

> And ever he rode hinderest of the rout.

In this manner he has jumbled his dumb dollies° together, and is praised by his equals for it; for both himself and his friend° are equally masters of Chaucer's language. They both think that the Wife of Bath is a young beautiful blooming damsel; and H——° says that she is the Fair Wife of Bath, and that the spring appears in her cheeks. Now hear what Chaucer has made her say of herself, who is no modest one:

> But Lord when it remembereth me
> Upon my youth and on my jollity,
> It tickleth me about the heart root.
> Unto this day it doth my heart boot,
> That I have had my world as in my time;
> But age, alas, that all will envenime,
> Hath me bireft, my beauty and my pith
> Let go; farewell: the devil go therewith,
> The flower is gone, there is no more to tell.
> The bran, as best, I can, I now mote sell;
> And yet, to be right merry, will I fond,
> Now forth to tell of my fourth husband.

She has had four husbands, a fit subject for this painter; yet the painter ought to be very much offended with his friend H——, who has called his 'a common scene', 'and very ordinary forms'—which is the truest part of all, for it is so, and very wretchedly so indeed. What merit can there be in a picture of which such words are spoken with truth?

But the prospectus says that the painter has represented Chaucer himself as a knave, who thrusts himself among honest people to make game of and laugh at them—though I must do justice to the painter, and say that he has made him look more like a fool than a knave. But it appears, in all the writings of Chaucer, and particularly in his *Canterbury Tales*, that he was very devout, and paid respect to true enthusiastic superstition. He has laughed at his knaves and fools as I do now. But he has respected his true pilgrims, who are a majority of his company, and are not thrown together in the random manner that Mr. S——° has done. Chaucer has nowhere called the Plowman old, worn out with age and labour, as the prospectus has represented him, and says that the picture has done so too. He is worn down with labour, but not with age. How spots of brown and yellow, smeared about at random, can be either young or old, I cannot see. It may be an old man; it may be a young one; it may be any thing that a prospectus pleases. But I know that where there are no lineaments there can be no character. And what connoisseurs call touch,° I know by experience must be the destruction of all character and expression, as it is of every lineament.

The scene of Mr. S——'s picture is by Dulwich Hills, which was not the way to Canterbury; but perhaps the painter thought he would give them a ride round about, because they were a burlesque set of scarecrows, not worth any man's respect or care.

But the painter's thoughts being always upon gold, he has introduced a character that Chaucer has not, namely, a Goldsmith°— for so the prospectus tells us. Why he has introduced a Goldsmith, and what is the wit of it, the prospectus does not explain. But it takes care to mention the reserve and modesty of the painter; this makes a good epigram enough:

> The fox, the owl, the spider, and the mole,
> By sweet reserve and modesty get fat.

But the prospectus tells us that the painter has introduced a Sea Captain. Chaucer has a Shipman, a sailor, a trading master of a vessel, called by courtesy Captain, as every master of a boat is; but this does not make him a sea captain. Chaucer has purposely omitted such a personage, as it only exists in certain periods: it is the soldier by sea. He who would be a soldier in inland nations is a sea captain in commercial nations.

All is misconceived, and its mis-execution is equal to its misconception. I have no objection to Rubens and Rembrandt being employed, or even to their living in a palace; but it shall not be at the expense of Rafael and Michelangelo living in a cottage, and in contempt and derision. I have been scorned long enough by these fellows, who owe to me all that they have; it shall be so no longer.

> I found them blind; I taught them how to see
> And now they know me not, nor yet themselves.

The Bard, from Gray°

> On a rock, whose haughty brow
> Frown'd o'er old Conway's foaming flood,
> Robed in the sable garb of woe,
> With haggard eyes the Poet stood,
> Loose his beard, and hoary hair
> Stream'd like a meteor to the troubled air.
>
> Weave the warp, and weave the woof,
> The winding sheet of Edward's race.

Weaving the winding sheet of Edward's race by means of sounds of spiritual music, and its accompanying expressions of articulate speech, is a bold, and daring, and most masterly conception, that the public have embraced and approved with avidity. Poetry consists in these conceptions; and shall painting be confined to the sordid drudgery of facsimile representations of merely mortal and perishing substances and not be, as poetry and music are, elevated into its own proper sphere of invention and visionary conception? No, it shall not be so! Painting, as well as poetry and music, exists and exults in immortal thoughts. If Mr. B.'s Canterbury Pilgrims had been done by any other power than that of the poetic visionary, it would have been as dull as his adversary's.

The spirits of the murdered bards assist in weaving the deadly woof:

> With me in dreadful harmony they join,
> And weave, with bloody hands, the tissue of thy line.

The connoisseurs and artists who have made objections to Mr. B.'s mode of representing spirits with real bodies would do well to consider that the Venus, the Minerva, the Jupiter, the Apollo, which they admire in Greek statues, are all of them representations of spiritual existences of gods immortal to the mortal perishing organ of sight; and yet they are embodied and organized in solid marble. Mr. B. requires the same latitude, and all is well. The Prophets describe what they saw in vision as real and existing men whom they saw with their imaginative and immortal organs; the Apostles the same. The clearer the organ the more distinct the object. A spirit and a vision are not, as the modern philosophy supposes, a cloudy vapour or a nothing; they are organized and minutely articulated beyond all that the mortal and perishing nature can produce. He who does not imagine in stronger and better lineaments, and in stronger and better light than his perishing mortal eye can see, does not imagine at all. The painter of this work asserts that all his imaginations appear to him infinitely more perfect, and more minutely organized, than anything seen by his mortal eye. Spirits are organized men. Moderns wish to draw figures without lines, and with great and heavy shadows; are not shadows more unmeaning than lines, and more heavy? Oh who can doubt this?

King Edward and his Queen Elenor are prostrated, with their horses, at the foot of a rock on which the Bard stands: prostrated by the terrors of his harp on the margin of the river Conway, whose waves bear up a corpse of a slaughtered bard at the foot of the rock. The armies of Edward are seen winding among the mountains:

> He wound with toilsome march his long array.

Mortimer and Gloucester lie spellbound behind their king. The execution of this picture is also in water colours, or fresco.

The Ancient Britons

In the last battle of King Arthur only three Britons escaped; these were the Strongest Man, the Beautifullest Man, and the Ugliest Man.° These three marched through the field unsubdued, as Gods; and the sun of Britain set, but shall arise again with tenfold splendour when Arthur shall awake from sleep, and resume his dominion over earth and ocean.

The three general classes of men who are represented by the most Beautiful, the most Strong, and the most Ugly, could not be represented by any historical facts but those of our own country, the ancient Britons', without violating costume. The Britons (say historians) were naked civilized men: learned, studious, abstruse in thought and contemplation, naked, simple, plain in their acts and manners, wiser than after-ages. They were overwhelmed by brutal arms, all but a small remnant. Strength, Beauty, and Ugliness escaped the wreck, and remain for ever unsubdued, age after age.

The British antiquities are now in the artist's hands: all his visionary contemplations relating to his own country and its ancient glory, when it was as it again shall be, the source of learning and inspiration. (Arthur was a name for the constellation Arcturus, or Boötes,° the keeper of the North Pole.) And all the fables of Arthur and his round table; of the warlike naked Britons; of Merlin; of Arthur's conquest of the whole world; of his death, or sleep, and promise to return again; of the Druid monuments, or temples; of the pavement of Watling Street;° of London Stone;° of the caverns in Cornwall, Wales, Derbyshire, and Scotland; of the giants of Ireland and Britain; of the elemental beings, called by us by the general name of Fairies; and of these three who escaped, namely, Beauty, Strength, and Ugliness. Mr. B. has in his hands poems of the highest antiquity. Adam was a Druid, and Noah; also Abraham was called to succeed the Druidical age, which began to turn allegoric and mental signification into corporeal command, whereby human sacrifice would have depopulated the earth. All these things are written in Eden. The artist is an inhabitant of that happy country; and if everything goes on as it has begun, the world of vegetation and generation may expect to be opened again to Heaven, through Eden, as it was in the beginning.

The Strong man represents the human sublime. The Beautiful man represents the human pathetic, which was in the wars of Eden divided into male and female. The Ugly man represents the human reason. They were originally one man, who was fourfold; he was self-divided, and his real humanity slain on the stems of generation; and the form of the fourth was like the Son of God. How he became divided is a subject of great sublimity and pathos. The artist has written it° under inspiration and will, if God please, publish it; it is voluminous, and contains the ancient history of Britain, and the world of Satan and of Adam.

In the mean time he has painted this picture, which supposes that in the reign of that British prince,° who lived in the fifth century, there were remains of those naked heroes in the Welsh mountains. They are there now; Gray saw them in the person of his bard on Snowdon. There they dwell in naked simplicity. Happy is he who can see and converse with them above the shadows of generation and death. The giant Albion was patriarch of the Atlantic; he is the Atlas of the Greeks, one of those the Greeks called Titans. The stories of Arthur are the acts of Albion, applied to a prince of the fifth century, who conquered Europe, and held the empire of the world in the dark age, which the Romans never again recovered. In this picture, believing with Milton the ancient British history,° Mr. B. has done, as all the ancients did, and as all the moderns who are worthy of fame: given the historical fact in its poetical vigour so as it always happens, and not in that dull way that some historians pretend, who, being weakly organized themselves, cannot see either miracle or prodigy. All is to them a dull round of probability and possibilities; but the history of all times and places is nothing else but improbabilities and impossibilities—what we should say was impossible if we did not see it always before our eyes.

The antiquities of every nation under Heaven is no less sacred than that of the Jews. They are the same thing as Jacob Bryant° and all antiquaries have proved. How other antiquities came to be neglected and disbelieved, while those of the Jews are collected and arranged, is an enquiry worthy of both the antiquarian and the divine. All had originally one language, and one religion; this was the religion of Jesus, the everlasting gospel. Antiquity preaches the gospel of Jesus. The reasoning historian, turner and twister of causes and consequences, such as Hume,° Gibbon and Voltaire, cannot with all their artifice turn or twist one fact or disarrange self-evident action and reality. Reasons and opinions concerning acts are not history. Acts themselves alone are history, and these are neither the exclusive property of Hume, Gibbon nor Voltaire, Echard,° Rapin,° Plutarch nor Herodotus. Tell me the acts, O historian, and leave me to reason upon them as I please; away with your reasoning and your rubbish. All that is not action is not worth reading. Tell me the What; I do not want you to tell me the Why, and the How. I can find that out myself, as well as you can, and I will not be fooled by you into opinions that you please to impose, to disbelieve what you think improbable or impossible. His opinion who does not see spiritual agency, is not worth any man's reading. He who rejects a fact because it is improbable, must reject all history and retain doubts only.

It has been said to the artist, 'Take the Apollo for the model of your Beautiful man and the Hercules for your Strong man, and the Dancing Faun for your Ugly man.' Now he comes to his trial. He knows that what he does is not inferior to the grandest antiques. Superior they cannot be, for human power cannot go beyond either what he does, or what they have done; it is the gift of God; it is inspiration and vision. He had resolved to emulate those precious remains of antiquity. He has done so, and the result you behold. His ideas of strength and beauty have not been greatly different. Poetry as it exists now on earth in the various remains of ancient authors, music as it exists in old tunes or melodies, painting and sculpture as it exists in the remains of antiquity and in the works of more modern genius, is inspiration, and cannot be surpassed; it is perfect and eternal. Milton, Shakespeare, Michelangelo, Rafael, the finest specimens of ancient sculpture and painting and architecture, Gothic, Grecian, Hindu and Egyptian, are the extent of the human mind. The human mind cannot go beyond the gift of God, the Holy Ghost. To suppose that art can go beyond the finest specimens of art that are now in the world is not knowing what art is; it is being blind to the gifts of the spirit.

It will be necessary for the painter to say something concerning his ideas of Beauty, Strength and Ugliness.

The Beauty that is annexed and appended to folly is a lamentable accident and error of the mortal and perishing life. It does but seldom happen, but with this unnatural mixture the sublime artist can have nothing to do; it is fit for the burlesque. The Beauty proper for sublime art is lineaments, or forms and features that are capable of being the receptacles of intellect. Accordingly the painter has given in his beautiful man his own idea of intellectual beauty. The face and limbs that deviates or alters least, from infancy to old age, is the face and limbs of greatest Beauty and perfection.

The Ugly likewise, when accompanied and annexed to imbecility and disease, is a subject for burlesque and not for historical grandeur. The artist has imagined his Ugly man one approaching to the beast in features and form: his forehead small, without frontals;° his jaws large; his nose high on the ridge, and narrow; his chest and the stamina of his make comparatively little, and his joints and his extremities large; his eyes with scarce any whites, narrow and cunning—and everything tending toward what is truly Ugly: the incapability of intellect.

The artist has considered his Strong man as a receptacle of wisdom, a sublime energizer. His features and limbs do not spindle out into length, without strength, nor are they too large and unwieldy for his brain and bosom. Strength consists in accumulation of power to the principal seat, and from thence a regular gradation and subordination; strength is compactness, not extent nor bulk.

The Strong man acts from conscious superiority, and marches on in fearless dependence on the divine decrees, raging with the inspirations of a prophetic mind. The Beautiful man acts from duty, and anxious solicitude for the fates of those for whom he combats. The Ugly man acts from love of carnage, and delight in the savage barbarities of war, rushing with sportive precipitation into the very teeth of the affrighted enemy.

The Roman soldiers rolled together in a heap before them ('like the rolling thing before the whirlwind')° each show a different character, and a different expression of fear, or revenge, or envy, or blank horror, or amazement, or devout wonder and unresisting awe.

The dead and the dying, Britons naked, mingled with armed Romans, strew the field beneath. Among these the last of the bards who were capable of attending warlike deeds is seen falling, outstretched among the dead and the dying, singing to his harp in the pains of death.

Distant among the mountains are Druid temples, similar to Stonehenge. The sun sets behind the mountains, bloody with the day of battle.

The flush of health in flesh exposed to the open air, nourished by the spirits of forests and floods, in that ancient happy period which history has recorded, cannot be like the sickly daubs of Titian or Rubens. Where will the copier of Nature, as it now is, find a civilized man who has been accustomed to go naked? Imagination only can furnish us with colouring appropriate, such as is found in the frescos of Rafael and Michelangelo; the disposition of forms always directs colouring in works of true art. As to a modern man stripped from his load of clothing, he is like a dead corpse. Hence Rubens, Titian, Correggio, and all of that class, are like leather and chalk; their men are like leather, and their women like chalk, for the disposition of their forms will not admit of grand colouring. In Mr. B.'s Britons the blood is seen to circulate in their limbs; he defies competition in colouring.

The Spiritual Preceptor, an experiment picture

This subject is taken from the visions of Emanuel Swedenborg (Universal Theology, No. 623).° The Learned, who strive to ascend into Heaven by means of learning, appear to children like dead horses, when repelled by the celestial spheres. The works of this visionary are well worthy the attention of painters and poets; they are foundations for grand things. The reason they have not been more attended to is because corporeal demons have gained a predominance; who the leaders of these are, will be shown below. Unworthy men who gain fame among men continue to govern mankind after death, and in their spiritual bodies oppose the spirits of those who worthily are famous; and, as Swedenborg observes, by entering into disease and excrement, drunkenness and concupiscence, they possess themselves of the bodies of mortal men, and shut the doors of mind and of thought, by placing learning above inspiration. O Artist! you may disbelieve all this, but it shall be at your own peril.

Satan Calling up his Legions, from Milton's Paradise Lost: *a composition for a more perfect picture,*° *afterward executed for a lady of high rank. An experiment picture*

This picture was likewise painted at intervals, for experiment on colours, without an oily vehicle. It may be worthy of attention, not only on account of its composition, but of the great labour which has been bestowed on it, that is, three or four times as much as would have finished a more perfect picture. The labour has destroyed the lineaments; it was with difficulty brought back again to a certain effect, which it had at first, when all the lineaments were perfect.

These pictures, among numerous others painted for experiment, were the result of temptations and perturbations, labouring to destroy imaginative power by means of that infernal machine, called Chiaroscuro,° in the hands of Venetian and Flemish demons—whose enmity to the painter himself, and to all artists who study in the Florentine and Roman Schools, may be removed by an exhibition and exposure of their vile tricks. They cause that everything in art shall become a machine. They cause that the execution shall be all blocked up with brown shadows. They put the original artist in fear and doubt of his own original conception. The spirit of Titian was particularly active in raising doubts concerning the possibility of executing without a model, and when once he had raised the doubt it became easy for him to snatch away the vision time after time. For when the artist took his pencil to execute his ideas, his power of imagination weakened so much, and darkened, that memory of Nature and of pictures of the various schools possessed his mind, instead of appropriate execution resulting from the inventions—like walking in another man's style, or speaking or looking in another man's style and manner, unappropriate and repugnant to your own individual character, tormenting the true artist, till he leaves the Florentine and adopts the Venetian practice, or does as Mr. B has done: has the courage to suffer poverty and disgrace, till he ultimately conquers.

Rubens is a most outrageous demon, and by infusing the remembrances of his pictures and style of execution hinders all power of individual thought: so that the man who is possessed by this demon loses all admiration of any other artist but Rubens, and those who were his imitators and journeymen. He causes to the Florentine and Roman artist fear to execute; and though the original conception was all fire and animation, he loads it with hellish brownness, and blocks up all its gates of light, except one, and that one he closes with iron bars, till the victim is obliged to give up the Florentine and Roman practice, and adopt the Venetian and Flemish.

Correggio is a soft, and effeminate, and consequently a most cruel demon, whose whole delight is to cause endless labour to whoever suffers him to enter his mind. The story that is told in all lives of the painters, about Correggio being poor and but badly paid for his pictures, is altogether false:° he was a petty prince in Italy, and employed numerous journeymen in manufacturing (as Rubens and Titian did) the pictures that go under his name. The manual labour in these pictures of Correggio is immense, and was paid for originally at the immense prices that those who keep manufactories of art always charge to their employers, while they themselves pay their journeymen little enough. But though Correggio was not poor, he will make any true artist so who permits him to enter his mind, and take possession of his affections. He infuses a love of soft and even tints without boundaries, and of endless reflected lights, that confuse one another, and hinder all correct drawing from appearing to be correct. For if one of Rafael or Michelangelo's figures was to be traced, and Correggio's reflections and refractions to be added to it, there would soon be an end of proportion and strength, and it would be weak, and pappy, and lumbering, and thick-headed, like his own works. But then it would have softness and evenness, by a twelve month's labour, where a month would with judgement have finished it better and higher; and the poor wretch who executed it would be the Correggio that the life-writers have written of: a drudge and a miserable man, compelled to softness by poverty. I say again, O Artist, you may disbelieve all this, but it shall be at your own peril.

Note. These experiment pictures have been bruised and knocked about, without mercy, to try all experiments.

The Bramins. A Drawing

The subject is Mr. Wilkin translating the *Geeta*: an ideal design,° suggested by the first publication of that part of the Hindu scriptures, translated by Mr. Wilkin.° I understand that my costume is incorrect, but in this I plead the authority of the ancients, who often deviated from the habits, to preserve the manners: as in the instance of Laocoon who, though a priest, is represented naked.°

The above four drawings° the artist wishes were in fresco on an enlarged scale, to ornament the altars of churches, and to make England like Italy: respected by respectable men of other countries on account of art. It is not the want of genius that can hereafter be laid to our charge. The artist who has done these pictures and drawings will take care of that; let those who govern the nation take care of the other. The times require that every one should speak out boldly; England expects that every man should do his duty° in arts, as well as in arms, or in the senate.

Ruth. A Drawing

This design is taken from that most pathetic passage in the Book of Ruth where—Naomi having taken leave of her daughters-in-law with intent to return to her own country—Ruth cannot leave her, but says, 'Whither thou goest I will go; and where thou lodgest I will lodge, thy people shall be my people, and thy God my God: where thou diest I will die, and there will I be buried; God do so to me and more also, if ought but death part thee and me.'°

The distinction that is made in modern times between a painting and a drawing proceeds from ignorance of art. The merit of a picture is the same as the merit of a drawing. The dauber daubs his drawings; he who draws his drawings draws his pictures. There is no difference between Rafael's cartoons and his frescos, or pictures, except that the frescos, or pictures, are more finished. When Mr. B. formerly painted in oil colours his pictures were shown to certain painters and connoisseurs, who said that they were very admirable drawings on canvas, but not pictures: but they said the same of Rafael's pictures.

Mr. B. thought this the greatest of compliments, though it was meant otherwise. If losing and obliterating the outline constitutes a picture, Mr. B. will never be so foolish as to do one. Such art of losing the outlines is the art of Venice and Flanders: it loses all character, and leaves what some people call expression. But this is a false notion of expression; expression cannot exist without character as its stamina, and neither character nor expression can exist without firm and determinate outline. Fresco painting is susceptible of higher finishing than drawing on paper, or than any other method of painting. But he must have a strange organization of sight who does not prefer a drawing on paper to a daubing in oil by the same master, supposing both to be done with equal care.

The great and golden rule of art, as well as of life, is this: that the more distinct, sharp, and wiry the bounding line, the more perfect the work of art; and the less keen and sharp, the greater is the evidence of weak imitation, plagiarism, and bungling. Great inventors, in all ages, knew this: Protogenes and Apelles° knew each other by this line. Rafael, and Michelangelo, and Albert Durer are known by this and this alone. The want of this determinate and bounding form evidences the want of idea in the artist's mind, and the pretence of the plagiary° in all its branches. How do we distinguish the oak from the beech, the horse from the ox, but by the bounding outline? How do we distinguish one face or countenance from another, but by the bounding line and its infinite inflexions° and movements? What is it that builds a house and plants a garden, but the definite and determinate? What is it that distinguishes honesty from knavery, but the hard and wiry line of rectitude and certainty in the actions and intention? Leave out this line and you leave out life itself; all is chaos again, and the line of the Almighty° must be drawn out upon it before man or beast can exist. Talk no more then of Correggio, or Rembrandt, or any other of those plagiaries of Venice or Flanders. They were but the lame imitators of lines drawn by their predecessors, and their works prove themselves contemptible dis-arranged imitations and blundering misapplied copies.

The Penance of Jane Shore° in Saint Paul's Church. A Drawing

This drawing was done above thirty years ago, and proves to the author (and he thinks will prove to any discerning eye) that the productions of our youth and of our maturer age are equal in all essential points. If a man is master of his profession, he cannot be ignorant that he is so; and if he is not employed by those who pretend to encourage art, he will employ himself, and laugh in secret at the pretences of the ignorant—while he has every night dropped into his shoe,° as soon as he puts it off, and puts out the candle, and gets into bed, a reward for the labours of the day such as the world cannot give, and patience and time await to give him all that the world can give.

FINIS

From 'Public Address' (c.1811)

If men of weak capacities have alone the power of execution in art Mr. B. has now put to the test. If to invent and to draw well hinders the executive power in art, and his strokes are still to be condemned because they are unlike those of artists who are unacquainted with drawing, is now to be decided by the public. Mr. B.'s inventive powers and his scientific knowledge of drawing is on all hands acknowledged. It only remains to be certified whether physiognomic strength and power is to give place to imbecility. In a work of art it is not fine tints that are required, but fine forms. Fine tints without fine forms are always the subterfuge of the blockhead.

> Rafael sublime, majestic, graceful, wise,
> His executive power must I despise?
> Rubens loud, vulgar, stupid, ignorant,
> His power of execution I must grant?
> Learn the laborious stumble of a fool,
> And from an idiot's actions form my rule?
> Go send your children to the slobbering° school!

I account it a public duty respectfully to address myself to the Chalcographic Society,° and to express to them my opinion (the result of the incessant practice and experience of many years) that engraving as an art is lost in England, owing to an artfully propagated opinion that drawing spoils an engraver. I request the society to inspect my print,° of which drawing is the foundation and indeed the superstructure; it is drawing on copper, as painting ought to be drawing on canvas or any other surface, and nothing else. I request likewise that the Society will compare the prints of Bartolozzi, Woollett, Strange° with the old English portraits (that is, compare the modern art with the art as it existed previous to the entrance of Vandyke and Rubens into this country—since which English engraving is lost) and I am sure the result of this comparison will be that the Society must be of my opinion that engraving, by losing drawing, has lost all character and all expression, without which the art is lost.

In this plate Mr. B. has resumed the style with which he set out in life, of which Heath° and Stothard were the awkward imitators at that time. It is the style of Albert Durer's Histories° and the old engravers, which cannot be imitated by anyone who does not understand drawing, and which, according to Heath and Stothard, Flaxman and even Romney, spoils an engraver—for each of these men have repeatedly asserted this absurdity to me in condemnation of my work and approbation of Heath's lame imitation (Stothard being such a fool as to suppose that his blundering blurs can be made out and delineated by any engraver who knows how to cut dots and lozenges equally well with those little prints which I engraved after him five and twenty years ago,° and by which he got his reputation as a draughtsman).

The manner in which my character both as an artist and a man has been blasted these thirty years may be seen particularly in a Sunday paper called the *Examiner*,° published in Beaufort Buildings (we all know that editors of newspapers trouble their heads very little about art and science, and that they are always paid for what they put in upon these ungracious subjects) and the manner in which I have routed out the nest of villains will be seen in a poem° concerning my three years Herculean labours at Felpham which I will soon publish. Secret calumny and open professions of friendship are common enough all the world over, but have never been so good an occasion of poetic imagery. When a base man means to be your enemy he always begins with being your friend.

Flaxman cannot deny that one of the very first monuments he did I gratuitously designed for him: at the same time he was blasting my character as an artist to Macklin° my employer, as Macklin told me at the time. How much of his Homer and Dante he will allow to be mine I do not know, as he went far enough off to publish them, even to Italy;° but the public will know and posterity will know.

Many people are so foolish as to think that they can wound Mr Fuseli over my shoulder.° They will find themselves mistaken; they could not wound even Mr Barry so.

A certain portrait painter said to me in a boasting way: 'Since I have practised painting I have lost all idea of drawing.' Such a man must know that I looked upon him with contempt. He did not care for this any more than West° did, who hesitated and equivocated with me upon the same subject: at which time he asserted that Woollett's prints were superior to Basire's° because they had more labour and care. Now this is contrary to the truth. Woollett did not know how to put so much labour into a head or a foot as Basire did. He did not know how to draw the leaf of a tree; all his study was clean strokes and mossy tints. How then shall he be able to make use of either labour or care, unless the labour and care of imbecility? The life's labour of mental weakness scarcely equals one hour of the labour of ordinary capacity, like the full gallop of the gouty man to the ordinary walk of youth and health. I allow that there is such a thing as high finished ignorance—as there may be a fool or a knave in an embroidered coat—but I say that the embroidery of the ignorant finisher is not like a coat made by another, but is an emanation from ignorance itself, and its finishing is like its master: the life's labour of five hundred idiots (for he never does the work himself).

What is called the English style of engraving such as proceeded from the toilettes of Woollett and Strange (for theirs were fribbles' toilettes)° can never produce character and expression. I knew the men intimately from their intimacy with Basire my master and knew them both to be heavy lumps of cunning and ignorance, as their works show to all the Continent—who laugh at the contemptible pretences of Englishmen to improve art before they even know the first beginnings of art. I hope this print will redeem my country from this coxcomb situation, and show that it is only some Englishmen, and not all, who are thus ridiculous in their pretences. Advertisements in newspapers are no proofs of popular approbation, but often the contrary. A man who pretends to improve fine art does not know what fine art is. Ye English engravers must come down from your high flights; ye must condescend to study Marc Antonio° and Albert Durer. Ye must begin before you attempt to finish or improve, and when you have begun you will know better than to think of improving what cannot be improved. It is very true what you have said for these thirty-two years: I am mad or else you are so; both of us cannot be in our right senses. Posterity will judge by our works. Woollett's and Strange's works are like those of Titian and Correggio: the life's labour of ignorant journeymen, suited to the purposes of commerce no doubt, for commerce cannot endure individual merit—its insatiable maw must be fed by what all can do equally well (at least it is so in England, as I have found to my cost these forty years).

Woollett's best works were etched by Jack Browne;° Woollett etched very bad himself. Strange's prints were, when I knew him, all done by Aliamet° and his French journeyman, whose name I forget.

'The Cottagers' and 'Jocund Peasants', the 'Views in Kew Gardens', 'Foots Cray' and 'Diana and Acteon', and in short all that are called Woollett's, were etched by Jack Browne, and in Woollett's works the etching is all—though in these a single leaf of a tree is never correct.

Such prints as Woollett and Strange produced will do for those who choose to purchase the life's labour of ignorance and imbecility in preference to the inspired moments of genius and animation. I also knew something of Tom Cook,° who engraved after Hogarth. Cook wished to give to Hogarth what he could take from Rafael, that is, outline and mass and colour, but he could not.

I do not pretend to paint better than Rafael, or Michelangelo, or Julio Romano, or Albert Durer, but I do pretend to paint finer than Rubens, or Rembrandt, or Correggio, or Titian. I do not pretend to engrave finer than Albert Durer, Goltzius,° Sadeler° or Edelinck,° but I do pretend to engrave finer than Strange, Woollett, Hall,° or Bartolozzi— and all because I understand drawing, which they understood not.

I do not condemn Rubens, Rembrandt, or Titian because they did not understand drawing, but because they did not understand colouring. How long shall I be forced to beat this into men's ears? I do not condemn Strange or Woollett because they did not understand drawing, but because they did not understand graving. I do not condemn Pope or Dryden because they did not understand imagination, but because they did not understand verse. Their colouring, graving and verse can never be applied to art. That is not either colouring, graving or verse which is unappropriate to the subject. He who makes a design must know the effect and colouring proper to be put to that design, and will never take that of Rubens, Rembrandt or Titian to turn that which is soul and life into a mill or machine.

They say there is no straight line in Nature; this is a lie, like all that they say, for there is every line in Nature. But I will tell them what is not in Nature. An even tint is not in Nature; it produces heaviness. Nature's shadows are ever-varying, and a ruled sky° that is quite even never can produce a natural sky. The same with every object in a picture: its spots are its beauties. Now Gentlemen Critics, how do you like this? You may rage, but what I say I will prove by such practice, and have already done so, that you will rage to your own destruction. Woollett I knew very intimately by his intimacy with Basire, and I knew him to be one of the most ignorant fellows that I ever knew. A machine is not a man nor a work of art; it is destructive of humanity and of art.

Woollett I know did not know how to grind his graver. I know this; he has often proved his ignorance before me at Basire's by laughing at Basire's knife, tools and ridiculing the forms of Basire's other gravers— till Basire was quite dashed and out of conceit° with what he himself knew. But his impudence had a contrary effect on me. Englishmen have been so used to journeymen's undecided bungling that they cannot bear the firmness of a master's touch. Every line is the line of beauty;° it is only fumble and bungle which cannot draw a line; this only is ugliness. That is not a line which doubts and hesitates in the midst of its course.

Commerce is so far from being beneficial to arts or to empire that it is destructive of both, as all their history shows—for the above reason of individual merit being its great hatred. Empires flourish till they become commercial, and then they are scattered abroad to the four winds.

In this manner the English public have been imposed upon for many years, under the impression that engraving and painting are somewhat else besides drawing. Painting is drawing on canvas, and engraving is drawing on copper and nothing else, and he who pretends to be either painter or engraver without being a master of drawing is an impostor. We may be clever as pugilists,° but as artists we are and have long been the contempt of the Continent. Gravelot° once said to my master Basire, 'De English may be very clever in deir own opinions, but dey do not draw de draw.'

Resentment for personal injuries has had some share in this Public Address, but love to my art and zeal for my country a much greater.

Men think they can copy Nature as correctly as I copy imagination. This they will find impossible, and all the copies or pretended copiers of Nature from Rembrandt to Reynolds prove that Nature becomes to its victim nothing but blots and blurs. Why are copiers of Nature incorrect while copiers of imagination are correct? This is manifest to all.

The originality of this production makes it necessary to say a few words. While the works of Pope and Dryden are looked upon as the same art with those of Milton and Shakespeare, while the works of Strange and Woollett are looked upon as the same art with those of Rafael and Albert Durer, there can be no art in a nation but such as is subservient to the interest of the monopolizing trader. Englishmen, rouse yourselves from the fatal slumber into which booksellers and trading dealers have thrown you, under the artfully propagated pretence that a translation of a copy of any kind can be as honourable to a nation as an original—belying the English character in that well-known saying, 'Englishmen improve what others invent'. This even Hogarth's works prove a detestable falsehood. No man can improve an original invention. Nor can an original invention exist without execution, organized and minutely delineated and articulated either by God or Man.

I do not mean smoothed up, and niggled, and poco piud° and all the beauties picked out and blurred and blotted, but drawn with a firm hand and decided at once, like Fuseli and Michelangelo, Shakespeare and Milton.

BLAKE'S APOLOGY FOR HIS CATALOGUE

Having given great offence by writing in prose
I'll write in verse as soft as Bartelloze.°
Some blush at what others can see no harm in,
But nobody sees any harm in rhyming.
Dryden in rhyme cries 'Milton only planned'°—
Every fool shook his bells throughout the land.
Tom Cook cut Hogarth down with his clean graving—
Thousands of connoisseurs with joy ran raving.
Thus Hayley on his toilette° seeing his soap,
Cries 'Homer is very much improved by Pope'.°
Some say I've given great provision to my foes,
And that now I lead my false friends by the nose.
Flaxman and Stothard smelling a sweet savour
Cry 'Blakified drawing spoils painter and engraver'.
While I, looking up to my umbrella,
Resolved to be a very contrary fellow,
Cry looking quite from skumference to centre,
'No one can finish so high as the original inventor'.
Thus poor Schiavonetti° died on the Cromek,
A thing that is tied around the *Examiner*'s neck.
This is my sweet apology to my friends,
That I may put them in mind of their latter ends.

I have heard many people say, 'Give me the ideas. It is no matter what words you put them into', and others say, 'Give me the design; it is no matter for the execution'. These people know enough of artifice, but nothing of art. Ideas cannot be given but in their minutely appropriate words; nor can a design be made without its minutely appropriate execution. The unorganized blots and blurs of Rubens and Titian are not art; nor can their method ever express ideas or imaginations any more than Pope's metaphysical° jargon° of rhyming.

Unappropriate execution is the most nauseous of all affectation and foppery.

He who copies does not execute; he only imitates what is already executed. Execution is only the result of invention.

Whoever looks at any of the great and expressive works of engraving that have been published by English traders must feel a loathing and disgust, and accordingly most Englishmen have a contempt for art, which is the greatest curse that can fall upon a nation.

He who could represent Christ uniformly like a drayman° must have queer conceptions: consequently his execution must have been as queer, and those must be queer fellows who give great sums for such nonsense and think it fine art.

The modern chalcographic connoisseurs and amateurs° admire only the work of the journeyman: picking out of whites and blacks in what is called tints.° They despise drawing, which despises them in return. They see only whether everything is covered down but one spot of light.

Mr. B. submits to a more severe tribunal: he invites the admirers of old English portraits to look at his print.

I do not know whether Homer is a liar, and that there is no such thing as generous contention.° I know that all those with whom I have contended in art have strove, not to excel, but to starve me out by calumny and the arts of trading combination.°

It is nonsense for noblemen and gentlemen to offer premiums° for the encouragement of art, when such pictures as these can be done without premiums. Let them encourage what exists already, and not endeavour to counteract by tricks. Let it no more be said that empires encourage arts, for it is arts that encourage empires. (Arts and artists are spiritual, and laugh at mortal contingencies.) It is in their power to hinder instruction, but not to instruct—just as it is in their power to murder a man, but not to make a man.

Let us teach Buonaparte, and whomsoever else it may concern, that it is not arts that follow and attend upon empire, but empire that attends upon and follows the arts.

No man of sense can think that an imitation of the objects of Nature is the art of painting, or that such imitation (which anyone may easily perform) is worthy of notice—much less that such an art should be the glory and pride of a nation. The Italians laugh at English connoisseurs, who are most of them such silly fellows as to believe this.

A man sets himself down with colours and with all the articles of painting. He puts a model before him, and he copies that so neat as to make it a deception. Now let any man of sense ask himself one question: 'Is this art? Can it be worthy of admiration to anybody of understanding? Who could not do this? What man who has eyes and an ordinary share of patience cannot do this neatly? Is this art? Or is it glorious to a nation to produce such contemptible copies?' Countrymen, countrymen! do not suffer yourselves to be disgraced.

The English artist may be assured that he is doing an injury and injustice to his country while he studies and imitates the effects of Nature. England will never rival Italy while we servilely copy what the wise Italians Rafael and Michelangelo scorned, nay, abhorred—as Vasari tells us.°

> Call that the 'public voice' which is their error—
> Like as a monkey peeping in a mirror
> Admires all his colours brown and warm,
> And never once perceives his ugly form.

What kind of intellects° must he have who sees only the colours of things and not the forms of things?

A jockey that is anything of a jockey will never buy a horse by the colour, and a man who has got any brains will never buy a picture by the colour.

When I tell any truth it is not for the sake of convincing those who do not know it, but for the sake of defending those who do.

No man of sense ever supposes that copying from Nature is the art of painting. If the art is no more than this it is no better than any other's manual labour; anybody may do it, and the fool often will do it best as it is a work of no mind.

The greatest part of what are called in England old pictures are oil colour copies from fresco originals;° the comparison is easily made and the copy detected. Note I mean fresco easel or cabinet pictures, on canvas and wood and copper etc.

The painter hopes that his friends Anytus, Melitus and Lycon° will perceive that they are not now in ancient Greece, and though they can use the poison of calumny the English public will be convinced that such a picture as this could never be painted by a madman, or by one in a state of outrageous manners—as these bad men both print and publish by all the means in their power. The painter begs public protection, and all will be well.

I wonder who can say, 'Speak no ill of the dead', when it is asserted in the Bible that the name of the wicked shall rot.° It is deistical virtue I suppose, but as I have none of this I will pour aquafortis° on the name of the wicked, and turn it into an ornament and an example—to be avoided by some and imitated by others if they please.

Columbus discovered America but Americus Vesputius° finished and smoothed it over, like an English engraver, or Correggio and Titian.

What man of sense will lay out his money upon the life's labours of imbecility and imbecility's journeymen, or think to educate a fool how to build a universe with farthing balls? The contemptible idiots who have been called great men of late years ought to rouse the public indignation of men of sense in all professions.

There is not, because there cannot be, any difference of effect in the pictures of Rubens and Rembrandt; when you have seen one of their pictures you have seen all. It is not so with Rafael, Julio Romano, Albert Durer, Michalangelo. Every picture of theirs has a different and appropriate effect.

Yet I do not shrink from the comparison in either relief° or strength of colour with either Rembrandt or Rubens. On the contrary, I court the comparison and fear not the result—but not in a dark corner. Their effects are in every picture the same; mine are in every picture different.

I hope my countrymen will excuse me if I tell them a wholesome truth. Most Englishmen when they look at a picture immediately set about searching for points of light, and clap the picture into a dark corner; this when done by grand works is like looking for epigrams in Homer. A point of light is a witticism. Many are destructive of all art; one is an epigram only, and no grand work can have them. They produce system and monotony.

Rafael, Michelangelo, Albert Durer, Julio Romano are accounted ignorant of that epigrammatic wit in art, because they avoid it as a destructive machine—as it is.

That vulgar epigram in art, Rembrandt's hundred guilders,° has entirely put an end to all genuine and appropriate effect. All, both morning and night, is now a dark cavern; it is the fashion.

When you view a collection of pictures painted since Venetian art was the fashion, or go into a modern exhibition, with a very few exceptions every picture has the same effect: a piece of machinery of points of light, to be put into a dark hole.

Who that has eyes cannot see that Rubens and Correggio must have been very weak and vulgar fellows—and we are to imitate their execution! This is like what Sir Francis Bacon says:° that a healthy child should be taught and compelled to walk like a cripple, while the cripple must be taught to walk like healthy people. Oh rare wisdom!

There is just the same science in Lebrun, or Rubens, or even Vanloo° that there is in Rafael or Michalangelo, but not the same genius. Science is soon got; the other never can be acquired, but must be born.

Let a man who has made a drawing go on and on, and he will produce a picture or painting; but if he chooses to leave off before he has spoiled it he will do a better thing.

Mr. B. repeats that there is not one character or expression in this print which could be produced with the execution of Titian, Rubens, Correggio, Rembrandt or any of that class. Character and expression can only be expressed by those who feel them. Even Hogarth's execution cannot be copied or improved. Gentlemen of fortune who give great prices for pictures should consider the following: Ruben's Luxembourg Gallery° is confessed on all hands to be the work of a blockhead; it bears this evidence in its face. How can its execution be any other than the work of a blockhead? Bloated gods, Mercury, Juno, Venus and the rattletraps° of mythology, and the lumber of an awkward French palace, are thrown together around clumsy and ricketty princes and princesses higgledy piggledy. On the contrary, Julio Romano's Palace of T at Mantua is allowed on all hands to be the production of a man of the most profound sense and genius, and yet his execution is pronounced by English connoisseurs (and Reynolds their doll) to be unfit for the study of the painter. Can I speak with too great contempt of such contemptible fellows? If all the princes in Europe, like Louis XIV and Charles the First, were to patronize such blockheads I, William Blake, a mental prince, should decollate and hang their souls as guilty of mental high treason.

Princes appear to me to be fools; Houses of Commons and Houses of Lords appear to me to be fools. They seem to me to be something else besides human life.

I am really sorry to see my countrymen trouble themselves about politics. If men were wise, the most arbitrary princes could not hurt them. If they are not wise, the freest government is compelled to be a tyranny.

The wretched state of the arts in this country and in Europe— originating in the wretched state of political science, which is the science of sciences—demands a firm and determinate conduct on the part of artists: to resist the contemptible counterarts established by such contemptible politicians as Louis XIV, and originally set on foot by Venetian picture traders, music traders, and rhyme traders, to the destruction of all true art as it is this day.

An example of these contrary arts is given us in the characters of Milton and Dryden as they are written in a poem signed with the name of Nat Lee, which perhaps he never wrote, and perhaps he wrote in a paroxysm of insanity: in which it is said that Milton's poem is a rough unfinished piece, and Dryden has finished it.° Now let Dryden's *Fall* and Milton's *Paradise* be read, and I will assert that everybody of understanding must cry out 'shame' on such niggling and poco piu as Dryden has degraded Milton with; but at the same time I will allow that stupidity will prefer Dryden, because it is in rhyme and monotonous sing-song, sing-song from beginning to end. Such are Bartolozzi, Woollett and Strange.

To recover art has been the business of my life—to the Florentine original, and if possible to go beyond that original. This I thought the only pursuit worthy of a man. To imitate I abhor. I obstinately adhere to the true style of art, such as Michelangelo, Rafael, Julio Romano, Albert Durer left it. I demand therefore of the amateurs of art the encouragement which is my due. If they continue to refuse theirs is the loss, not mine—and theirs is the contempt of posterity. I have enough in the approbation of fellow labourers; this is my glory and exceeding great reward.° I go on and nothing can hinder my course;

> And in melodious accents I
> Will sit me down and cry I, I.

The painters of England are unemployed in public work, while the sculptors have continual and superabundant employment. Our churches and abbeys are treasures of their producing for ages back, while painting is excluded. Painting, the principal art, has no place among our almost only public works. Yet it is more adapted to solemn ornament than marble can be, as it is capable of being placed on any heighth, and indeed would make a noble finish placed above the great public monuments in Westminster, St Paul's, and other cathedrals. To the Society for Encouragement of Arts° I address myself with respectful duty, requesting their consideration of my plan as a great public means of advancing fine art in Protestant communities. Monuments to the dead painted by historical and poetical artists like Barry and Mortimer—I forbear to name living artists though equally worthy—I say monuments so painted must make England what Italy is: an envied storehouse of intellectual riches.

It has been said of late years, the English public have no taste for painting. This is a falsehood. The English are as good judges of painting as of poetry, and they prove it in their contempt for great collections of all the rubbish of the Continent brought here by ignorant picture dealers. An Englishman may well say, 'I am no judge of painting', when he is shown these smears and daubs at an immense price, and told that such is the art of painting. I say the English public are true encouragers of great art, while they discourage and look with contempt on false art.

In a commercial nation impostors are abroad in all professions; these are the greatest enemies of genius. In the art of painting these impostors sedulously propagate an opinion that great inventors cannot execute. This opinion is as destructive of the true artists as it is false by all experience. Even Hogarth cannot be either copied or improved. Can Anglus° never discern perfection but in the journeyman's labour?

I know my execution is not like anybody else. I do not intend it should be so. None but blockheads copy one another. My conception and invention are on all hands allowed to be superior. My execution will be found so too. To what is it that gentlemen of the first rank, both in genius and fortune, have subscribed their names? To my inventions. The executive part they never disputed.

The lavish praise I have received from all quarters for invention and drawing has generally been accompanied by this: 'He can conceive, but he cannot execute'. This absurd assertion has done me, and may still do me, the greatest mischief. I call for public protection against these villains. I am like others: just equal in invention and in execution, as my works show. I in my own defence challenge a competition with the finest engravings, and defy the most critical judge to make the comparison honestly: asserting in my own defence that this print is the finest that has been done, or is likely to be done, in England—where drawing, its foundation, is contemned, and absurd nonsense about dots and lozenges and clean strokes° made to occupy the attention to the neglect of all real art. I defy any man to cut cleaner strokes than I do, or rougher where I please, and assert that he who thinks he can engrave or paint either, without being a master of drawing, is a fool. Painting is drawing on canvas, and engraving is drawing on copper and nothing else. Drawing is execution and nothing else, and he who draws best must be the best artist. To this I subscribe my name as a public duty,

                                                        WILLIAM BLAKE

On Homer's Poetry (c.1821)

Every poem must necessarily be a perfect unit, but why Homer's is peculiarly so I cannot tell: he has told the story of Bellerophon and omitted the Judgement of Paris, which is not only a part, but a principal part, of Homer's subject.°

But when a work has unity it is as much in a part as in the whole; the Torso is as much a unity as the Laocoon.

As unity is the cloak of folly, so goodness is the cloak of knavery. Those who will have unity exclusively in Homer come out with a moral like a sting in the tail. Aristotle says characters are either good or bad; now goodness or badness has nothing to do with character. An apple-tree, a pear-tree, a horse, a lion, are characters—but a good apple-tree or a bad is an apple-tree still. A horse is not more a lion for being a bad horse; that is its character. Its goodness or badness is another consideration.

It is the same with the moral of a whole poem as with the moral goodness of its parts. Unity and morality are secondary considerations, and belong to philosophy and not to poetry, to exception and not to rule, to accident and not to substance. The ancients called it eating of the Tree of Good and Evil.

The classics, it is the classics!—and not Goths nor monks—that desolate Europe with wars.

On Virgil (c.1821)

Sacred truth has pronounced that Greece and Rome, as Babylon and Egypt, so far from being parents of arts and sciences as they pretend, were destroyers of all art. Homer, Virgil and Ovid confirm this opinion, and make us reverence the word of God: the only light of antiquity that remains unperverted by war. Virgil in the *Aeneid* Book VI, line 848 says, 'Let others study art; Rome has somewhat better to do, namely, war and dominion.'

Rome and Greece swept art into their maw and destroyed it. A warlike state never can produce art. It will rob, and plunder, and accumulate into one place, and translate, and copy, and buy, and sell, and criticise, but not make. Grecian is mathematic form. Gothic is living form. Mathematic form is eternal in the reasoning memory. Living form is eternal existence.

From the annotations to Wordsworth's 1815 Poems (1826)

I see in Wordsworth the natural man rising up against the spiritual man continually, and then he is no poet but a heathen philosopher: at enmity against all true poetry or inspiration.

There is no such thing as natural piety,° because the natural man is at enmity with God.

This° is all in the highest degree imaginative and equal to any poet, but not superior. I cannot think that real poets have any competition. None are greatest in the Kingdom of Heaven;° it is so in poetry.

Natural objects always did, and now do, weaken, deaden and obliterate imagination° in me. Wordsworth must know that what he writes valuable is not to be found in Nature.

Read Michelangelo's sonnet,° Vol. II, p. 179.

I do not know who wrote these prefaces;° they are very mischievous, and direct contrary to Wordsworth's own practice.

I believe both Macpherson and Chatterton: that what they say is ancient is so.

I own myself an admirer of Ossian equally with any other poet whatever—Rowley° and Chatterton also.

EARLY VISIONARY AND NARRATIVE WRITINGS

(All before 1783)

To Spring

O thou, with dewy locks, who lookest down
Through the clear windows of the morning: turn°
Thine angel eyes upon our western isle,
Which in full choir hails thy approach, O Spring!

The hills tell each other, and the list'ning
Valleys hear; all our longing eyes are turned
Up to thy bright pavilions. Issue forth,
And let thy holy feet visit our clime.

Come o'er the eastern hills, and let our winds
Kiss thy perfumed garments; let us taste   10
Thy morn and evening breath; scatter thy pearls
Upon our love-sick land that mourns for thee.

Oh deck her forth with thy fair fingers; pour
Thy soft kisses on her bosom; and put
Thy golden crown upon her languished head,
Whose modest tresses were bound up for thee!

To Summer

O thou, who passest through our valleys in
Thy strength, curb thy fierce steeds, allay the heat
That flames from their large nostrils! Thou, O Summer,
Oft pitched'st here thy golden tent, and oft
Beneath our oaks hast slept, while we beheld
With joy thy ruddy limbs and flourishing hair.

Beneath our thickest shades we oft have heard
Thy voice, when noon upon his fervid car
Rode o'er the deep of heaven; beside our springs
Sit down, and in our mossy valleys; on   10
Some bank beside a river clear throw thy
Silk draperies off, and rush into the stream.
Our vallies love the Summer in his pride.

Our bards are famed who strike the silver wire;
Our youth are bolder than the southern swains;
Our maidens fairer in the sprightly dance;
We lack not songs, nor instruments of joy,
Nor echoes sweet, nor waters clear as heaven,
Nor laurel wreaths against the sultry heat.

To Autumn

O Autumn, laden with fruit, and stained
With the blood of the grape, pass not, but sit
Beneath my shady roof; there thou may'st rest,
And tune thy jolly voice to my fresh pipe;
And all the daughters of the year shall dance!
Sing now the lusty song of fruits and flowers.

'The narrow bud opens her beauties to
The sun, and love runs in her thrilling veins;
Blossoms hang round the brows of morning, and
Flourish down the bright cheek of modest eve,
Till clust'ring Summer breaks forth into singing,   10
And feathered clouds strew flowers round her head.

'The spirits of the air live on the smells
Of fruit; and joy, with pinions light, roves round
The gardens, or sits singing in the trees.'
Thus sang the jolly Autumn as he sat;
Then rose, girded himself, and o'er the bleak
Hills fled from our sight; but left his golden load.

To Winter

O Winter! bar thine adamantine doors;
The north is thine; there hast thou built thy dark
Deep-founded habitation. Shake not thy roofs,
Nor bend thy pillars with thine iron car.

He hears me not, but o'er the yawning deep
Rides heavy; his storms are unchained; sheathed
In ribbed steel, I dare not lift mine eyes;
For he hath reared his sceptre o'er the world.

Lo! now the direful monster, whose skin clings
To his strong bones, strides o'er the groaning rocks.   10
He withers all in silence, and his hand
Unclothes the earth, and freezes up frail life.

He takes his seat upon the cliff; the mariner
Cries in vain. Poor little wretch! that deal'st
With storms; till heaven smiles, and the monster
Is driv'n yelling to his caves beneath Mount Hecla.°

To the Evening Star

Thou fair-haired angel of the evening,
Now, whilst the sun rests on the mountains, light
Thy bright torch of love; thy radiant crown
Put on, and smile upon our evening bed!
Smile on our loves; and, while thou drawest the
Blue curtains of the sky, scatter thy silver dew
On every flower that shuts its sweet eyes
In timely sleep. Let thy west wind sleep on
The lake; speak silence with thy glimmering eyes,
And wash the dusk with silver. Soon, full soon,   10
Dost thou withdraw; then the wolf rages wide,
And the lion glares through the dun forest.
The fleeces of our flocks are covered with
Thy sacred dew; protect them with thine influence.

To Morning

O holy virgin! clad in purest white,
Unlock heav'n's golden gates, and issue forth;
Awake the dawn that sleeps in heaven; let light
Rise from the chambers of the east, and bring
The honied dew that cometh on waking day.
O radiant morning, salute the sun,
Roused like a huntsman to the chase; and, with
Thy buskined feet, appear upon our hills.°

Fair Elenor

The bell struck one, and shook the silent tower;
The graves give up their dead; fair Elenor
Walked by the castle gate, and looked in.
A hollow groan ran through the dreary vaults.

She shrieked aloud, and sunk upon the steps,
On the cold stone her pale cheek. Sickly smells
Of death issue as from a sepulchre,
And all is silent but the sighing vaults.

Chill death withdraws his hand, and she revives;
Amazed, she finds herself upon her feet,   10
And, like a ghost, through narrow passages
Walking, feeling the cold walls with her hands.

Fancy returns, and now she thinks of bones,
And grinning skulls, and corruptible death,
Wrapped in his shroud; and, now, fancies she hears
Deep sighs, and sees pale sickly ghosts gliding.

At length, no fancy, but reality,
Distracts her. A rushing sound, and the feet
Of one that fled, approaches—Ellen stood,
Like a dumb statue, froze to stone with fear.   20

The wretch approaches, crying, 'The deed is done;
Take this, and send it by whom thou wilt send;
It is my life—send it to Elenor—
He's dead, and howling after me for blood!

'Take this,' he cried; and thrust into her arms
A wet napkin, wrapped about; then rushed
Past, howling. She received into her arms
Pale death, and followed on the wings of fear.

They passed swift through the outer gate; the wretch,
Howling, leaped o'er the wall into the moat,   30
Stifling in mud. Fair Ellen passed the bridge,
And heard a gloomy voice cry, 'Is it done?'

As the deer wounded Ellen flew over
The pathless plain, as the arrows that fly°
By night; destruction flies, and strikes in darkness;
She fled from fear, till at her house arrived.

Her maids await her; on her bed she falls,
That bed of joy, where erst her lord hath pressed.
'Ah, woman's fear!' she cried, 'Ah, cursed duke!
Ah, my dear lord! Ah, wretched Elenor!   40

'My lord was like a flower upon the brows
Of lusty May! Ah, life as frail as flower!
O ghastly death! withdraw thy cruel hand,
Seek'st thou that flower to deck thy horrid temples?

'My lord was like a star, in highest heav'n
Drawn down to earth by spells and wickedness;
My lord was like the opening eyes of day,
When western winds creep softly o'er the flowers.

'But he is darkened; like the summer's noon,
Clouded; fall'n like the stately tree, cut down;   50
The breath of heaven dwelt among his leaves.
O Elenor, weak woman, filled with woe!'

Thus having spoke she raised up her head,
And saw the bloody napkin by her side,
Which in her arms she brought; and now, tenfold
More terrified, saw it unfold itself.

Her eyes were fixed; the bloody cloth unfolds,
Disclosing to her sight the murdered head
Of her dear lord, all ghastly pale, clotted
With gory blood; it groaned, and thus it spake: 60

'O Elenor, behold thy husband's head,
Who, sleeping on the stones of yonder tower,°
Was 'reft of life by the accursed duke!
A hired villain turned my sleep to death!

'O Elenor, beware the cursed duke,
O give not him thy hand, now I am dead;
He seeks thy love; who, coward, in the night,
Hired a villain to bereave my life.'

She sat with dead cold limbs, stiffened to stone;
She took the gory head up in her arms;   70
She kissed the pale lips; she had no tears to shed;
She hugged it to her breast, and groaned her last.

To the Muses

Whether on Ida's shady brow,°
    Or in the chambers of the East,
The chambers of the sun, that now
    From ancient melody have ceased;°

Whether in Heav'n ye wander fair,
    Or the green corners of the earth,
Or the blue regions of the air,
    Where the melodious winds have birth;

Whether on crystal rocks ye rove,
    Beneath the bosom of the sea   10
Wand'ring in many a coral grove,
    Fair Nine, forsaking poetry!

How have you left the ancient love
   That bards of old enjoyed in you!
The languid strings do scarcely move!
   The sound is forced, the notes are few!

Gwin, King of Norway

Come, kings, and listen to my song,
   When Gwin, the son of Nore,
Over the nations of the north
   His cruel sceptre bore.

The nobles of the land did feed°
   Upon the hungry poor;
They tear the poor man's lamb, and drive
   The needy from their door!

The land is desolate; our wives
   And children cry for bread;   10
Arise, and pull the tyrant down;
   Let Gwin be humbled.

Gordred the giant roused himself
   From sleeping in his cave;
He shook the hills, and in the clouds
   The troubled banners wave.

Beneath them rolled, like tempests black,
   The num'rous sons of blood;
Like lions' whelps, roaring abroad,
   Seeking their nightly food.   20

Down Bleron's hills they dreadful rush,
   Their cry ascends the clouds;
The trampling horse, and clanging arms
   Like rushing mighty floods!

Their wives and children, weeping loud,
    Follow in wild array,
Howling like ghosts, furious as wolves,
    In the bleak wintry day.

'Pull down the tyrant to the dust,
    Let Gwin be humbled,'   30
They cry, 'and let ten thousand lives
    Pay for the tyrant's head.'

From tower to tower the watchmen cry,
    'O Gwin, the son of Nore,
Arouse thyself! the nations black,
    Like clouds, come rolling o'er!'

Gwin reared his shield, his palace shakes,
    His chiefs come rushing round,
Each, like an awful thunder cloud,
    With voice of solemn sound.   40

Like reared stones around a grave
    They stand around the King;
Then suddenly each seized his spear,
    And clashing steel does ring.

The husbandman does leave his plough,
    To wade through fields of gore;
The merchant binds his brows in steel,
    And leaves the trading shore;

The shepherd leaves his mellow pipe,
    And sounds the trumpet shrill;   50
The workman throws his hammer down
    To heave the bloody bill.

Like the tall ghost of Barraton,
    Who sports in stormy sky,
Gwin leads his host as black as night,
    When pestilence does fly,

With horses and with chariots—
   And all his spearmen bold
March to the sound of mournful song,
   Like clouds around him rolled.   60

Gwin lifts his hand—the nations halt;
  'Prepare for war,' he cries,
'Gordred appears!—his frowning brow
  Troubles our northern skies.'

The armies stand, like balances
  Held in th' Almighty's hand;
'Gwin, thou hast filled thy measure up;
  Thou'rt swept from out the land.'

And now the raging armies rushed,
  Like warring mighty seas;   70
The heav'ns are shook with roaring war;
  The dust ascends the skies!

Earth smokes with blood, and groans, and shakes,
  To drink her children's gore,
A sea of blood; nor can the eye
  See to the trembling shore!

And on the verge of his wild sea
  Famine and death doth cry;
The cries of women and of babes
  Over the field doth fly.   80

The King is seen raging afar,
  With all his men of might;
Like blazing comets, scattering death
  Through the red fev'rous night.

Beneath his arm like sheep they die,
  And groan upon the plain;
The battle faints, and bloody men
  Fight upon hills of slain.

Now death is sick, and riven men
  Labour and toil for life;   90
Steed rolls on steed, and shield on shield,
  Sunk in this sea of strife!

The god of war is drunk with blood,
  The earth doth faint and fail;
The stench of blood makes sick the heav'ns;
  Ghosts glut the throat of hell!

Oh what have Kings to answer for,
  Before that awful throne!
When thousand deaths for vengeance cry,
  And ghosts accusing groan!   100

Like blazing comets in the sky,
  That shake the stars of light,
Which drop like fruit unto the earth,
  Through the fierce burning night;

Like these did Gwin and Gordred meet,
  And the first blow decides;
Down from the brow unto the breast
  Gordred his head divides!

Gwin fell; the sons of Norway fled,
  All that remained alive;   110
The rest did fill the vale of death:
  For them the eagles strive.

The river Dorman rolled their blood
  Into the northern sea;
Who mourned his sons, and overwhelmed
  The pleasant south country.

An Imitation of Spenser

Golden Apollo, that through heaven wide
    Scatter'st the rays of light, and truth's beams!
In lucent words my darkling verses dight,
    And wash my earthy mind in thy clear streams,
    That wisdom may descend in fairy dreams:
All while the jocund hours in thy train
Scatter their fancies at thy poet's feet;
    And when thou yields to night thy wide domain,
Let rays of truth enlight his sleeping brain.

For brutish Pan in vain might thee assay°   10
    With tinkling sounds to dash thy nervous verse,°
Sound without sense; yet in his rude affray
    (For ignorance is folly's leesing nurse,°
    And love of folly needs none other curse)
Midas the praise hath gained of lengthened ears,
    For which himself might deem him ne'er the worse
    To sit in council with his modern peers,
And judge of tinkling rhymes, and elegances terse.

And thou, Mercurius, that with winged brow
    Dost mount aloft into the yielding sky,   20
And through Heav'n's halls thy airy flight dost throw,
Entering with holy feet to where on high
Jove weighs the counsel of futurity;
    Then, laden with eternal fate, dost go
Down, like a falling star, from autumn sky,
And o'er the surface of the silent deep dost fly.

If thou arrivest at the sandy shore,
Where nought but envious hissing adders dwell,
    Thy golden rod, thrown on the dusty floor,
Can charm to harmony with potent spell;   30
Such is sweet eloquence, that does dispel
    Envy and hate that thirst for human gore,
And cause in sweet society to dwell
Vile savage minds that lurk in lonely cell.

    O Mercury, assist my lab'ring sense,
That round the circle of the world would fly!
    As the winged eagle scorns the towery fence
Of Alpine hills round his high aery,
And searches through the corners of the sky,
    Sports in the clouds to hear the thunder's sound,   40
And see the winged lightnings as they fly,
    Then, bosomed in an amber cloud, around
Plumes his wide wings, and seeks Sol's palace high.

And thou, O warrior, maid invincible,
Armed with the terrors of almighty Jove!
    Pallas, Minerva, maiden terrible,
Lov'st thou to walk the peaceful solemn grove,
    In solemn gloom of branches interwove?
Or bear'st thy aegis o'er the burning field,
    Where, like the sea, the waves of battle move?   50
Or have thy soft piteous eyes beheld
    The weary wanderer through the desert rove?
Or does th' afflicted man thy heav'nly bosom move?

King Edward the Third

PERSONS

King Edward	*Sir Walter Manny*
The Black Prince	*Lord Audley*
Queen Philippa	*Lord Percy*
Duke of Clarence	*Bishop*
Sir John Chandos	*William*, Dagworth's man
Sir Thomas Dagworth	*Peter Blunt*, a common soldier

SCENE I

The coast of France. King Edward and nobles. The army

KING. O thou, to whose fury the nations are
But as dust! maintain thy servant's right.
Without thine aid the twisted mail, and spear,
And forged helm, and shield of seven times beaten brass,
Are idle trophies of the vanquisher.
When confusion rages, when the field is in a flame,
When the cries of blood tear horror from heav'n,
And yelling death runs up and down the ranks,
Let Liberty, the chartered right of Englishmen,
Won by our fathers in many a glorious field,   10
Enerve my soldiers; let Liberty
Blaze in each countenance, and fire the battle.
The enemy fight in chains, invisible chains, but heavy;
Their minds are fettered; then how can they be free,
While, like the mounting flame,
We spring to battle o'er the floods of death?
And these fair youths, the flow'r of England,
Vent'ring their lives in my most righteous cause,
Oh sheathe their hearts with triple steel, that they
May emulate their fathers' virtues.   20
And thou, my son, be strong; thou fightest for a crown
That death can never ravish from thy brow,
A crown of glory; but from thy very dust
Shall beam a radiance, to fire the breasts
Of youth unborn! Our names are written equal
In fame's wide trophied hall; 'tis ours to gild
The letters, and to make them shine with gold

That never tarnishes: whether Third Edward,
Or the Prince of Wales, or Montacute, or Mortimer,
Or ev'n the least by birth, shall gain the brightest fame,   30
Is in his hand to whom all men are equal.
The world of men are like the num'rous stars,
That beam and twinkle in the depth of night,
Each clad in glory according to his sphere;
But we, that wander from our native seats,
And beam forth lustre on a darkling world,
Grow larger as we advance! And some perhaps
The most obscure at home, that scarce were seen
To twinkle in their sphere, may so advance
That the astonished world, with up-turned eyes,   40
Regardless of the moon, and those that once were bright,
Stand only for to gaze upon their splendour!
            [*He here knights the Prince, and other young nobles*
Now let us take a just revenge for those°
Brave lords, who fell beneath the bloody axe
At Paris. Thanks, noble Harcourt, for 'twas
By your advice we landed here in Brittany°—
A country not yet sown with destruction,°
And where the fiery whirlwind of swift war
Has not yet swept its desolating wing.
Into three parties we divide by day,   50
And separate march, but join again at night;
Each knows his rank, and Heav'n marshall all.

                                    [*Exeunt*

SCENE II

English Court. Lionel Duke of Clarence, Queen Philippa, lords, bishop, etc

CLARENCE. My Lords, I have, by the advice of her
Whom I am doubly bound to obey, my parent
And my sovereign, called you together.
My task is great, my burden heavier than
My unfledged years;
Yet, with your kind assistance, Lords, I hope
England shall dwell in peace; that while my father
Toils in his wars, and turns his eyes on this
His native shore—and sees Commerce fly round
With his white wings, and sees his golden London   10

And her silver Thames thronged with shining spires
And corded ships, her merchants buzzing round
Like summer bees, and all the golden cities
In his land overflowing with honey—
Glory may not be dimmed with clouds of care.
Say, Lords, should not our thoughts be first to commerce?
My Lord Biship, you would recommend us agriculture?

BISHOP. Sweet Prince! the arts of peace are great,
And no less glorious than those of war,
Perhaps more glorious in the philosophic mind.   20
When I sit at my home, a private man,
My thoughts are on my gardens, and my fields,
How to employ the hand that lacketh bread.
If Industry is in my diocese,
Religion will flourish; each man's heart
Is cultivated, and will bring forth fruit:
This is my private duty and my pleasure.
But as I sit in council with my prince,
My thoughts take in the gen'ral good of the whole,
And England is the land favoured by Commerce;   30
For Commerce, though the child of Agriculture,
Fosters his parent, who else must sweat and toil,
And gain but scanty fare. Then, my dear Lord,
Be England's trade our care; and we, as tradesmen,
Looking to the gain of this our native land.

CLAR. O my good Lord, true wisdom drops like honey
From your tongue, as from a worshipped oak!
Forgive, my Lords, my talkative youth, that speaks
Not merely what my narrow observation has
Picked up, but what I have concluded from your lessons.
Now, by the Queen's advice, I ask your leave   40
To dine to-morrow with the Mayor of London.
If I obtain your leave, I have another boon
To ask, which is, the favour of your company.
I fear Lord Percy will not give me leave.

PERCY. Dear Sir, a prince should always keep his state,
And grant his favours with a sparing hand,
Or they are never rightly valued.
These are my thoughts, yet it were best to go;
But keep a proper dignity, for now   50
You represent the sacred person of

Your father; 'tis with princes as 'tis with the sun,
If not sometimes o'er-clouded, we grow weary
Of his officious glory.

CLAR. Then you will give me leave to shine sometimes,
My Lord?

LORD. Thou hast a gallant spirit, which I fear
Will be imposed on by the closer sort!   [*Aside*

CLAR. Well, I'll endeavour to take
Lord Percy's advice; I have been used so much   60
To dignity, that I'm sick on't.

QUEEN PHILIPPA. Fie, fie, Lord Clarence; you proceed not to business,
But speak of your own pleasures.
I hope their Lordships will excuse your giddiness.

CLAR. My Lords, the French have fitted out many
Small ships of war, that, like to ravening wolves,
Infest our English seas, devouring all
Our burdened vessels, spoiling our naval flocks.
The merchants do complain, and beg our aid.

PERCY. The merchants are rich enough;   70
Can they not help themselves?

BISH. They can, and may; but how to gain their will,
Requires our countenance and help.

PERCY. When that they find they must, my Lord, they will.
Let them but suffer awhile, and you shall see
They will bestir themselves.

BISH. Lord Percy cannot mean that we should suffer
This disgrace; if so, we are not sovereigns
Of the sea: our right, that Heaven gave
To England, when at the birth of Nature   80
She was seated in the deep, the ocean ceased
His mighty roar, and, fawning, played around
Her snowy feet, and owned his awful queen.
Lord Percy, if the heart is sick, the head
Must be aggrieved; if but one member suffer,
The heart doth fail. You say, my Lord, the merchants
Can, if they will, defend themselves against
These rovers; this is a noble scheme,
Worthy the brave Lord Percy, and as worthy
His generous aid to put it into practice.   90

PERCY. Lord Bishop, what was rash in me, is wise

In you; I dare not own the plan. 'Tis not
Mine. Yet will I, if you please,
Quickly to the Lord Mayor, and work him onward
To this most glorious voyage; on which cast
I'll set my whole estate,
But we will bring these Gallic rovers under.
QUEEN. Thanks, brave Lord Percy; you have the thanks
Of England's Queen, and will, ere long, of England.

[*Exeunt*

SCENE III

At Crécy. Sir Thomas Dagworth and Lord Audley, meeting

AUDLEY. Good morrow, brave Sir Thomas; the bright morn
Smiles on our army, and the gallant sun
Springs from the hills like a young hero
Into the battle, shaking his golden locks
Exultingly; this is a promising day.
DAGWORTH. Why, my Lord Audley, I don't know.
Give me your hand, and now I'll tell you what
I think you do not know—Edward's afraid of Philip.°
AUD. Ha ha, Sir Thomas! you but joke;
Did you e'er see him fear? At Blanchetaque,   10
When almost singly he drove six thousand
French from the ford, did he fear then?
DAGW. Yes, fear; that made him fight so.
AUD. By the same reason I might say, 'tis fear
That makes you fight.
DAGW. Mayhap you may; look upon Edward's face—
No one can say he fears. But when he turns
His back, then I will say it to his face,
He is afraid; he makes us all afraid.
I cannot bear the enemy at my back.   20
Now here we are at Crécy; where, to-morrow,
To-morrow we shall know. I say, Lord Audley,
That Edward runs away from Philip.
AUD. Perhaps you think the Prince too is afraid?
DAGW. No; God forbid! I'm sure he is not—
He is a young lion. Oh I have seen him fight,
And give command, and lightning has flashed
From his eyes across the field; I have seen him

Shake hands with death, and strike a bargain for
The enemy; he has danced in the field   30
Of battle, like the youth at morris play.
I'm sure he's not afraid, nor Warwick, nor none,
None of us but me; and I am very much afraid.

AUD. Are you afraid too, Sir Thomas?
  I believe that as much as I believe
  The King's afraid; but what are you afraid of?

DAGW. Of having my back laid open; we turn
  Our backs to the fire, till we shall burn our skirts.

AUD. And this, Sir Thomas, you call fear? Your fear
  Is of a different kind then from the King's;   40
  He fears to turn his face, and you to turn your back.
  I do not think, Sir Thomas, you know what fear is.

Enter Sir John Chandos

CHANDOS. Good morrow, Generals; I give you joy.
  Welcome to the fields of Crécy. Here we stop,
  And wait for Philip.

DAGW. I hope so.

AUD. There, Sir Thomas; do you call that fear?

DAGW. I don't know; perhaps he takes it by fits.
  Why noble Chandos, look you here—
  One rotten sheep spoils the whole flock;   50
  And if the bell-wether is tainted, I wish
  The Prince may not catch the distemper too.

CHAND. Distemper, Sir Thomas! what distemper?
  I have not heard.

DAGW. Why, Chandos, you are a wise man,
  I know you understand me; a distemper°
  The King caught here in France of running away.

AUD. Sir Thomas, you say, you have caught it too.

DAG. And so will the whole army; 'tis very catching,
  For when the coward runs, the brave man totters.   60
  Perhaps the air of the country is the cause.
  I feel it coming upon me, so I strive against it;
  You yet are whole, but after a few more
  Retreats we all shall know how to retreat
  Better than fight. To be plain, I think retreating
  Too often takes away a soldier's courage.

CHAND. Here comes the King himself; tell him your thoughts
   Plainly, Sir Thomas.
DAGW. I've told him before, but his disorder
   Makes him deaf.   70

Enter King Edward and Black Prince

KING. Good morrow, Generals; when English courage fails,
   Down goes our right to France;
   But we are conquerors everywhere; nothing
   Can stand our soldiers; each man is worthy
   Of a triumph. Such an army of heroes
   Ne'er shouted to the heav'ns, nor shook the field.
   Edward, my son, thou art
   Most happy, having such command; the man
   Were base who were not fired to deeds
   Above heroic, having such examples.   80
PRINCE. Sire! with respect and deference I look
   Upon such noble souls, and wish myself
   Worthy the high command that Heaven and you
   Have given me. When I have seen the field glow,
   And in each countenance the soul of war,
   Curbed by the manliest reason, I have been winged
   With certain victory; and 'tis my boast,
   And shall be still my glory, I was inspired
   By these brave troops.
DAGW. Your Grace had better make   90
   Them all generals.
KING. Sir Thomas Dagworth, you must have your joke,
   And shall, while you can fight as you did at
   The ford.
DAGW. I have a small petition to your Majesty.
KING. What can Sir Thomas Dagworth ask, that Edward
   Can refuse?
DAGW. I hope your Majesty cannot refuse so great
   A trifle; I've gilt your cause with my best blood,°
   And would again, were I not forbid   100
   By him whom I am bound to obey; my hands
   Are tied up, my courage shrunk and withered,
   My sinews slackened, and my voice scarce heard.
   Therefore I beg I may return to England.

KING. I know not what you could have asked, Sir Thomas,
    That I would not have sooner parted with°
    Than such a soldier as you have been, and such a friend;
    Nay, I will know the most remote particulars
    Of this your strange petition, that, if I can,
    I still may keep you here.   110

DAGW. Here on the fields of Crécy we are settled,
    Till Philip springs the tim'rous covey again.
    The wolf is hunted down by causeless fear;
    The lion flees, and fear usurps his heart,
    Startled, astonished at the clam'rous cock;
    The eagle, that doth gaze upon the sun,
    Fears the small fire that plays about the fen;
    If, at this moment of their idle fear,
    The dog doth seize the wolf, the forester the lion,
    The negro in the crevice of the rock   120
    Doth seize the soaring eagle, undone by flight
    They tame submit; such the effect flight has
    On noble souls. Now hear its opposite:
    The tim'rous stag starts from the thicket wild;
    The fearful crane springs from the splashy fen;
    The shining snake glides o'er the bending grass.
    The stag turns head! and bays the crying hounds;
    The crane, o'ertaken, fighteth with the hawk;
    The snake doth turn, and bite the padding foot;
    And, if your Majesty's afraid of Philip,   130
    You are more like a lion than a crane.
    Therefore I beg I may return to England.

KING. Sir Thomas, now I understand your mirth,
    Which often plays with wisdom for its pastime,
    And brings good counsel from the breast of laughter,
    I hope you'll stay, and see us fight this battle,
    And reap rich harvest in the fields of Crécy;
    Then go to England, tell them how we fight,
    And set all hearts on fire to be with us.
    Philip is plumed, and thinks we flee from him,   140
    Else he would never dare to attack us. Now,
    Now the quarry's set! and Death doth sport
    In the bright sunshine of this fatal day.

DAGW. Now my heart dances, and I am as light
    As the young bridegroom going to be married.

Now must I to my soldiers, get them ready,
Furbish our armours bright, new plume our helms,
And we will sing, like the young housewives busied
In the dairy; my feet are winged, but not
For flight, an please your grace.   150
KING. If all my soldiers are as pleased as you,
    'Twill be a gallant thing to fight or die;
    Then I can never be afraid of Philip.
DAGW. A raw-boned fellow t'other day passed by me;
    I told him to put off his hungry looks;
    He answered me, 'I hunger for another battle'.
    I saw a little Welshman with a fiery face;
    I told him he looked like a candle half°
    Burned out; he answered, he was 'pig enough°
    To light another pattle.' Last night beneath   160
    The moon I walked abroad, when all had pitched
    Their tents, and all were still;
    I heard a blooming youth singing a song
    He had composed, and at each pause he wiped
    His dropping eyes. The ditty was, 'If he
    Returned victorious, he should wed a maiden
    Fairer than snow, and rich as midsummer.'
    Another wept, and wished health to his father.
    I chid them both, but gave them noble hopes.
    These are the minds that glory in the battle,   170
    And leap and dance to hear the trumpet sound.
KING. Sir Thomas Dagworth, be thou near our person;
    Thy heart is richer than the vales of France.
    I will not part with such a man as thee.
    If Philip came armed in the ribs of death,°
    And shook his mortal dart against my head,°
    Thoud'st laugh his fury into nerveless shame!
    Go now, for thou art suited to the work,
    Throughout the camp; enflame the timorous,
    Blow up the sluggish into ardour, and   180
    Confirm the strong with strength; the weak inspire,
    And wing their brows with hope and expectation.
    Then to our tent return, and meet to council. [*Exit Dagworth*
CHAND. That man's a hero in his closet, and more°
    A hero to the servants of his house,
    Than to the gaping world; he carries windows

In that enlarged breast of his, that all
May see what's done within.

PRINCE. He is a genuine Englishman, my Chandos,
And hath the spirit of liberty within him.   190
Forgive my prejudice, Sir John; I think
My Englishmen the bravest people on
The face of the earth.

CHAND. Courage, my Lord, proceeds from self-dependence;
Teach man to think he's a free agent,
Give but a slave his liberty, he'll shake
Off sloth, and build himself a hut, and hedge
A spot of ground; this he'll defend; 'tis his
By right of nature. Thus set in action,
He will still move onward to plan conveniences,   200
'Till glory fires his breast to enlarge his castle,
While the poor slave drudges all day, in hope
To rest at night.

KING. O Liberty, how glorious art thou!
I see thee hov'ring o'er my army, with
Thy wide-stretched plumes; I see thee
Lead them on to battle;
I see thee blow thy golden trumpet, while
Thy sons shout the strong shout of victory!
O noble Chandos! think thyself a gardener,   210
My son a vine, which I commit unto
Thy care; prune all extravagant shoots, and guide
Th' ambitious tendrils in the paths of wisdom;
Water him with thy advice, and Heav'n
Rain fresh'ning dew upon his branches. And,
O Edward, my dear son! learn to think lowly of
Thyself, as we may all each prefer other.
'Tis the best policy, and 'tis our duty.   [Exit King Edward

PRINCE. And may our duty, Chandos, be our pleasure—
Now we are alone, Sir John, I will unburden,   220
And breathe my hopes into the burning air,
Where thousand deaths are posting up and down,
Commissioned to this fatal field of Crécy;
Methinks I see them arm my gallant soldiers,
And gird the sword upon each thigh, and fit
Each shining helm, and string each stubborn bow,
And dance to the neighing of our steeds.

Methinks the shout begins, the battle burns;
Methinks I see them perch on English crests,
And roar the wild flame of fierce war upon   230
The thronged enemy! In truth, I am too full;
It is my sin to love the noise of war.
Chandos, thou seest my weakness; strong nature
Will bend or break us; my blood, like a spring-tide,
Does rise so high, to overflow all bounds
Of moderation; while Reason, in his
Frail bark, can see no shore or bound for vast
Ambition. Come, take the helm, my Chandos,
That my full-blown sails overset me not
In the wild tempest; condemn my 'ventrous youth,   240
That plays with danger, as the innocent child,
Unthinking, plays upon the viper's den.
I am a coward, in my reason, Chandos.
CHAND. You are a man, my prince, and a brave man,
If I can judge of actions; but your heat
Is the effect of youth, and want of use;
Use makes the armed field and noisy war
Pass over as a summer cloud, unregarded,
Or but expected as a thing of course.
Age is contemplative; each rolling year   250
Brings forth fruit to the mind's treasure-house;
While vacant youth doth crave and seek about
Within itself, and findeth discontent;
Then, tired of thought, impatient takes the wing,
Seizes the fruits of time, attacks experience,
Roams round vast Nature's forest, where no bounds
Are set, the swiftest may have room, the strongest
Find prey—till, tired at length, sated and tired
With the changing sameness, old variety,
We sit us down, and view our former joys   260
With distaste and dislike.
PRINCE. Then if we must tug for experience,
Let us not fear to beat round Nature's wilds,
And rouse the strongest prey; then if we fall,
We fall with glory. I know the wolf
Is dangerous to fight, not good for food,
Nor is the hide a comely vestment; so
We have our battle for our pains. I know

That youth has need of age to point fit prey,
And oft the stander-by shall steal the fruit   270
Of th' others' labour. This is philosophy;
These are the tricks of the world; but the pure soul
Shall mount on native wings, disdaining
Little sport, and cut a path into the heaven of glory,
Leaving a track of light for men to wonder at.
I'm glad my father does not hear me talk;
You can find friendly excuses for me, Chandos.
But do you not think, Sir John, that if it please
Th' Almighty to stretch out my span of life,
I shall with pleasure view a glorious action,   280
Which my youth mastered?
CHAND. Considerate age, my Lord, views motives,
And not acts. When neither warbling voice
Nor trilling pipe is heard, nor pleasure sits
With trembling age, the voice of Conscience then,
Sweeter than music in a summer's eve,
Shall warble round the snowy head, and keep
Sweet symphony to feathered angels, sitting
As guardians round your chair; then shall the pulse
Beat slow, and taste, and touch, and sight, and sound, and smell, 290
That sing and dance round Reason's fine-wrought throne,
Shall flee away, and leave him all forlorn;
Yet not forlorn if Conscience is his friend.

                                                        [*Exeunt*

SCENE IV

Sir Thomas Dagworth's Tent. Dagworth and William his man

DAGWORTH. Bring hither my armour, William;
    Ambition is the growth of ev'ry clime.
WILLIAM. Does it grow in England, Sir?
DAGW. Aye, it grows most in lands most cultivated.
WILL. Then it grows most in France; the vines here
    Are finer than any we have in England.
DAGW. Aye, but the oaks are not.
WILL. What is the tree you mentioned? I don't think
    I ever saw it.
DAGW. Ambition.
WILL. Is it a little creeping root that grows in ditches?

DAGW. Thou dost not understand me, William.
It is a root that grows in every breast;
Ambition is the desire or passion that one man
Has to get before another, in any pursuit after glory;
But I don't think you have any of it.

WILL. Yes, I have; I have a great ambition to know every thing, Sir.

DAGW. But when our first ideas are wrong, what follows must all be wrong of course; 'tis best to know a little, and to know that little aright.

WILL. Then, Sir, I should be glad to know if it was not ambition that brought over our king to France to fight for his right?

DAGW. Though the knowledge of that will not profit thee much, yet I will tell you that it was ambition.

WILL. Then if ambition is a sin, we are all guilty in coming with him, and in fighting for him.

DAGW. Now, William, thou dost thrust the question home; but I must tell you, that guilt being an act of the mind, none are guilty but those whose minds are prompted by that same ambition.

WILL. Now I always thought, that a man might be guilty of doing wrong, without knowing it was wrong.

DAGW. Thou art a natural philosopher,° and knowest truth by instinct; while reason runs aground, as we have run our argument. Only remember, William, all have it in their power to know the motives of their own actions, and 'tis a sin to act without some reason.

WILL. And whoever acts without reason, may do a great deal of harm without knowing it.

DAGW. Thou art an endless moralist.

WILL. Now there's a story come into my head, that I will tell your honour, if you'll give me leave.

DAGW. No, William, save it till another time; this is no time for story-telling; but here comes one who is as entertaining as a good story.

Enter Peter Blunt

PETER. Yonder's a musician going to play before the King; it's a new song about the French and English, and the Prince has made the minstrel a squire, and given him I don't know what, and I can't tell whether he don't mention us all one by one; and he is to write another about all us that are to die, that we may be remembered in Old England, for all our blood and bones are in France; and a great deal more than we shall all hear by and by; and I came to tell your honour, because you love to hear war-songs.

DAGW. And who is this minstrel, Peter, do'st know?

PETER. Oh aye, I forgot to tell that; he has got the same name as Sir John Chandos, that the prince is always with—the wise man, that knows us all as well as your honour, only e'nt so good natured.

DAGW. I thank you, Peter, for your information, but not for your compliment, which is not true; there's as much difference between him and me, as between glittering sand and fruitful mould; or shining glass and a wrought diamond, set in rich gold, and fitted to the finger of an emperor: such is that worthy Chandos.

PETER. I know your honour does not think anything of yourself, but everybody else does.

DAGW. Go, Peter, get you gone; flattery is delicious, even from the lips of a babbler.   [*Exit Peter*

WILL. I never flatter your honour.

DAGW. I don't know that.

WILL. Why you know, Sir, when we were in England, at the tournament at Windsor, and the Earl of Warwick was tumbled over, you asked me if he did not look well when he fell? and I said, No, he looked very foolish; and you was very angry with me for not flattering you.

DAGW. You mean that I was angry with you for not flattering the Earl of Warwick.   [*Exeunt*

SCENE V

Sir Thomas Dagworth's Tent. Sir Thomas Dagworth. To him enters Sir Walter Manny

SIR WALTER. Sir Thomas Dagworth, I have been weeping
    Over the men that are to die to-day.

DAGW. Why, brave Sir Walter, you or I may fall.

SIR WALTER. I know this breathing flesh must lie and rot,
    Covered with silence and forgetfulness.
    Death wons in cities' smoke, and in still night,°
    When men sleep in their beds, walketh about!
    How many in walled cities lie and groan,
    Turning themselves upon their beds,
    Talking with Death, answering his hard demands!   10
    How many walk in darkness; terrors are round
    The curtains of their beds; destruction is
    Ready at the door! How many sleep
    In earth, covered with stones and deathy dust,

Resting in quietness, whose spirits walk
Upon the clouds of Heaven, to die no more!
Yet Death is terrible, though borne on angels' wings!
How terrible then is the field of Death,
Where he doth rend the vault of Heaven,
And shake the gates of Hell!   20
O Dagworth, France is sick! The very sky,
Though sunshine light it, seems to me as pale
As the pale fainting man on his death-bed,
Whose face is shown by light of sickly taper!
It makes me sad and sick at very heart,
Thousands must fall to-day!

DAGW. Thousands of souls must leave this prison-house,
To be exalted to those heavenly fields,
Where songs of triumph, palms of victory,
Where peace, and joy, and love, and calm content   30
Sit singing in the azure clouds, and strew
Flowers of Heaven's growth over the banquet-table.
Bind ardent hope upon your feet like shoes,
Put on the robe of preparation;
The table is prepared in shining Heaven;
The flowers of immortality are blown.
Let those that fight, fight in good stedfastness,
And those that fall shall rise in victory.

SIR WALTER. I've often seen the burning field of war,
And often heard the dismal clang of arms;   40
But never, till this fatal day of Crécy,
Has my soul fainted with these views of Death!
I seem to be in one great charnel-house,°
And seem to scent the rotten carcases!
I seem to hear the dismal yells of Death,
While the black gore drops from his horrid jaws:
Yet I not fear the monster in his pride.
But oh the souls that are to die to-day!

DAGW. Stop, brave Sir Walter; let me drop a tear,
Then let the clarion of war begin.   50
I'll fight and weep, 'tis in my country's cause;
I'll weep and shout for glorious liberty.
Grim War shall laugh and shout, decked in tears,
And blood shall flow like streams across the meadows,
That murmur down their pebbly channels, and

Spend their sweet lives to do their country service.
Then shall England's verdure shoot, her fields shall smile,
Her ships shall sing across the foaming sea,
Her mariners shall use the flute and viol,
And rattling guns, and black and dreary war,   60
Shall be no more.
SIR WALTER. Well; let the trumpet sound, and the drum beat;
Let war stain the blue heavens with bloody banners,
I'll draw my sword, nor ever sheath it up,
'Till England blow the trump of victory,
Or I lay stretched upon the field of Death!

[*Exeunt*

SCENE VI

*In the Camp. Several of the warriors met at the King's tent with a minstrel,
who sings the following song:*

O sons of Trojan Brutus, clothed in war,°
Whose voices are the thunder of the field,
Rolling dark clouds o'er France, muffling the sun
In sickly darkness like a dim eclipse,
Threatening as the red brow of storms, as fire
Burning up nations in your wrath and fury!

Your ancestors came from the fires of Troy,
(Like lions roused by lightning from their dens,
Whose eyes do glare against the stormy fires)
Heated with war, filled with the blood of Greeks,   10
With helmets hewn, and shields covered with gore,
In navies black, broken with wind and tide!°

They landed in firm array upon the rocks
Of Albion; they kissed the rocky shore;
'Be thou our mother, and our nurse,' they said,°
'Our children's mother; and thou shalt be our grave,
The sepulchre of ancient Troy, from whence
Shall rise cities, and thrones, and arms, and awful powers.'

Our fathers swarm from the ships. Giant voices
Are heard from the hills, the enormous sons   20
Of ocean run from rocks and caves: wild men,
Naked and roaring like lions, hurling rocks,
And wielding knotty clubs, like oaks entangled
Thick as a forest, ready for the axe.

Our fathers move in firm array to battle;
The savage monsters rush like roaring fire.
Like as a forest roars with crackling flames,
When the red lightning, borne by furious storms,
Lights on some woody shore; the parched heavens
Rain fire into the molten raging sea!   30

The smoking trees are strewn upon the shore,
Spoiled of their verdure! Oh how oft have they
Defied the storm that howled o'er their heads!
Our fathers, sweating, lean on their spears, and view
The mighty dead: giant bodies, streaming blood,
Dread visages, frowning in silent death!

Then Brutus spoke, inspired; our fathers sit
Attentive on the melancholy shore.
Hear ye the voice of Brutus. 'The flowing waves
Of time come rolling o'er my breast,' he said,   40
'And my heart labours with futurity.
Our sons shall rule the empire of the sea;

'Their mighty wings shall stretch from east to west;
Their nest is in the sea, but they shall roam
Like eagles for the prey. Nor shall the young
Crave or be heard, for plenty shall bring forth;
Cities shall sing, and vales in rich array
Shall laugh, whose fruitful laps bend down with fulness.

'Our sons shall rise from thrones in joy,
Each one buckling on his armour; morning   50
Shall be prevented by their swords' gleaming,°
And evening hear their song of victory!
Their towers shall be built upon the rocks;
Their daughters shall sing, surrounded with shining spears!

'Liberty shall stand upon the cliffs of Albion,
Casting her blue eyes over the green ocean;
Or, tow'ring, stand upon the roaring waves,
Stretching her mighty spear o'er distant lands;
While, with her eagle wings, she covereth
Fair Albion's shore, and all her families.'   60

Prologue, Intended for a Dramatic Piece of King Edward the Fourth

Oh for a voice like thunder, and a tongue°
To drown the throat of war! When the senses
Are shaken, and the soul is driven to madness,
Who can stand? When the souls of the oppressed°
Fight in the troubled air that rages, who can stand?
When the whirlwind of fury comes from the
Throne of God, when the frowns of his countenance
Drive the nations together, who can stand?
When Sin claps his broad wings over the battle,
And sails rejoicing in the flood of death;   10
When souls are torn to everlasting fire,
And fiends of Hell rejoice upon the slain,
Oh who can stand? Oh who hath caused this?
Oh who can answer at the throne of God?
The kings and nobles of the land have done it!
Hear it not, Heaven, thy ministers have done it!

Prologue to King John

Justice hath heaved a sword to plunge in Albion's breast; for Albion's sins are crimson dyed, and the red scourge follows her desolate sons! Then Patriot rose; full oft did Patriot rise, when Tyranny hath stained fair Albion's breast with her own children's gore. Round his majestic feet deep thunders roll; each heart does tremble, and each knee grows slack. The stars of heaven tremble; the roaring voice of war, the trumpet, calls to battle! Brother in brother's blood must bathe, rivers of death! O land, most hapless! O beauteous island, how forsaken! Weep from thy silver fountains; weep from thy gentle rivers! The angel of the island weeps! Thy widowed virgins weep beneath thy shades! Thy aged fathers gird themselves for war! The sucking infant lives to die in battle; the weeping mother feeds him for the slaughter! The husbandman doth leave his bending harvest! Blood cries afar! The land doth sow itself!° The glittering youth of courts must gleam in arms! The aged senators their ancient swords assume! The trembling sinews of old age must work the work of death against their progeny; for Tyranny hath stretched his purple arm, and 'Blood,' he cries, 'the chariots and the horses, the noise of shout, and dreadful thunder of the battle heard afar!' Beware, O proud! thou shalt be humbled; thy cruel brow, thine iron heart is smitten, though lingering fate is slow. Oh yet may Albion smile again, and stretch her peaceful arms, and raise her golden head, exultingly! Her citizens shall throng about her gates, her mariners shall sing upon the sea, and myriads shall to her temples crowd! Her sons shall joy as in the morning! Her daughters sing as to the rising year!

A War Song to Englishmen

Prepare, prepare, the iron helm of war;°
Bring forth the lots, cast in the spacious orb.
Th' angel of fate turns them with mighty hands,
And casts them out upon the darkened earth!

       Prepare, prepare.

Prepare your hearts for death's cold hand! prepare
Your souls for flight, your bodies for the earth!
Prepare your arms for glorious victory!
Prepare your eyes to meet a holy God!

       Prepare, prepare. 10

Whose fatal scroll is that? Methinks 'tis mine!°
Why sinks my heart, why faltereth my tongue?
Had I three lives, I'd die in such a cause,
And rise, with ghosts, over the well-fought field.

       Prepare, prepare.

The arrows of almighty God are drawn!
Angels of death stand in the low'ring heavens!
Thousands of souls must seek the realms of light,
And walk together on the clouds of heaven!

       Prepare, prepare. 20

Soldiers, prepare! Our cause is Heaven's cause;
Soldiers, prepare! Be worthy of our cause:
Prepare to meet our fathers in the sky:
Prepare, O troops, that are to fall to-day!

                        Prepare, prepare.

Alfred shall smile, and make his harp rejoice;
The Norman William, and the learned clerk,°
And Lion Heart, and black-browed Edward, with
His loyal queen shall rise, and welcome us!

                        Prepare, prepare.   30

The Couch of Death

The veiled evening walked solitary down the western hills, and silence reposed in the valley; the birds of day were heard in their nests, rustling in brakes and thickets; and the owl and bat flew round the darkening trees: all is silent when Nature takes her repose. In former times, on such an evening, when the cold clay breathed with life, and our ancestors, who now sleep in their graves, walked on the steadfast globe, the remains of a family of the tribes of earth, a mother and a sister, were gathered to the sick-bed of a youth. Sorrow linked them together, leaning on one another's necks alternately like lilies; dropping tears in each other's bosom, they stood by the bed, like reeds bending over a lake when the evening drops trickle down. His voice was low as the whisperings of the woods when the wind is asleep, and the visions of Heaven unfold their visitation. 'Parting is hard, and death is terrible; I seem to walk through a deep valley, far from the light of day, alone and comfortless! The damps of death fall thick upon me! Horrors stare me in the face! I look behind, there is no returning; Death follows after me; I walk in regions of Death, where no tree is; without a lantern to direct my steps, without a staff to support me.' Thus he laments through the still evening, till the curtains of darkness were drawn! Like the sound of a broken pipe, the aged woman raised her voice. 'O my son, my son, I know but little of the path thou goest! But lo, there is a God, who made the world; stretch out thy hand to Him.' The youth replied, like a voice heard from a sepulchre: 'My hand is feeble, how should I stretch it out? My ways are sinful, how should I raise mine eyes? My voice hath used deceit, how should I call on Him who is truth? My breath is loathsome, how should He not be offended? If I lay my face in the dust, the grave opens its mouth for me; if I lift up my head, sin covers me as a cloak!° O my dear friends, pray ye for me! Stretch forth your hands, that my helper may come! Through the void space I walk between the sinful world and eternity! Beneath me burns eternal fire! Oh for a hand to pluck me forth!' As the voice of an omen heard in the silent valley, when the few inhabitants cling trembling together; as the voice of the angel of death, when the thin beams of the moon give a faint light, such was this young man's voice to his friends! Like the bubbling waters of the brook in the dead of night, the aged woman raised her cry, and said: 'O voice, that dwellest in my breast, can I not cry, and lift my eyes to Heaven? Thinking of this, my spirit is turned within me into confusion! O my child, my child! is thy breath infected? So is mine. As the deer, wounded by the brooks of water, so the arrows of sin stick in my flesh; the poison hath entered into my marrow.' Like rolling waves, upon a desert shore, sighs succeeded sighs; they covered their faces, and wept! The youth lay silent; his mother's arm was under his head; he was like a cloud tossed by the winds, till the sun shine, and the drops of rain glisten, the yellow harvest breathes, and the thankful eyes of the villagers are turned up in smiles. The traveller that hath taken shelter under an oak, eyes the distant country with joy! Such smiles were seen upon the face of the youth! A visionary hand wiped away his tears,° and a ray of light beamed around his head! All was still. The moon hung not out her lamp,° and the stars faintly glimmered in the summer sky; the breath of night slept among the leaves of the forest; the bosom of the lofty hill drank in the silent dew, while on his majestic brow the voice of angels is heard, and stringed sounds ride upon the wings of night. The sorrowful pair lift up their heads; hovering angels are around them; voices of comfort are heard over the Couch of Death, and the youth breathes out his soul with joy into eternity.

Contemplation

Who is this, that with unerring step dares tempt the wilds, where only Nature's foot hath trod? 'Tis Contemplation, daughter of the grey morning! Majestical she steppeth, and with her pure quill on every flower writeth Wisdom's name. Now lowly bending, whispers in mine ear, 'O man, how great, how little thou! O man, slave of each moment, lord of eternity! seest thou where Mirth sits on the painted cheek? Doth it not seem ashamed of such a place, and grow immoderate to brave it out? Oh what an humble garb true Joy puts on! Those who want

Happiness must stoop to find it; it is a flower that grows in every vale. Vain foolish man, that roams on lofty rocks! where, 'cause his garments are swoln with wind, he fancies he is grown into a giant! Lo, then, Humility; take it, and wear it in thine heart, lord of thyself; thou then art lord of all. Clamour brawls along the streets, and Destruction hovers in the city's smoke; but on these plains, and in these silent woods, true joys descend. Here build thy nest; here fix thy staff. Delights blossom around; numberless beauties blow; the green grass springs in joy, and the nimble air kisses the leaves; the brook stretches its arms along the velvet meadow; its silver inhabitants sport and play; the youthful sun joys like a hunter roused to the chase; he rushes up the sky, and lays hold on the immortal coursers of day; the sky glitters with the jingling trappings! Like a triumph, season follows season, while the airy music fills the world with joyful sounds.' I answered: 'Heavenly goddess! I am wrapped in mortality; my flesh is a prison, my bones the bars of death; Misery builds over our cottage roofs, and Discontent runs like a brook. Even in childhood, Sorrow slept with me in my cradle; he followed me up and down in the house; when I grew up he was my school-fellow; thus he was in my steps and in my play, till he became to me as my brother. I walked through dreary places with him, and in church-yards; and I oft found myself sitting by Sorrow on a tomb-stone!'

Samson

Samson, the strongest of the children of men, I sing:° how he was foiled by woman's arts, by a false wife brought to the gates of death! O Truth, that shinest with propitious beams, turning our earthly night to heavenly day from presence of the Almighty Father! thou visitest our darkling world with blessed feet, bringing good news of Sin and Death destroyed! O white-robed angel, guide my timorous hand to write as on a lofty rock with iron pens the words of truth, that all who pass may read. Now Night, noon-tide of damned spirits, over the silent earth spreads her pavilion, while in dark council sat Philista's lords;° and where strength failed, black thoughts in ambush lay. Their helmed youth and aged warriors in dust together lie, and Desolation spreads his wings over the land of Palestine; from side to side the land groans, her prowess lost, and seeks to hide her bruised head under the mists of night, breeding dark plots. For Dalila's fair arts have long been tried in vain; in vain she wept in many a treacherous tear. 'Go on, fair traitress; do thy guileful work; ere once again the changing moon her circuit hath performed, thou shalt overcome, and conquer him by force unconquerable, and wrest his secret from him. Call thine alluring arts and honest-seeming brow, the holy kiss of love, and the transparent tear; put on fair linen, that with the lily vies, purple and silver; neglect thy hair, to seem more lovely in thy loose attire; put on thy country's pride, deceit; and eyes of love decked in mild sorrow, and sell thy lord for gold.' For now, upon her sumptuous couch reclined, in gorgeous pride, she still entreats, and still she grasps his vigorous knees with her fair arms. 'Thou lov'st me not! Thou'rt war, thou art not love! O foolish Dalila! O weak woman! it is death clothed in flesh thou lovest, and thou hast been encircled in his arms! Alas, my lord, what am I calling thee? Thou art my God! To thee I pour my tears for sacrifice morning and evening. My days are covered with sorrow! shut up, darkened. By night I am deceived! Who says that thou wast born of mortal kind? Destruction was thy father; a lioness suckled thee;° thy young hands tore human limbs, and gorged human flesh! Come hither, Death; art thou not Samson's servant? 'Tis Dalila that calls, thy master's wife. No, stay, and let thy master do the deed; one blow of that strong arm would ease my pain; then should I lay at quiet, and have rest. Pity forsook thee at thy birth! O Dagon furious, and all ye gods of Palestine, withdraw your hand! I am but a weak woman. Alas, I am wedded to your enemy! I will go mad, and tear my crisped hair; I'll run about, and pierce the ears o'th'gods! O Samson, hold me not; thou lovest me not! Look not upon me with those deathful eyes! Thou wouldst my death, and death approaches fast.' Thus, in false tears, she bathed his feet, and thus she day by day oppressed his soul. He seemed a mountain, his brow among the clouds; she seemed a silver stream, his feet embracing. Dark thoughts rolled to and fro in his mind, like thunder clouds, troubling the sky; his visage was troubled; his soul was distressed. 'Though I shall tell her all my heart, what can I fear? Though I should tell this secret of my birth, the utmost may be warded off as well when told as now.' She saw him moved, and thus resumes her wiles. 'Samson, I'm thine; do with me what thou wilt; my friends are enemies; my life is death; I am a traitor to my nation, and despised; my joy is given into the hands of him who hates me, using deceit to the wife of his bosom. Thrice hast thou mocked me, and grieved my soul. Didst thou not tell me with green withes to bind thy nervous arms, and after that, when I had found thy falsehood, with new ropes to bind thee fast? I knew thou didst but mock me. Alas, when in thy sleep I bound thee with them to try thy truth, I cried, "The Philistines be upon thee, Samson!" Then did suspicion wake thee; how didst thou rend the feeble ties! Thou fearest nought, what shouldst thou fear? Thy power is more than mortal; none can hurt thee; thy bones are brass; thy sinews are iron! Ten thousand spears are like the summer grass; an army of mighty men are as flocks in the vallies; what canst thou fear? I drink my tears like water; I live upon sorrow! O worse than wolves and tigers, what canst thou give when such a trifle is denied me? But oh at last thou mockest me to shame my over-fond inquiry! Thou toldest me to weave thee to the beam by thy strong hair; I did even that to try thy truth; but when I cried, "The Philistines be upon thee", then didst thou leave me to bewail that Samson loved me not.' He sat, and inward grieved; he saw and loved the beauteous suppliant, nor could conceal aught that might appease her; then, leaning on her bosom, thus he spoke: 'Hear, O Dalila! Doubt no more of Samson's love; for that fair breast was made the ivory palace of my inmost heart, where it shall lie at rest; for sorrow is the lot of all of woman born. For care was I brought forth, and labour is my lot; nor matchless might, nor wisdom, nor every gift enjoyed, can from the heart of man hide sorrow. Twice was my birth foretold from Heaven,° and twice a sacred vow enjoined me that I should drink no wine, nor eat of any unclean thing, for holy unto Israel's God I am, a Nazarite even from my mother's womb. Twice was it told, that it might not be broken. "Grant me a son, kind Heaven", Manoa cried; but Heaven refused! Childless he mourned, but thought his God knew best. In solitude, though not obscure, in Israel he lived, till venerable age came on; his flocks increased, and plenty crowned his board: beloved, revered of man! But God hath other joys in store. Is burdened Israel his grief? The son of his old age shall set it free! The venerable sweetener of his life receives the promise first from Heaven. She saw the maidens play, and blessed their innocent mirth; she blessed each new-joined pair; but from her the long-wished deliverer shall spring. Pensive, alone she sat within the house, when busy day was fading, and calm evening, time for contemplation, rose from the forsaken east, and drew the curtains of heaven; pensive she sat, and thought on Israel's grief, and silent prayed to Israel's God. When, lo, an angel from the fields of light entered the house! His form was manhood in the prime, and from his spacious brow shot terrors through the evening shade! But mild he hailed her: "Hail, highly favoured!"° said he, "for, lo, thou shalt conceive, and bear a son, and Israel's strength shall be upon his shoulders, and he shall be called Israel's Deliverer! Now therefore drink no wine, and eat not any unclean thing, for he shall be a Nazarite to God." Then, as a neighbour when his evening tale is told departs, his blessing leaving, so seemed he to depart. She wondered with exceeding joy, nor knew he was an angel. Manoa left his fields to sit in the house, and take his evening's rest from labour, the sweetest time that God has allotted mortal man. He sat, and heard with joy, and praised God who Israel still doth keep. The time rolled on, and Israel groaned oppressed. The sword was bright, while the plough-share rusted, till hope grew feeble, and was ready to give place to doubting; then prayed Manoa; "O Lord, thy flock is scattered on the hills! The wolf teareth them; oppression stretches his rod over our land; our country is ploughed with swords, and reaped in blood! The echoes of slaughter reach from hill to hill! Instead of peaceful pipe, the shepherd bears a sword; the ox-goad is turned into a spear! Oh when shall our deliverer come? The Philistine riots on our flocks; our vintage is gathered by bands of enemies! Stretch forth thy hand, and save." Thus prayed Manoa. The aged woman walked into the field, and, lo, again the angel came! clad as a traveller fresh risen on his journey; she ran and called her husband, who came and talked with him. "O man of God," said he, "thou comest from far! Let us detain thee while I make ready a kid, that thou mayest sit and eat, and tell us of thy name and warfare: that when thy sayings come to pass, we may honour thee." The angel answered, "My name is wonderful;° enquire not after it, seeing it is a secret; but, if thou wilt, offer an offering unto the Lord." '

[The Vision of Pride]

then she° bore pale Desire, father of Curiosity, a virgin ever young, and after leaden Sloth: from whom came Ignorance, who brought forth Wonder. These are the Gods which came from Fear, for Gods like these nor male nor female are but single pregnate or, if they list, together mingling bring forth mighty powers. She knew them not, yet they all war with Shame and strengthen her weak arm.

Pride awoke, nor knew that Joy was born, and, taking poisonous seed from her own bowel, in the monster Shame infused. Forth came Ambition, crawling like a toad. Pride rears it in her bosom and the Gods all bow to it. So great its power that Pride, inspired by it, prophetic saw the kingdoms of the world and all their glory,° giants of mighty arm before the flood,° Cain's city° built with murder. Then Babel mighty reared him to the skies, Babel with thousand tongues. Confusion° it was called, and given to Shame. This Pride observing inly grieved, but knew not that the rest was given to Shame as well as this.

Then Nineveh, and Babylon, and costly Tyre, and ev'n Jerusalem was shown, the holy city. Then Athens' learning and the pride of Greece, and further from the rising sun was Rome seated on seven hills, the mistress of the world, emblem of Pride. She saw the arts their treasures bring, and Luxury his bounteous table spread. But now a cloud o'ercast, and back to th' east, to Constantine's great city, Empire fled; ere long to bleed and die—sacrifice done by a priestly hand. So once the sun his chariot drew back to prolong a good king's life.°

The cloud o'erpassed and Rome now shone again, mitered and crowned with triple crown.° Then Pride was better pleased. She saw the world fall down in adoration. But now full to the setting sun a sun arose out of the sea;° it rose and shed sweet influence o'er the earth. Pride feared for her city, but not long, for, looking steadfastly, she saw that Pride reigned here.

Now direful pains accost her and, still pregnant, so Envy came and Hate, twin progeny. Envy hath a serpent's head° of fearful bulk hissing with hundred tongues; her poisonous breath breeds Satire, foul contagion from which none are free. O'erwhelmed by ever-during thirst she swalloweth her own poison, which consumes her nether parts; from whence a river springs. Most black and loathsome through the land it runs, rolling with furious noise, but at the last it settles in a lake called Oblivion. 'Tis at this river's fount where every mortal's cup is mixed. My cup is filled with Envy's rankest draught; a miracle no less can set me right. Desire still pines but for one cooling drop and 'tis denied, while others in Contentment's downy nest do sleep. It is the cursed thorn wounding my breast that makes me sing. However sweet 'tis Envy that inspires my song. Pricked by the fame of others, how I mourn; and my complaints are sweeter than their joys, but, oh, could I at Envy shake my hands, my notes should rise to meet the new-born day.

Hate, meagre hag, sets Envy on, unable to do aught herself but worn away, a bloodless demon. The Gods all serve her at her will; so great her power is, like fabled Hecate° she doth bind them to her law. Far in a direful cave she lives unseen, closed from the eye of day, to the hard rock transfixed by fate, and here she works her witcheries: that when she groans she shakes the solid ground. Now Envy she controls with numbing trance and Melancholy sprung from her dark womb.

There is a Melancholy, oh how lovely 'tis, whose heaven is in the heavenly mind; for she from Heaven came, and where she goes Heaven still doth follow her. She brings true Joy, once fled, and Contemplation is her daughter. Sweet Contemplation! She brings Humility to Man. 'Take her', she says, 'and wear her in thine heart; lord of thyself, thou then art lord of all'. 'Tis Contemplation teacheth Knowledge, truly how to know, and reinstates him on his throne, once lost; how lost I'll tell.

But stop the motley song. I'll show how Conscience came from Heaven. But, oh, who listens to his voice? 'Twas Conscience who brought Melancholy down. Conscience was sent a guard to Reason: Reason, once fairer than the light, till fouled in Knowledge's dark prison house. For Knowledge drove sweet Innocence away, and Reason would have followed but fate suffered not. Then down came Conscience with his lovely band.

The eager song goes on, telling how Pride against her father° warred and overcame. Down his white beard the silver torrents roll, and swelling sighs burst forth; his children all in arms appear to tear him from his throne. Black was the deed, most black. Shame in a mist sat round his troubled head, and filled him with confusion. Fear as a torrent wild roared round his throne; the mighty pillars shake. Now all the Gods in blackening ranks appear, like a tempestuous thunder cloud. Pride leads them on. Now they surround the God and bind him fast. Pride bound him, then usurped o'er all the Gods. She rode upon the swelling wind, and scattered all who durst t'oppose but Shame, opposing fierce and hovering over her in the dark'ning storm. She brought forth Rage, and Shame bore Honour and made league with Pride.

Meanwhile Strife, mighty prince, was born. Envy in direful pains him bore; then Envy brought forth Care. Care sitteth in the wrinkled brow. Strife shapeless sitteth under thrones of kings, like smould'ring fire, or in the buzz of cities flies abroad.° Care brought forth Covet, eyeless and prone to th'earth, and Strife brought forth Revenge. Hate brooding in her dismal den grew pregnant, and bore Scorn and Slander. Scorn waits on Pride, but Slander flies around the world to do the work of Hate, her drudge and elf. But Policy doth drudge for Hate as well as Slander, and oft makes use of her—Policy, son of Shame. Indeed Hate controls all the Gods at will. Policy brought forth Guile and Fraud. These Gods last named live in the smoke of cities, on dusky wing breathing forth clamour and destruction. Alas, in cities where's the man whose face is not a mask unto his heart?

Pride made a Goddess fair—or image rather,° till Knowledge animated it; 'twas called Self-love. The Gods admiring loaded her with gifts, as once Pandora.° She 'mongst men was sent and worse ills° attended her by far. She was a Goddess powerful, and bore Conceit and Emulation; and Policy doth dwell with her, by whom she had a son called Suspicion. Go see the city, friends joined hand in hand. Go see the natural tie of flesh and blood. Go see, more strong, the ties of marriage love. Thou scarce shall find but Self-love stands between.°

' "Woe!" cried the Muse'

'Woe!' cried the Muse.° Tears started at the sound. Grief perched upon my brow and Thought embraced her. 'What does this mean,' I cried, 'when all around Summer hath spread her plumes and tunes her notes, when buxom Joy doth fan his wings, and golden Pleasures beam around my head? Why, Grief, dost thou accost me?' The Muse then struck her deepest string, and Sympathy came forth. She spread her awful wings and gave me up. My nerves with trembling curdle all my blood, and ev'ry piece of flesh doth cry out, 'Woe'. How soon the winds sing round the dark'ning storm erewhile so fair! And now they fall and beg the skies will weep. A day like this laid Elfrid° in the dust, sweet Elfrid, fairer than the beaming sun. Oh, soon cut off in th' morning of her days, 'twas the rude thunder stroke that closed her eyes, and laid her lilied beauties on the green!

The dance was broke—the circle just begun. The flower was plucked and yet it was not blown. 'But what art thou?' I could no more, till mute Attention struck my listening ear. It spoke: 'I come, my friend, to take my last farewell, sunk by the hand of death in wat'ry tomb. O'er yonder lake, swift as the nightly blast that blights the infant bud, the winds their sad complainings bear for comrade lost, untimely lost—thy comrade once. When living, thee I loved ev'n unto death; now, dead, I'll guard thee from approaching ill. Farewell, my time is gone.' It said no more, but vanished ever from my sight.

SEPTENARY VERSE OF THE
FRENCH REVOLUTIONARY PERIOD

Tiriel (c.1789)

I

And aged Tiriel stood before the gates of his beautiful palace°
With Myratana, once the queen of all the western plains;°
But now his eyes were darkened, and his wife fading in death.
They stood before their once delightful palace, and thus the voice
Of aged Tiriel arose, that his sons might hear in their gates:

'Accursed race of Tiriel, behold your father.
Come forth and look on her that bore you; come you accursed sons!
In my weak arms I here have borne your dying mother.
Come forth, sons of the curse, come forth, see the death of
    Myratana!'

His sons ran from their gates and saw their aged parents stand,   10
And thus the eldest son of Tiriel raised his mighty voice:

'Old man, unworthy to be called the father of Tiriel's race
(For every one of those thy wrinkles, each of those grey hairs,
Are cruel as death, and as obdurate as the devouring pit),
Why should thy sons care for thy curses, thou accursed man?
Were we not slaves till we rebelled? Who cares for Tiriel's curse?
His blessing was a cruel curse; his curse may be a blessing.'

He ceased. The aged man raised up his right hand to the heavens:
His left supported Myratana shrinking in pangs of death.
The orbs of his large eyes he opened, and thus his voice went forth: 20

'Serpents, not sons, wreathing around the bones of Tiriel!°
Ye worms of death feasting upon your aged parents' flesh,
Listen and hear your mother's groans. No more accursed sons
She bears; she groans not at the birth of Heuxos or Yuva.°
These are the groans of death, ye serpents, these are the groans of
    death.
Nourished with milk, ye serpents, nourished with mother's tears
    and cares,
Look at my eyes, blind as the orbless skull among the stones.
Look at my bald head! Hark, listen, ye serpents! Listen!
What, Myratana? What, my wife? O soul, O spirit, O fire!
What, Myratana, art thou dead? Look here, ye serpents look!   30

The serpents sprung from her own bowels have drained her dry as
    this.
Curse on your ruthless heads, for I will bury her even here.'°

So saying be began to dig a grave with his aged hands,
But Heuxos called a son of Zazel to dig their mother a grave:°

'Old cruelty, desist, and let us dig a grave for thee.
Thou hast refused our charity, thou hast refused our food,
Thou has refused our clothes, our beds, our houses for thy
    dwelling,
Choosing to wander like a son of Zazel in the rocks.
Why dost thou curse? Is not the curse now come upon your head?
Was it not you enslaved the sons of Zazel, and they have cursed,   40
And now you feel it? Dig a grave and let us bury our mother.'

'There take the body, cursed sons, and may the heavens rain wrath
As thick as northern fogs around your gates to choke you up,
That you may lie, as now your mother lies, like dogs cast out,
The stink of your dead carcases annoying man and beast,
Till your white bones are bleached with age for a memorial.
No! your remembrance shall perish; for when your carcases
Lie stinking on the earth the buriers shall arise from the east,
And not a bone of all the sons of Tiriel remain.
Bury your mother; but you cannot bury the curse of Tiriel.'   50

He ceased, and darkling o'er the mountains sought his pathless
    way.

II

He wandered day and night. To him both day and night were dark;
The sun he felt, but the bright moon was now a useless globe.
O'er mountains, and through vales of woe, the blind and aged man
Wandered, till he that leadeth all led him to the vales of Har;°

And Har and Heva like two children sat beneath the oak.°
Mnetha, now aged, waited on them, and brought them food and
    clothing,°
But they were as the shadow of Har, and as the years forgotten.
Playing with flowers, and running after birds, they spent the day,
And in the night like infants slept delighted with infant dreams.   60

Soon as the blind wanderer entered the pleasant gardens of Har
They ran weeping like frighted infants for refuge in Mnetha's
　　arms.
The blind man felt his way and cried: 'Peace to these open doors!
Let no one fear, for poor blind Tiriel hurts none but himself.
Tell me, O friends, where am I now, and in what pleasant place?'

'This is the valley of Har,' said Mnetha, 'and this the tent of Har.
Who art thou, poor blind man, that takest the name of Tiriel on
　　thee?
Tiriel is king of all the west. Who art thou? I am Mnetha,
And this is Har and Heva, trembling like infants by my side.'

'I know Tiriel is king of the west, and there he lives in joy.   70
No matter who I am, O Mnetha, if thou hast any food
Give it to me, for I cannot stay; my journey is far from hence.'

Then Har said: 'O my mother Mnetha, venture not so near him,
For he is the king of rotten wood, and of the bones of death.
He wanders without eyes, and passes through thick walls and
　　doors.
Thou shalt not smite my mother Mnetha, O thou eyeless man!'

'A wanderer, I beg for food. You see I cannot weep.
I cast away my staff, the kind companion of my travel,
And I kneel down that you may see I am a harmless man.'

He kneeled down, and Mnetha said: 'Come, Har and Heva, rise.   80
He is an innocent old man, and hungry with his travel.'

Then Har arose, and laid his hand upon old Tiriel's head.

'God bless thy poor bald pate. God bless thy hollow winking eyes.
God bless thy shrivelled beard. God bless thy many-wrinkled
　　forehead.
Thou hast no teeth, old man; and thus I kiss thy sleek bald head.
Heva, come kiss his bald head, for he will not hurt us, Heva.'

Then Heva came, and took old Tiriel in her mother's arms.

'Bless thy poor eyes, old man, and bless the old father of Tiriel.°
Thou art my Tiriel's old father. I know thee through thy wrinkles,
Because thou smellest like the fig tree; thou smellest like ripe figs.   90
How didst thou lose thy eyes, old Tiriel? Bless thy wrinkled face.'

Mnetha said: 'Come in, aged wanderer; tell us of thy name.
Why shouldest thou conceal thyself from those of thine own flesh?'

'I am not of this region,' said Tiriel dissemblingly.
'I am an aged wanderer, once father of a race
Far in the north, but they were wicked and were all destroyed,
And I their father sent an outcast. I have told you all;
Ask me no more, I pray, for grief hath sealed my precious sight.'

'O Lord,' said Mnetha, 'how I tremble! Are there then more
    people,
More human creatures on this earth, beside the sons of Har?'   100

'No more,' said Tiriel, 'but I, remain on all this globe;
And I remain an outcast. Hast thou anything to drink?'

Then Mnetha gave him milk and fruits, and they sat down
    together.

III

They sat and ate, and Har and Heva smiled on Tiriel.

'Thou art a very old, old man, but I am older than thou.
How came thine hair to leave thy forehead? How came thy face so
    brown?
My hair is very long. My beard doth cover all my breast.
God bless thy piteous face! To count the wrinkles in thy face
Would puzzle Mnetha. Bless thy face, for thou art Tiriel.'

'Tiriel I never saw but once. I sat with him and ate.   110
He was as cheerful as a prince, and gave me entertainment;
But long I stayed not at his palace, for I am forced to wander.'

'What! Wilt thou leave us too?' said Heva. 'Thou shalt not leave us
    too.
For we have many sports to show thee, and many songs to sing;
And after dinner we will walk into the cage of Har,
And thou shalt help us to catch birds, and gather them ripe
    cherries.
Then let thy name be Tiriel, and never leave us more.'

'If thou dost go,' said Har, 'I wish thine eyes may see thy folly.
My sons have left me; did thine leave thee? Oh, 'twas very cruel!'° 120

'No, venerable man,' said Tiriel, 'Ask me not such things;
For thou dost make my heart to bleed. My sons were not like thine,
But worse. Oh, never ask me more, or I must flee away.'

'Thou shalt not go,' said Heva, 'till thou hast seen our singing
    birds,
And heard Har sing in the great cage, and slept upon our fleeces.
Go not, for thou are so like Tiriel that I love thine head—
Though it is wrinkled, like the earth parched with the summer
    heat.'

Then Tiriel rose up from the seat, and said: 'God bless these tents.
My journey is o'er rocks and mountains, not in pleasant vales.
I must not sleep nor rest, because of madness and dismay.'

And Mnetha said: 'Thou must not go to wander dark, alone, 130
But dwell with us and let us be to thee instead of eyes;
And I will bring thee food, old man, till death shall call thee hence.'

Then Tiriel frowned, and answered: 'Did I not command you,
    saying
Madness and deep dismay possess the heart of the blind man,
The wanderer who seeks the woods leaning upon his staff?'

Then Mnetha trembling at his frowns led him to the tent door,
And gave to him his staff and blessed him. He went on his way.

But Har and Heva stood and watched him till he entered the wood,
And then they went and wept to Mnetha; but they soon forgot
    their tears.

IV

Over the weary hills the blind man took his lonely way. 140
To him the day and night alike was dark and desolate.
But far he had not gone when Ijim, from his woods come down,°
Met him at entrance of the forest in a dark and lonely way.

'Who art thou, eyeless wretch, that thus obstruct'st the lion's path?
Ijim shall rend thy feeble joints, thou tempter of dark Ijim.
Thou hast the form of Tiriel, but I know thee well enough.
Stand from my path, foul fiend. Is this the last of thy deceits:
To be a hypocrite, and stand in shape of a blind beggar?'

The blind man heard his brother's voice, and kneeled down on his
    knee.
'O brother Ijim—if it is thy voice that speaks to me—   150
Smite not thy brother Tiriel, though weary of his life.
My sons have smitten me already, and if thou smitest me
The curse that rolls over their heads will rest itself on thine.
'Tis now seven years since in my palace I beheld thy face.'

'Come, thou dark fiend, I dare thy cunning! Know that Ijim scorns
To smite thee in the form of helpless age and eyeless policy.
Rise up, for I discern thee, and I dare thy eloquent tongue!
Come, I will lead thee on thy way, and use thee as a scoff.'

'O brother Ijim, thou beholdest wretched Tiriel.
Kiss me, my brother, and then leave me to wander desolate.'   160

'No, artful fiend; but I will lead thee. Dost thou want to go?
Reply not, lest I bind thee with the green flags of the brook.
Ay, now thou art discovered I will use thee like a slave.'

When Tiriel heard the words of Ijim he sought not to reply.
He knew 'twas vain, for Ijim's words were as the voice of fate.

And they went on together, over hills, through woody dales,
Blind to the pleasures of the sight, and deaf to warbling birds.
All day they walked, and all the night beneath the pleasant moon,
Westwardly journeying, till Tiriel grew weary with his travel.

'O Ijim, I am faint and weary, for my knees forbid   170
To bear me further. Urge me not, lest I should die with travel.
A little rest I crave, a little water from a brook,
Or I shall soon discover that I am a mortal man,
And you will lose your once-loved Tiriel. Alas, how faint I am!'

'Impudent fiend,' said Ijim, 'Hold thy glib and eloquent tongue!
Tiriel is a king, and thou the tempter of dark Ijim.
Drink of this running brook, and I will bear thee on my shoulders.'

He drank, and Ijim raised him up and bore him on his shoulders.
All day he bore him and, when evening drew her solemn curtain,
Entered the gates of Tiriel's palace, and stood and called aloud: 180

'Heuxos, come forth! I here have brought the fiend that troubles
   Ijim.
Look! Know'st thou aught of this grey beard, or of these blinded
   eyes?'

Heuxos and Lotho ran forth at the sound of Ijim's voice,°
And saw their aged father borne upon his mighty shoulders.
Their eloquent tongues were dumb, and sweat stood on their
   trembling limbs.
They knew 'twas vain to strive with Ijim; they bowed and silent
   stood.

'What, Heuxos! Call thy father, for I mean to sport tonight.
This is the hypocrite that sometimes roars a dreadful lion.°
Then I have rent his limbs, and left him rotting in the forest
For birds to eat; but I have scarce departed from the place   190
But like a tiger he would come, and so I rent him too.
Then like a river he would seek to drown me in his waves,
But soon I buffeted the torrent; anon like to a cloud
Fraught with the swords of lightning, but I braved the vengeance
   too.

Then he would creep like a bright serpent, till around my neck,
While I was sleeping, he would twine; I squeezed his pois'nous
   soul.
Then, like a toad or like a newt, would whisper in my ears,°
Or like a rock stood in my way, or like a pois'nous shrub.
At last I caught him in the form of Tiriel, blind and old,
And so I'll keep him. Fetch your father, fetch forth Myratana!'°   200

They stood confounded, and thus Tiriel raised his silver voice:

'Serpents, not sons, why do you stand? Fetch hither Tiriel,
Fetch hither Myratana; and delight yourselves with scoffs.
For poor blind Tiriel is returned, and this much injured head
Is ready for your bitter taunts. Come forth, sons of the curse!'

Meantime the other sons of Tiriel ran around their father.
Confounded at the terrible strength of Ijim, they knew 'twas vain;
Both spear and shield were useless, and the coat of iron mail.
When Ijim stretched his mighty arm the arrow from his limbs
Rebounded, and the piercing sword broke on his naked flesh.   210

'Then is it true, Heuxos, that thou hast turned thy aged parent
To be the sport of wintry winds?' said Ijim. 'Is this true?
It is a lie, and I am like the tree torn by the wind.
Thou eyeless fiend, and you dissemblers! Is this Tiriel's house?
It is as false as Matha, and as dark as vacant Orcus.°
Escape, ye fiends, for Ijim will not lift his hand against ye.'

So saying, Ijim gloomy turned his back, and silent sought
The secret forests, and all night wandered in desolate ways.

                                    v

And aged Tiriel stood and said: 'Where does the thunder sleep?
Where doth he hide his terrible head? And his swift and fiery
     daughters,   220
Where do they shroud their fiery wings and the terrors of their
     hair?
Earth, thus I stamp thy bosom. Rouse the earthquake from his
     den,
To raise his dark and burning visage through the cleaving ground,
To thrust these towers with his shoulders. Let his fiery dogs
Rise from the centre, belching flames and roarings, dark smoke.°
Where art thou, pestilence that bathest in fogs and standing lakes?°
Rise up thy sluggish limbs, and let the loathsomest of poisons
Drop from thy garments as thou walkest wrapped in yellow clouds.
Here take thy seat, in this wide court; let it be strewn with dead,
And sit and smile upon these cursed sons of Tiriel.   230
Thunder and fire and pestilence, hear you not Tiriel's curse?'

He ceased; the heaving clouds confused rolled round the lofty
     towers,
Discharging their enormous voices. At the father's curse
The earth trembled, fires belched from the yawning clefts,
And when the shaking ceased a fog possessed the accursed clime.

The cry was great in Tiriel's palace. His five daughters ran
And caught him by the garments, weeping with cries of bitter woe.

'Ay, now you feel the curse, you cry; but may all ears be deaf
As Tiriel's, and all eyes as blind as Tiriel's to your woes!
May never stars shine on your roofs, may never sun nor moon   240
Visit you, but eternal fogs hover around your walls!
Hela, my youngest daughter, you shall lead me from this place,°
And let the curse fall on the rest and wrap them up together!'

He ceased, and Hela led her father from the noisome place.
In haste they fled, while all the sons and daughters of Tiriel,
Chained in thick darkness, uttered cries of mourning all the night.
And in the morning, lo, an hundred men in ghastly death,
The four daughters stretched on the marble pavement, silent all,
Fall'n by the pestilence. The rest moped round in guilty fears;
And all the children in their beds were cut off in one night.°   250
Thirty of Tiriel's sons remained to wither in the palace:
Desolate, loathed, dumb, astonished, waiting for black death.

VI

And Hela led her father through the silent of the night,°
Astonished, silent, till the morning beams began to spring.

'Now, Hela, I can go with pleasure and dwell with Har and Heva,
Now that the curse shall clean devour all those guilty sons.
This is the right and ready way; I know it by the sound
That our feet make. Remember, Hela, I have saved thee from
     death.
Then be obedient to thy father, for the curse is taken off thee.
I dwelt with Myratana five years in the desolate rock,   260
And all that time we waited for the fire to fall from heaven,
Or for the torrents of the sea to overwhelm you all.
But now my wife is dead and all the time of grace is past.
You see the parents' curse. Now lead me where I have com-
     manded.'

'O leagued with evil spirits, thou accursed man of sin!
True, I was born thy slave. Who asked thee to save me from death?
'Twas for thyself, thou cruel, because thou wantest eyes.'

'True, Hela: this is the desert of all those cruel ones.
Is Tiriel cruel? Look! His daughter—and his youngest daughter—
Laughs at affection, glories in rebellion, scoffs at love.   270
I have not eat these two days. Lead me to Har and Heva's tent,
Or I will wrap thee up in such a terrible father's curse
That thou shalt feel worms in thy marrow creeping through thy
    bones.
Yet thou shalt lead me. Lead me, I command, to Har and Heva.'

'O cruel! O destroyer! O consumer! O avenger!
To Har and Heva I will lead thee then. Would that they would
    curse!
Then would they curse as thou hast cursed. But they are not like
    thee.
Oh, they are holy and forgiving, filled with loving mercy,
Forgetting the offences of their most rebellious children!
Or else thou wouldest not have lived to curse thy helpless
    children.'   280

'Look on my eyes, Hela, and see, for thou hast eyes to see.
The tears swell from my stony fountains; wherefore do I weep?
Wherefore from my blind orbs art thou not seized with pois'nous
    stings?
Laugh, serpent, youngest venomous reptile of the flesh of Tiriel,
Laugh! For thy father Tiriel shall give thee cause to laugh,
Unless thou lead me to the tent of Har, child of the curse.'

'Silence thy evil tongue, thou murderer of thy helpless children!
I lead thee to the tent of Har: not that I mind thy curse,
But that I feel they will curse thee, and hang upon thy bones
Fell shaking agonies, and in each wrinkle of that face   290
Plant worms of death, to feast upon the tongue of terrible curses.'

'Hela, my daughter, listen! Thou art the daughter of Tiriel.
Thy father calls. Thy father lifts his hand unto the heavens,
For thou hast laughed at my tears, and cursed thy aged father.
Let snakes rise from thy bedded locks and laugh among thy curls!'

He ceased; her dark hair upright stood, while snakes enfolded
    round
Her madding brows. Her shrieks appalled the soul of Tiriel.

'What have I done, Hela, my daughter? Fear'st thou now the curse,
Or wherefore dost thou cry? Ah, wretch to curse thy aged father!
Lead me to Har and Heva, and the curse of Tiriel   300
Shall fail. If thou refuse, howl in the desolate mountains!'

VII

She howling led him over mountains and through frighted vales,
Till to the caves of Zazel they approached at eventide.

Forth from their caves old Zazel and his sons ran, when they saw
Their tyrant prince blind, and his daughter howling and leading him.

They laughed and mocked. Some threw dirt and stones as they passed by.
But when Tiriel turned around and raised his awful voice
Some fled away, but Zazel stood still and thus began:
'Bald tyrant, wrinkled cunning, listen to Zazel's chains!
'Twas thou that chained thy brother Zazel. Where are now thine eyes?   310
Shout, beautiful daughter of Tiriel! Thou singest a sweet song.
Where are you going? Come and eat some roots and drink some water.
Thy crown is bald, old man; the sun will dry thy brains away,
And thou wilt be as foolish as thy foolish brother Zazel.'

The blind man heard, and smote his breast and trembling passed on.
They threw dirt after them, till to the covert of a wood
The howling maiden led her father, where wild beasts resort,
Hoping to end her woes; but from her cries the tigers fled.
All night they wandered through the wood, and when the sun arose
They entered on the mountains of Har. At noon the happy tents   320
Were frighted by the dismal cries of Hela on the mountains,

But Har and Heva slept, fearless as babes on loving breasts.
Mnetha awoke; she ran and stood at the tent door, and saw
The aged wanderer led towards the tents. She took her bow
And chose her arrows, then advanced to meet the terrible pair.

VIII

And Mnetha hasted and met them at the gate of the lower garden.

'Stand still, or from my bow receive a sharp and winged death!'

Then Tiriel stood, saying: 'What soft voice threatens such bitter
    things?
Lead me to Har and Heva. I am Tiriel, King of the west.'

And Mnetha led them to the tent of Har, and Har and Heva
Ran to the door. When Tiriel felt the ankles of aged Har
He said: 'O weak mistaken father of a lawless race!
Thy laws O Har, and Tiriel's wisdom, end together in a curse.
Why is one law given to the lion and the patient ox,
And why men bound beneath the heavens in a reptile form:
A worm of sixty winters creeping on the dusky ground?
The child springs from the womb, the father ready stands to form
The infant head, while the mother idle plays with her dog on her
    couch.
The young bosom is cold for lack of mother's nourishment, and
    milk°
Is cut off from the weeping mouth, With difficulty and pain   340
The little lids are lifted and the little nostrils opened.
The father forms a whip to rouse the sluggish senses to act,
And scourges off all youthful fancies from the newborn man.
And when the drone has reached his crawling length
Black berries appear that poison all around him. Such was Tiriel:
Compelled to pray repugnant, and to humble the immortal
    spirit°—
Till I am subtle as a serpent in a paradise,
Consuming all, both flowers and fruits, insects and warbling birds.
And now my paradise is fall'n, and a drear sandy plain
Returns my thirsty hissings in a curse on thee, O Har,°   350
Mistaken father of a lawless race. My voice is past.'

He ceased, outstretched at Har and Heva's feet in awful death.

The Book of Thel (c.*1789*)

THEL'S° MOTTO

Does the eagle know what is in the pit?
Or wilt thou go ask the mole?
Can wisdom be put in a silver rod,
Or love in a golden bowl?°

I

The daughters of Mne Seraphim led round their sunny flocks,°
All but the youngest. She in paleness sought the secret air,
To fade away like morning beauty from her mortal day.
Down by the river of Adona her soft voice is heard,°
And thus her gentle lamentation falls like morning dew:

'O life of this our spring, why fades the lotus of the water?   10
Why fade these children of the spring, born but to smile and fall?
Ah! Thel is like a wat'ry bow, and like a parting cloud,°
Like a reflection in a glass, like shadows in the water,°
Like dreams of infants, like a smile upon an infant's face,
Like the dove's voice, like transient day, like music in the air.
Ah! gentle may I lay me down, and gentle rest my head,°
And gentle sleep the sleep of death, and gentle hear the voice
Of him that walketh in the garden in the evening time.'°

The Lily of the valley breathing in the humble grass
Answered the lovely maid, and said: 'I am a wat'ry weed,°   20
And I am very small, and love to dwell in lowly vales:
So weak, the gilded butterfly scarce perches on my head.°
Yet I am visited from heaven, and he that smiles on all
Walks in the valley, and each morn over me spreads his hand,
Saying: "Rejoice, thou humble grass, thou new-born lily flower,
Thou gentle maid of silent valleys and of modest brooks;
For thou shalt be clothed in light and fed with morning manna,
Till summer's heat melts thee beside the fountains and the springs,
To flourish in eternal vales." Then why should Thel complain?
Why should the mistress of the vales of Har utter a sigh?'   30

She ceased and smiled in tears, then sat down in her silver shrine.

Thel answered: 'O thou little virgin of the peaceful valley,
Giving to those that cannot crave, the voiceless, the o'ertired—
Thy breath doth nourish the innocent lamb; he smells thy milky
    garments;
He crops thy flowers, while thou sittest smiling in his face,
Wiping his mild and meekin mouth from all contagious taints.°
Thy wine doth purify the golden honey; thy perfume,
Which thou dost scatter on every little blade of grass that springs,
Revives the milked cow and tames the fire-breathing steed.
But Thel is like a faint cloud kindled at the rising sun;   40
I vanish from my pearly throne, and who shall find my place?'

'Queen of the vales', the Lily answered, 'ask the tender cloud,
And it shall tell thee why it glitters in the morning sky,
And why it scatters its bright beauty through the humid air.
Descend, O little Cloud, and hover before the eyes of Thel.'

The Cloud descended, and the Lily bowed her modest head,
And went to mind her numerous charge among the verdant grass.

II

'O little Cloud', the virgin said, 'I charge thee, tell to me
Why thou complainest not, when in one hour thou fade away—
Then we shall seek thee but not find. Ah! Thel is like to thee: 50
I pass away. Yet I complain, and no one hears my voice.'

The Cloud then showed his golden head, and his bright form
    emerged,
Hovering and glittering on the air before the face of Thel:

'O virgin, know'st thou not our steeds drink of the golden springs
Where Luvah doth renew his horses? Look'st thou on my youth,°
And fearest thou because I vanish and am seen no more,°
Nothing remains? O maid, I tell thee, when I pass away
It is to tenfold life, to love, to peace, and raptures holy.
Unseen descending weigh my light wings upon balmy flowers,
And court the fair-eyed dew to take me to her shining tent.   60
The weeping virgin trembling kneels before the risen sun,
Till we arise linked in a golden band and never part,
But walk united, bearing food to all our tender flowers.'

'Dost thou, O Little Cloud? I fear that I am not like thee;
For I walk through the vales of Har, and smell the sweetest
 flowers,
But I feed not the little flowers. I hear the warbling birds,
But I feed not the warbling birds; they fly and seek their food.
But Thel delights in these no more, because I fade away,
And all shall say: "Without a use this shining woman lived;
Or did she only live to be at death the food of worms?"' 70

The Cloud reclined upon his airy throne and answered thus:

'Then if thou art the food of worms, O virgin of the skies,
How great thy use, how great thy blessing! Every thing that lives
Lives not alone, nor for itself. Fear not, and I will call
The weak worm from its lowly bed, and thou shalt hear its voice.
Come forth, Worm of the silent valley, to thy pensive queen.'

The helpless Worm arose, and sat upon the Lily's leaf,
And the bright Cloud sailed on to find his partner in the vale.

III

Then Thel astonished viewed the Worm upon its dewy bed.

'Art thou a Worm? Image of weakness, art thou but a Worm?° 80
I see thee like an infant wrapped in the Lily's leaf.
Ah, weep not little voice! Thou canst not speak, but thou canst
 weep.
It this a Worm? I see thee lay helpless and naked, weeping,
And none to answer, none to cherish thee with mother's smiles.'

The Clod of Clay heard the Worm's voice and raised her pitying
 head;
She bowed over the weeping infant and her life exhaled
In milky fondness: then on Thel she fixed her humble eyes.

'O beauty of the vales of Har, we live not for ourselves.
Thou seest me, the meanest thing, and so I am indeed;
My bosom of itself is cold, and of itself is dark. 90
But he that loves the lowly pours his oil upon my head,
And kisses me, and binds his nuptial bands around my breast,
And says: "Thou mother of my children, I have loved thee,

And I have given thee a crown that none can take away".°
But how this is, sweet maid, I know not, and I cannot know;
I ponder, and I cannot ponder; yet I live and love.'

The daughter of beauty wiped her pitying tears with her white veil
And said: 'Alas! I knew not this, and therefore did I weep.
That God would love a worm I knew, and punish the evil foot
That wilful bruised its helpless form. But that he cherished it   100
With milk and oil I never knew, and therefore did I weep;
And I complained in the mild air, because I fade away,
And lay me down in thy cold bed, and leave my shining lot.'

'Queen of the vales', the matron Clay answered, 'I heard thy sighs,
And all thy moans flew o'er my roof, but I have called them down.
Wilt thou, O Queen, enter my house? 'Tis given thee to enter
And to return. Fear nothing; enter with thy virgin feet.'°

IV

The eternal gates' terrific porter lifted the northern bar.°
Thel entered in and saw the secrets of the land unknown.
She saw the couches of the dead, and where the fibrous roots   110
Of every heart on earth infixes deep its restless twists:
A land of sorrows and of tears, where never smile was seen.

She wandered in the land of clouds, through valleys dark,
    list'ning°
Dolours and lamentations. Waiting oft beside a dewy grave
She stood in silence, list'ning to the voices of the ground,
Till to her own grave plot she came, and there she sat down,
And heard this voice of sorrow breathed from the hollow pit:

'Why cannot the ear be closed to its own destruction,
Or the glist'ning eye to the poison of a smile?
Why are the eyelids stored with arrows ready drawn,   120
Where a thousand fighting men in ambush lie,
Or an eye of gifts and graces, show'ring fruits and coined gold?
Why a tongue impressed with honey from every wind?
Why an ear a whirlpool fierce to draw creations in?
Why a nostril wide inhaling terror, trembling and affright?
Why a tender curb upon the youthful burning boy?

Why a little curtain of flesh on the bed of our desire?'
The virgin started from her seat, and with a shriek
Fled back unhindered till she came into the vales of Har.

THE END

The French Revolution (*1791*)

BOOK THE FIRST

The dead brood over Europe; the cloud and vision descends over
    cheerful France.
O cloud well appointed! Sick, sick, the Prince on his couch,
    wreathed in dim°
And appalling mist, his strong hand outstretched; from his
    shoulder, down the bone,
Runs aching cold into the sceptre too heavy for mortal grasp—no
    more
To be swayed by visible hand, nor in cruelty bruise the mild
    flourishing mountains.

Sick the mountains, and all their vineyards weep, in the eyes of the
    kingly mourner;
Pale is the morning cloud in his visage. 'Rise, Necker! The ancient
    dawn calls us°
To awake from slumbers of five thousand years. I awake, but my
    soul is in dreams;
From my window I see the old mountains of France, like aged
    men, fading away.'

Troubled, leaning on Necker, descends the King to his chamber of   10
    council; shady mountains
In fear utter voices of thunder; the woods of France embosom the
    sound;
Clouds of wisdom prophetic reply, and roll over the palace roof
    heavy.
Forty men, each conversing with woes in the infinite shadows of
    his soul,
Like our ancient fathers in regions of twilight, walk, gathering
    round the King.°
Again the loud voice of France cries to the morning; the morning
    prophesies to its clouds.

For the Commons convene in the Hall of the Nation. France
    shakes! And the heavens of France
Perplexed vibrate round each careful countenance! Darkness of
    old times around them

Utters loud despair, shadowing Paris; her grey towers groan, and the Bastille trembles.

In its terrible towers the Governor stood, in dark fogs list'ning the horror;

A thousand his soldiers, old veterans of France, breathing red   20
clouds of power and dominion.

Sudden seized with howlings, despair, and black night, he stalked like a lion from tower

To tower; his howlings were heard in the Louvre. From court to court restless he dragged

His strong limbs; from court to court cursed the fierce torment unquelled,

Howling and giving the dark command. In his soul stood the purple plague,

Tugging his iron manacles, and piercing through the seven towers dark and sickly,

Panting over the prisoners like a wolf gorged— and the den named Horror held a man°

Chained hand and foot, round his neck an iron band, bound to the impregnable wall.

In his soul was the serpent coiled round his heart, hid from the light, as in a cleft rock—

And the man was confined for a writing prophetic. In the tower named Darkness was a man

Pinioned down to the stone floor, his strong bones scarce covered   30
with sinews. The iron rings

Were forged smaller as the flesh decayed; a mask of iron on his face hid the lineaments°

Of ancient kings, and the frown of the eternal lion was hid from the oppressed earth.

In the tower named Bloody a skeleton yellow remained in its chains on its couch

Of stone, once a man who refused to sign papers of abhorrence; the eternal worm

Crept in the skeleton. In the den named Religion a loathsome sick woman, bound down

To a bed of straw; the seven diseases of earth, like birds of prey, stood on the couch,

And fed on the body. She refused to be whore to the minister, and with a knife smote him.

In the tower named Order an old man, whose white beard covered the stone floor like weeds

On the margin of the sea, shrivelled up by heat of day and cold of
    night; his den was short
And narrow as a grave dug for a child, with spiders' webs wove,   40
    and with slime
Of ancient horrors covered. For snakes and scorpions are his
    companions; harmless they breathe
His sorrowful breath. He, by conscience urged, in the city of Paris
    raised a pulpit,
And taught wonders to darkened souls. In the den named Destiny
    a strong man sat,
His feet and hands cut off, and his eyes blinded; round his middle a
    chain and a band
Fastened into the wall. Fancy gave him to see an image of despair
    in his den,
Eternally rushing round, like a man on his hands and knees, day
    and night without rest.
He was friend to the favourite. In the seventh tower, named the
    Tower of God, was a man
Mad, with chains loose, which he dragged up and down; fed with
    hopes year by year, he pined
For liberty—vain hopes. His reason decayed, and the world of
    attraction in his bosom
Centred; and the rushing of chaos overwhelmed his dark soul. He   50
    was confined
For a letter of advice to a king, and his ravings in winds are heard
    over Versailles.

But the dens shook and trembled; the prisoners look up and assay
    to shout. They listen,
Then laugh in the dismal den, then are silent, and a light walks
    round the dark towers.

For the Commons convene in the Hall of the Nations. Like spirits
    of fire in the beautiful
Porches of the sun, to plant beauty in the desert craving abyss, they
    gleam
On the anxious city. All children new-born first behold them; tears
    are fled,
And they nestle in earth-breathing bosoms. So the city of Paris,
    their wives and children,
Look up to the morning Senate, and visions of sorrow leave
    pensive streets.

But heavy-browed jealousies lower o'er the Louvre, and terrors of ancient kings

Descend from the gloom and wander through the palace, and weep   60
round the King and his nobles.

While loud thunders roll, troubling the dead, kings are sick throughout all the earth.

The voice ceased; the nation sat; and the triple-forged fetters of times were unloosed.

The voice ceased; the nation sat; but ancient darkness and trembling wander through the palace.

As in day of havoc and routed battle, among thick shades of discontent

On the soul-skirting mountains of sorrow cold waving, the nobles fold round the King—

Each stern visage locked up as with strong bands of iron, each strong limb bound down as with marble,

In flames of red wrath burning, bound in astonishment a quarter of an hour.

Then the King glowed; his nobles fold round, like the sun of old time quenched in clouds.

In their darkness the King stood, his heart flamed, and uttered a with'ring heat, and these words burst forth:

'The nerves of five thousand years' ancestry tremble, shaking the   70
heavens of France.

Throbs of anguish beat on brazen war foreheads; they descend and look into their graves.

I see through darkness, through clouds rolling round me, the spirits of ancient kings

Shivering over their bleached bones; round them their counsellors look up from the dust,

Crying, "Hide from the living! Our bonds and our prisoners shout in the open field.

Hide in the nether earth! Hide in the bones! Sit obscured in the hollow skull.

Our flesh is corrupted, and we wear away. We are not numbered among the living. Let us hide

In stones, among roots of trees. The prisoners have burst their
    dens.
Let us hide; let us hide in the dust, and plague and wrath and
    tempest shall cease." '

He ceased, silent pond'ring, his brows folded heavy; his forehead
    was in affliction,
Like the central fire. From the window he saw his vast armies   80
    spread over the hills,
Breathing red fires from man to man, and from horse to horse.
    Then his bosom
Expanded like starry heaven; he sat down. His nobles took their
    ancient seats.

Then the ancientest peer, Duke of Burgundy, rose from the
    monarch's right hand, red as wines
From his mountains; an odour of war, like a ripe vineyard, rose
    from his garments,
And the chamber became as a clouded sky. O'er the council he
    stretched his red limbs,
Clothed in flames of crimson; as a ripe vineyard stretches over
    sheaves of corn,
The fierce duke hung over the council. Around him crowd,
    weeping in his burning robe,
A bright cloud of infant souls; his words fall like purple autumn on
    the sheaves.

'Shall this marble-built heaven become a clay cottage, this earth an
    oak stool, and these mowers
From the Atlantic mountains, mow down all this great starry   90
    harvest of six thousand years?°
And shall Necker, the hind of Geneva, stretch out his crooked
    sickle o'er fertile France?°
Till our purple and crimson is faded to russet, and the kingdoms of
    earth bound in sheaves,
And the ancient forests of chivalry hewn, and the joys of the
    combat burnt for fuel?
Till the power and dominion is rent from the pole, sword and
    sceptre from sun and moon,
The law and gospel from fire and air, and eternal reason and
    science

From the deep and the solid, and man lay his faded head down on
the rock

Of eternity, where the eternal lion and eagle remain to devour?

This to prevent, urged by cries in day, and prophetic dreams
hovering in night—

To enrich the lean earth that craves, furrowed with ploughs, whose
seed is departing from her—

Thy nobles have gathered thy starry hosts round this rebellious   100
city:

To rouse up the ancient forests of Europe with clarions of loud
breathing war,

To hear the horse neigh to the drum and trumpet, and the trumpet
and war shout reply.

Stretch the hand that beckons the eagles of heaven; they cry over
Paris, and wait

Till Fayette point his finger to Versailles. The eagles of heaven
must have their prey.'°

The King leaned on his mountains, then lifted his head and looked
on his armies, that shone

Through heaven, tinging morning with beams of blood; then,
turning to Burgundy troubled,

'Burgundy, thou wast born a lion! My soul is o'ergrown with
distress°

For the nobles of France, and dark mists roll round me and blot the
writing of God

Written in my bosom. Necker rise, leave the kingdom; thy life is
surrounded with snares.

We have called an assembly, but not to destroy; we have given   110
gifts, not to the weak.

I hear rushing of muskets, and bright'ning of swords, and visages
redd'ning with war,

Frowning and looking up from brooding villages and every
dark'ning city.

Ancient wonders frown over the kingdom, and cries of woman and
babes heard,

And tempests of doubt roll around me, and fierce sorrows, because
of the nobles of France.

Depart, answer not, for the tempest must fall, as in years that are
passed away.'

He ceased, and burned silent; red clouds roll round Necker, a
    weeping is heard o'er the palace.
Like a dark cloud Necker paused, and like thunder on the just
    man's burial day he paused°
(Silent sit the winds, silent the meadows, while the husbandman
    and woman of weakness,
And bright children, look after him into the grave, and water his
    clay with love,
Then turn towards pensive fields). So Necker paused, and his   120
    visage was covered with clouds.

Dropping a tear the old man his place left, and when he was gone
    out
He set his face toward Geneva to flee, and the women and children
    of the city
Kneeled round him and kissed his garments and wept. He stood a
    short space in the street,
Then fled; and the whole city knew he was fled to Geneva, and the
    Senate heard it.

But the nobles burned wrathful at Necker's departure, and
    wreathed their clouds and waters
In dismal volumes. As risen from beneath the Archbishop of Paris
    arose,
In the rushing of scales and hissing of flames and rolling of
    sulphurous smoke.

'Hearken, Monarch of France, to the terrors of Heaven, and let thy
    soul drink of my counsel.
Sleeping at midnight in my golden tower, the repose of the labours
    of men
Waved its solemn cloud over my head. I awoke; a cold hand passed   130
    over my limbs, and behold!
An aged form, white as snow, hov'ring in mist, weeping in the
    uncertain light.°
Dim the form, almost faded; tears fell down the shady cheeks. At
    his feet many, clothed
In white robes; strewn in air censers and harps; silent they lay
    prostrated.
Beneath, in the awful void, myriads descending and weeping
    through dismal winds;

Endless the shady train shiv'ring descended, from the gloom where the aged form wept.

At length, trembling, the vision, sighing in a low voice like the voice of the grasshopper, whispered:

"My groaning is heard in the abbeys, and God, so long worshipped, departs as a lamp

Without oil. For a curse is heard hoarse through the land, from a godless race

Descending to beasts; they look downward and labour and forget my holy law.

The sound of prayer fails from lips of flesh, and the holy hymn 140 from thickened tongues.

For the bars of Chaos are burst; her millions prepare their fiery way

Through the orbed abode of the holy dead, to root up and pull down and remove.

And nobles and clergy shall fail from before me, and my cloud and vision be no more;

The mitre become black, the crown vanish, and the sceptre and ivory staff

Of the ruler wither among bones of death. They shall consume from the thistly field,

And the sound of the bell, and voice of the sabbath, and singing of the holy choir,

Is turned into songs of the harlot in day, and cries of the virgin in night.

They shall drop at the plough and faint at the harrow, unredeemed, unconfessed, unpardoned;

The priest rot in his surplice by the lawless lover, the holy beside the accursed,

The king, frowning in purple, beside the grey ploughman, and 150 their worms embrace together."

The voice ceased; a groan shook my chamber. I slept, for the cloud of repose returned,

But morning dawned heavy upon me. I rose to bring my prince Heaven-uttered counsel.

Hear my counsel, O King, and send forth thy generals. The command of Heaven is upon thee;

Then do thou command, O King, to shut up this assembly in their final home.

Let thy soldiers possess this city of rebels, that threaten to bathe their feet

In the blood of nobility, trampling the heart and the head; let the
  Bastille devour
These rebellious seditious; seal them up, O Anointed, in everlast-
  ing chains.'

He sat down; a damp cold pervaded the nobles, and monsters of
  worlds unknown
Swam round them, watching to be delivered—when Aumont,
  whose chaos-born soul
Eternally wand'ring, a comet and swift-falling fire, pale entered   160
  the chamber.
Before the red council he stood, like a man that returns from
  hollow graves.

'Awe-surrounded, alone through the army a fear and a with'ring
  blight blown by the north:
The Abbé de Sieyès from the Nation's Assembly, O Princes and
  Generals of France,
Unquestioned, unhindered. Awe-struck are the soldiers; a dark
  shadowy man in the form
Of King Henry the Fourth walks before him in fires. The captains
  like men bound in chains°
Stood still as he passed. He is come to the Louvre, O King, with a
  message to thee.
The strong soldiers tremble, the horses their manes bow, and the
  guards of thy palace are fled.'
Up rose awful in his majestic beams Bourbon's strong Duke; his
  proud sword from his thigh
Drawn, he threw on the earth. The Duke of Bretagne and the Earl
  of Bourgogne
Rose inflamed, to and fro in the chamber, like thunder-clouds   170
  ready to burst.

'What, damp all our fires, O spectre of Henry,' said Bourbon, 'and
  rend the flames
From the head of our king! Rise, Monarch of France; command
  me, and I will lead
This army of superstition at large, that the ardour of noble souls
  quenchless
May yet burn in France, nor our shoulders be ploughed with the
  furrows of poverty.'

Then Orléans generous as mountains arose, and unfolded his robe, and put forth

His benevolent hand, looking on the Archbishop—who changed as pale as lead;

Would have risen but could not. His voice issued harsh grating; instead of words harsh hissings

Shook the chamber; he ceased abashed. Then Orléans spoke; all was silent.

He breathed on them, and said: 'O princes of fire, whose flames are for growth not consuming,

Fear not dreams, fear not visions, nor be you dismayed with   180
sorrows which flee at the morning!

Can the fires of nobility ever be quenched, or the stars by a stormy night?

Is the body diseased when the members are healthful? Can the man be bound in sorrow

Whose ev'ry function is filled with its fiery desire? Can the soul whose brain and heart

Cast their rivers in equal tides through the great Paradise, languish because the feet

Hands, head, bosom, and parts of love, follow their high-breathing joy?

And can nobles be bound when the people are free, or God weep when his children are happy?

Have you never seen Fayette's forehead, or Mirabeau's eyes, or the shoulders of Target,°

Or Bailly the strong foot of France, or Clermont the terrible voice, and your robes

Still retain their own crimson? Mine never yet faded, for fire delights in its form.

But go, merciless man, enter into the infinite labyrinth of another's   190
brain,

Ere thou measure the circle that he shall run. Go, thou cold recluse, into the fires

Of another's high flaming rich bosom, and return unconsumed, and write laws.

If thou canst not do this, doubt thy theories, learn to consider all men as thy equals,

Thy brethren, and not as thy foot or thy hand, unless thou first fearest to hurt them.'

The monarch stood up; the strong duke his sword to its golden
  scabbard returned;
The nobles sat round like clouds on the mountains, when the
  storm is passing away.

'Let the nation's ambassador come among nobles, like incense of
  the valley.'

Aumont went out and stood in the hollow porch, his ivory wand in
  his hand;
A cold orb of disdain revolved round him, and covered his soul
  with snows eternal.
Great Henry's soul shuddered, a whirlwind and fire tore furious   200
  from his angry bosom;
He indignant departed on horses of heav'n. Then the Abbé de
  Sieyès raised his feet
On the steps of the Louvre, like a voice of God following a storm.
  The Abbé followed
The pale fires of Aumont into the chamber; as a father that bows to
  his son,
Whose rich fields inheriting spread their old glory, so the voice of
  the people bowed
Before the ancient seat of the kingdom and mountains to be
  renewed.
'Hear O heavens of France, the voice of the people, arising from
  valley and hill,
O'erclouded with power. Hear the voice of valleys, the voice of
  meek cities,
Mourning oppressed on village and field, till the village and field is
  a waste.
For the husbandman weeps at blights of the fife, and blasting of
  trumpets consume°
The souls of mild France; the pale mother nourishes her child to   210
  the deadly slaughter.
When the heavens were sealed with a stone, and the terrible sun
  closed in an orb, and the moon
Rent from the nations, and each star appointed for watchers of
  night,
The millions of spirits immortal were bound in the ruins of
  sulphur heaven
To wander enslaved; black, depressed in dark ignorance, kept in
  awe with the whip,

To worship terrors, bred from the blood of revenge and breath of
desire

In bestial forms, or more terrible men, till the dawn of our peaceful
morning—

Till dawn, till morning, till the breaking of clouds, and swelling of
winds, and the universal voice—

Till Man raise his darkened limbs out of the cares of night, his eyes
and his heart

Expand. Where is space? Where, O sun, is thy dwelling? Where
thy tent, O faint slumb'rous moon?

Then the valleys of France shall cry to the soldier: "Throw down   220
thy sword and musket,

And run and embrace the meek peasant." Her nobles shall hear
and shall weep, and put off

The red robe of terror, the crown of oppression, the shoes of
contempt, and unbuckle

The girdle of war from the desolate earth. Then the priest in his
thund'rous cloud

Shall weep; bending to earth, embracing the valleys, and putting
his hand to the plough

Shall say: "No more I curse thee, but now I will bless thee; no
more in deadly black

Devour thy labour, nor lift up a cloud in thy heavens, O laborious
plough°—

That the wild raging millions, that wander in forests, and howl in
law-blasted wastes,

Strength maddened with slavery, honesty bound in the dens of
superstition,

May sing in the village, and shout in the harvest, and woo in
pleasant gardens

Their once savage loves, now beaming with knowledge, with   230
gentle awe adorned.

And the saw, and the hammer, the chisel, the pencil, the pen, and
the instruments

Of heavenly song sound in the wilds once forbidden, to teach the
laborious ploughman

And shepherd delivered from clouds of war, from pestilence, from
night-fear, from murder,

From falling, from stifling, from hunger, from cold, from slander,
discontent and sloth,

That walk in beasts and birds of night, driven back by the sandy
    desert
Like pestilent fogs round cities of men. And the happy earth sing
    in its course,
The mild peaceable nations be opened to Heav'n, and men walk
    with their fathers in bliss."
Then hear the first voice of the morning: "Depart, O clouds of
    night, and no more
Return. Be withdrawn, cloudy war; troops of warriors depart, nor
    around our peaceable city
Breathe fires, but ten miles from Paris; let all be peace, nor a soldier   240
    be seen." '

He ended. The wind of contention arose and the clouds cast their
    shadows—the princes
Like the mountains of France, whose aged trees utter an awful
    voice, and their branches
Are shattered. Till gradual a murmur is heard descending into the
    valley,
Like a voice in the vineyards of Burgundy, when grapes are shaken
    on grass,
Like the low voice of the labouring man, instead of the shout of
    joy.
And the palace appeared like a cloud driven abroad; blood ran
    down the ancient pillars.
Through the cloud a deep thunder, the Duke of Burgundy,
    delivers the King's command:
'Seest thou yonder dark castle that, moated around, keeps this city
    of Paris in awe.
Go command yonder tower, saying: "Bastille depart, and take thy
    shadowy course.
Overstep the dark river, thou terrible tower, and get thee up into   250
    the country ten miles.
And thou black southern prison, move along the dusky road to
    Versailles; there
Frown on the gardens." And if it obey and depart, then the king
    will disband
This war-breathing army; but if it refuse, let the Nation's
    Assembly thence learn
That this army of terrors, that prison of horrors, are the bands of
    the murmuring kingdom.'

Like the morning star arising above the black waves, when a
  shipwrecked soul sighs for morning,
Through the ranks, silent, walked the ambassador back to the
  Nation's Assembly, and told
The unwelcome message. Silent they heard; then a thunder rolled
  round loud and louder.
Like pillars of ancient halls, and ruins of times remote they sat.
Like a voice from the dim pillars Mirabeau rose; the thunders
  subsided away;
A rushing of wings around him was heard as he brightened, and   260
  cried out aloud:
'Where is the General of the Nation?' The walls re-echoed: 'Where
  is the General of the Nation?'
Sudden as the bullet wrapped in his fire, when brazen cannons
  rage in the field,
Fayette sprung from his seat saying, 'Ready!' Then, bowing like
  clouds, man toward man—the Assembly
Like a council of ardours seated in clouds, bending over the cities
  of men,°
And over the armies of strife where their children are marshalled
  together to battle—
They murmuring divide; while the wind sleeps beneath, and the
  numbers are counted in silence;
While they vote the removal of war, and the pestilence weighs his
  red wings in the sky.

So Fayette stood silent among the Assembly, and the votes were
  given and the numbers numbered;
And the vote was that Fayette should order the army to remove ten
  miles from Paris.

The aged sun rises appalled from dark mountains, and gleams a   270
  dusky beam
On Fayette, but on the whole army a shadow; for a cloud on the
  eastern hills
Hovered, and stretched across the city and across the army, and
  across the Louvre.
Like a flame of fire he stood before dark ranks, and before
  expecting captains.
On pestilent vapours around him flow frequent spectres of
  religious men weeping;

In winds driven out of the abbeys, their naked souls shiver in keen
    open air;
Driven out by the fiery cloud of Voltaire, and thund'rous rocks of
    Rousseau,
They dash like foam against the ridges of the army, uttering a faint
    feeble cry.
Gleams of fire streak the heavens, and of sulphur the earth, from
    Fayette as he lifted his hand;
But silent he stood, till all the officers rush round him like waves
Round the shore of France, in day of the British flag, when heavy   280
    cannons
Affright the coast, and the peasant looks over the sea and wipes a
    tear.
Over his head the soul of Voltaire shone fiery, and over the army
    Rousseau his white cloud
Unfolded, on souls of war-living terrors silent list'ning toward
    Fayette.
His voice loud inspired by liberty, and by spirits of the dead, thus
    thundered:

'The Nation's Assembly command that the Army remove ten
    miles from Paris,
Nor a soldier be seen in road or in field, till the nation command
    return.'

Rushing along iron ranks glittering, the officers each to his station
Depart, and the stern captain strokes his proud steed, and in front
    of his solid ranks
Waits the sound of trumpet. Captains of foot stand each by his
    cloudy drum.
Then the drum beats, and the steely ranks move, and trumpets   290
    rejoice in the sky.
Dark cavalry like clouds fraught with thunder ascend on the hills,
    and bright infantry, rank
Behind rank, to the soul-shaking drum and shrill fife along the
    roads glitter like fire.

The noise of trampling, the wind of trumpets, smote the palace
    walls with a blast.
Pale and cold sat the King in midst of his peers, and his noble heart
    sunk, and his pulses

Suspended their notion; a darkness crept over his eyelids, and chill
    cold sweat

Sat round his brows faded in faint death, his peers pale like
    mountains of the dead,

Covered with dews of night, groaning, shaking forests and floods.
    The cold newt,

And snake, and damp toad on the kingly foot crawl, or croak on the
    awful knee,

Shedding their slime; in folds of the robe the crowned adder builds
    and hisses

From stony brows. Shaken the forests of France, sick the kings of   300
    the nations,

And the bottoms of the world were opened, and the graves of arch-
    angels unsealed.

The enormous dead lift up their pale fires and look over the rocky
    cliffs.

A faint heat from their fires revived the cold Louvre; the frozen
    blood reflowed.

Awful up rose the King; him the peers followed. They saw the
    courts of the palace

Forsaken, and Paris without a soldier, silent; for the noise was gone
    up

And followed the army, and the Senate in peace sat beneath
    morning's beam.

Visions of the Daughters of Albion (1793)

The eye sees more than the heart knows.

THE ARGUMENT

I loved Theotormon,°
And I was not ashamed.
I trembled in my virgin fears,
And I hid in Leutha's vale!

I plucked Leutha's flower,
And I rose up from the vale;
But the terrible thunders tore
My virgin mantle in twain.

VISIONS

Enslaved, the daughters of Albion weep a trembling lamentation
Upon their mountains, in their valleys sighs toward America.   10
For the soft soul of America, Oothoon, wandered in woe
Along the vales of Leutha, seeking flowers to comfort her;
And thus she spoke to the bright marigold of Leutha's vale:

'Art thou a flower? Art thou a nymph? I see thee now a flower,
Now a nymph! I dare not pluck thee from thy dewy bed!'

The golden nymph replied: 'Pluck thou my flower, Oothoon the
    mild.
Another flower shall spring, because the soul of sweet delight
Can never pass away.' She ceased and closed her golden shrine.

Then Oothoon plucked the flower, saying, 'I pluck thee from thy
    bed,
Sweet flower, and put thee here to glow between my breasts,   20
And thus I turn my face to where my whole soul seeks.'

Over the waves she went in winged exulting swift delight,
And over Theotormon's reign took her impetuous course.

Bromion rent her with his thunders; on his stormy bed
Lay the faint maid, and soon her woes appalled his thunders
    hoarse.

Bromion spoke: 'Behold this harlot here on Bromion's bed,
And let the jealous dolphins sport around the lovely maid.°
Thy soft American plains are mine, and mine thy north and south.
Stamped with my signet are the swarthy children of the sun;°
They are obedient, they resist not, they obey the scourge;   30
Their daughters worship terrors and obey the violent.
Now thou may'st marry Bromion's harlot, and protect the child
Of Bromion's rage that Oothoon shall put forth in nine moons
 time.'

Then storms rent Theotormon's limbs; he rolled his waves around
And folded his black jealous waters round the adulterate pair.
Bound back to back in Bromion's caves, terror and meekness
 dwell.

At entrance Theotormon sits, wearing the threshold hard
With secret tears; beneath him sound like waves on a desert shore
The voice of slaves beneath the sun, the children bought with
 money,
That shiver in religious caves beneath the burning fires   40
Of lust, that belch incessant from the summits of the earth.

Oothoon weeps not; she cannot weep! Her tears are locked up;
But she can howl incessant, writhing her soft snowy limbs,
And calling Theotormon's eagles to prey upon her flesh.

'I call with holy voice! Kings of the sounding air,
Rend away this defiled bosom that I may reflect
The image of Theotormon on my pure transparent breast.'
The eagles at her call descend and rend their bleeding prey.
Theotormon severely smiles; her soul reflects the smile;   50
As the clear spring mudded with feet of beasts grows pure and smiles.

The daughters of Albion hear her woes, and echo back her sighs.

'Why does my Theotormon sit weeping upon the threshold,
And Oothoon hovers by his side, persuading him in vain?
I cry: "Arise, O Theotormon, for the village dog
Barks at the breaking day; the nightingale has done lamenting;
The lark does rustle in the ripe corn, and the eagle returns
From nightly prey and lifts his golden beak to the pure east,
Shaking the dust from his immortal pinions to awake

The sun that sleeps too long. Arise my Theotormon; I am pure,
Because the night is gone that closed me in its deadly black."   60
They told me that the night and day were all that I could see;
They told me that I had five senses to enclose me up,
And they enclosed my infinite brain into a narrow circle,
And sunk my heart into the abyss, a red round globe hot-burning,
Till all from life I was obliterated and erased.
Instead of morn arises a bright shadow, like an eye
In the eastern cloud; instead of night a sickly charnel-house,
That Theotormon hears me not! To him the night and morn
Are both alike: a night of sighs, a morning of fresh tears—
And none but Bromion can hear my lamentations.   70

'With what sense is it that the chicken shuns the ravenous hawk?°
With what sense does the tame pigeon measure out the expanse?
With what sense does the bee form cells? Have not the mouse and
    frog
Eyes and ears and sense of touch? Yet are their habitations
And their pursuits as different as their forms and as their joys!
Ask the wild ass why he refuses burdens, and the meek camel
Why he loves Man. Is it because of eye, ear, mouth or skin,
Or breathing nostrils? No; for these the wolf and tiger have.
Ask the blind worm the secrets of the grave, and why her spires
Love to curl round the bones of death; and ask the rav'nous snake   80
Where she gets poison, and the winged eagle why he loves the
    sun—
And then tell me the thoughts of Man that have been hid of old.

'Silent I hover all the night, and all day could be silent,
If Theotormon once would turn his loved eyes upon me.
How can I be defiled when I reflect thy image pure?
Sweetest the fruit that the worm feeds on, and the soul preyed on
    by woe,
The new-washed lamb tinged with the village smoke, and the
    bright swan
By the red earth of our immortal river. I bathe my wings,
And I am white and pure to hover round Theotormon's breast.'

Then Theotormon broke his silence, and he answered: 90

'Tell me what is the night or day to one o'erflowed with woe?
Tell me what is a thought, and of what substance is it made?

Tell me what is a joy, and in what gardens do joys grow?
And in what rivers swim the sorrows, and upon what mountains
Wave shadows of discontent? And in what houses dwell the
    wretched
Drunken with woe, forgotten and shut up from cold despair?

'Tell me where dwell the thoughts forgotten till thou call them
    forth?
Tell me where dwell the joys of old, and where the ancient loves?
And when will they renew again and the night of oblivion past?
That I might traverse times and spaces far remote, and bring   100
Comforts into a present sorrow and a night of pain.
Where goest thou, O thought? To what remote land is thy flight?
If thou returnest to the present moment of affliction
Wilt thou bring comforts on thy wings, and dews and honey and
    balm,
Or poison from the desert wilds, from the eyes of the envier?'

Then Bromion said, and shook the cavern with his lamentation:

'Thou knowest that the ancient trees seen by thine eyes have fruit,
But knowest thou that trees and fruits flourish upon the earth
To gratify senses unknown? Trees, beasts and birds unknown—
Unknown, not unperceived: spread in the infinite microscope,   110
In places yet unvisited by the voyager, and in worlds
Over another kind of seas, and in atmospheres unknown.
Ah! are there other wars, beside the wars of sword and fire?
And are there other sorrows, beside the sorrows of poverty?
And are there other joys, beside the joys of riches and ease?
And is there not one law for both the lion and the ox?
And is there not eternal fire, and eternal chains,
To bind the phantoms of existence from eternal life?'

Then Oothoon waited silent all the day, and all the night,
But when the morn arose, her lamentation renewed.   120
The daughters of Albion hear her woes, and echo back her sighs.

'O Urizen! Creator of men. Mistaken demon of Heaven:°
Thy joys are tears, thy labour vain, to form men to thine image.
How can one joy absorb another? Are not different joys
Holy, eternal, infinite! And each joy is a love.

'Does not the great mouth laugh at a gift, and the narrow eyelids
  mock
At the labour that is above payment? And wilt thou take the ape
For thy counsellor, or the dog for a schoolmaster to thy children?
Does he who contemns poverty, and he who turns with abhorrence
From usury, feel the same passion or are they moved alike?   130
How can the giver of gifts experience the delights of the merchant,
How the industrious citizen the pains of the husbandman?
How different far the fat-fed hireling with hollow drum,°
Who buys whole cornfields into wastes and sings upon the heath!
How different their eye and ear! How different the world to them!
With what sense does the parson claim the labour of the farmer?°
What are his nets and gins and traps, and how does he surround
  him
With cold floods of abstraction and with forests of solitude,
To build him castles and high spires, where kings and priests may
  dwell?
Till she who burns with youth, and knows no fixed lot, is bound   140
In spells of law to one she loathes. And must she drag the chain
Of life in weary lust? Must chilling murderous thoughts obscure
The clear heaven of her eternal spring?—to bear the wintry rage
Of a harsh terror, driv'n to madness, bound to hold a rod
Over her shrinking shoulders all the day—and all the night
To turn the wheel of false desire, and longings that wake her
  womb°
To the abhorred birth of cherubs in the human form,
That live a pestilence and die a meteor and are no more?
Till the child dwell with one he hates, and do the deed he loathes.
And the impure scourge force his seed into its unripe birth   150
Ere yet his eyelids can behold the arrows of the day.
Does the whale worship at thy footsteps as the hungry dog,
Or does he scent the mountain prey, because his nostrils wide
Draw in the ocean? Does his eye discern the flying cloud
As the raven's eye, or does he measure the expanse like the vulture?
Does the still spider view the cliffs where eagles hide their young?
Or does the fly rejoice, because the harvest is brought in?
Does not the eagle scorn the earth and despise the treasures
  beneath?
But the mole knoweth what is there, and the worm shall tell it thee.
Does not the worm erect a pillar in the mouldering churchyard,   160
And a palace of eternity in the jaws of the hungry grave?

Over his porch these words are written: "Take thy bliss, O Man!
And sweet shall be thy taste; and sweet thy infant joys renew!"

'Infancy, fearless, lustful, happy! nestling for delight
In laps of pleasure. Innocence! honest, open, seeking
The vigorous joys of morning light, open to virgin bliss.
Who taught thee modesty, subtle modesty, child of night and
    sleep?
When thou awakest, wilt thou dissemble all thy secret joys,
Or wert thou not awake when all this mystery was disclosed?
Then com'st thou forth a modest virgin, knowing to dissemble,   170
With nets found under thy night pillow to catch virgin joy,
And brand it with the name of whore, and sell it in the night,
In silence, ev'n without a whisper, and in seeming sleep.
Religious dreams and holy vespers light thy smoky fires:
Once were thy fires lighted by the eyes of honest morn.
And does my Theotormon seek this hypocrite modesty,
This knowing, artful, secret, fearful, cautious, trembling hypo-
    crite?
Then is Oothoon a whore indeed! And all the virgin joys
Of life are harlots; and Theotormon is a sick man's dream;
And Oothoon is the crafty slave of selfish holiness.   180

'But Oothoon is not so: a virgin filled with virgin fancies,
Open to joy and to delight wherever beauty appears.
If in the morning sun I find it, there my eyes are fixed
In happy copulation; if in evening mild, wearied with work,
Sit on a bank and draw the pleasures of this freeborn joy.

'The moment of desire! The moment of desire! The virgin
That pines for man shall awaken her womb to enormous joys
In the secret shadows of her chamber. The youth shut up from
The lustful joy shall forget to generate, and create an amorous
    image
In the shadows of his curtains and in the folds of his silent pillow.   190
Are not these the places of religion, the rewards of continence,
The self-enjoyings of self-denial? Why dost thou seek religion?
Is it because acts are not lovely that thou seekest solitude,
Where the horrible darkness is impressed with reflections of
    desire?

'Father of jealousy, be thou accursed from the earth!
Why hast thou taught my Theotormon this accursed thing?
Till beauty fades from off my shoulders, darkened and cast out,
A solitary wailing on the margin of non-entity.

'I cry, Love! Love! Love! Happy, happy Love! Free as the
    mountain wind!
Can that be Love that drinks another as a sponge drinks water,   200
That clouds with jealousy his nights, with weepings all the day—
To spin a web of age around him, grey and hoary, dark,
Till his eyes sicken at the fruit that hangs before his sight?
Such is self-love that envies all! A creeping skeleton
With lamplike eyes watching around the frozen marriage bed.

'But silken nets and traps of adamant will Oothoon spread,
And catch for thee girls of mild silver or of furious gold.
I'll lie beside thee on a bank and view their wanton play
In lovely copulation, bliss on bliss with Theotormon.
Red as the rosy morning, lustful as the first-born beam,   210
Oothoon shall view his dear delight, nor e'er with jealous cloud
Come in the heaven of generous love, nor selfish blightings bring.

'Does the sun walk in glorious raiment on the secret floor
Where the cold miser spreads his gold? Or does the bright cloud
    drop
On his stone threshold? Does his eye behold the beam that brings
Expansion to the eye of pity? Or will he bind himself
Beside the ox to thy hard furrow? Does not that mild beam blot
The bat, the owl, the glowing tiger, and the king of night?
The sea-fowl takes the wintry blast for a cov'ring to her limbs,
And the wild snake the pestilence to adorn him with gems and   220
    gold;
And trees and birds and beasts and men behold their eternal joy.
Arise, you little glancing wings, and sing your infant joy!
Arise and drink your bliss; for everything that lives is holy!'

Thus every morning wails Oothoon; but Theotormon sits
Upon the margined ocean, conversing with shadows dire.

The daughters of Albion hear her woes, and echo back her sighs.

THE END

Africa (1795)

I will sing you a song of Los, the eternal prophet;
He sung it to four harps at the tables of Eternity,°
    In heart-formed Africa.
Urizen faded! Ariston shuddered!°
    And thus the song began:

Adam stood in the garden of Eden,
And Noah on the mountains of Ararat;
They saw Urizen give his laws to the nations°
By the hands of the children of Los.

Adam shuddered! Noah faded! Black grew the sunny African,   10
When Rintrah gave abstract philosophy to Brahma in the east.
(Night spoke to the cloud:
'Lo, these human-formed spirits in smiling hypocrisy war
Against one another; so let them war on, slaves to the eternal
    elements'.)
Noah shrunk beneath the waters;
Abram fled in fires from Chaldea;°
Moses beheld upon Mount Sinai forms of dark delusion.

To Trismegistus Palamabron gave an abstract law,°
To Pythagoras, Socrates, and Plato.

Times rolled on o'er all the sons of Har; time after time   20
Orc on Mount Atlas howled, chained down with the chain of
    jealousy.°
Then Oothoon hovered over Judah and Jerusalem,
And Jesus heard her voice (a man of sorrows); he received°
A gospel from wretched Theotormon.

The human race began to wither, for the healthy built
Secluded places, fearing the joys of love,
And the diseased only propagated.
So Antamon called up Leutha from her valleys of delight,
And to Mahomet a loose Bible gave.°
But in the north to Odin Sotha gave a code of war,   30
Because of Diralada, thinking to reclaim his joy.

These were the churches, hospitals, castles, palaces,
Like nets and gins and traps to catch the joys of Eternity,
    And all the rest a desert;
Till like a dream Eternity was obliterated and erased.

Since that dread day when Har and Heva fled,
Because their brethren and sisters lived in war and lust;
And as they fled they shrunk
Into two narrow doleful forms,
Creeping in reptile flesh upon   40
The bosom of the ground,
And all the vast of Nature shrunk
Before their shrunken eyes.

Thus the terrible race of Los and Enitharmon gave
Laws and religions to the sons of Har, binding them more
And more to earth, closing and restraining,
Till a philosophy of five senses was complete.
Urizen wept and gave it into the hands of Newton and Locke.

Clouds roll heavy upon the Alps round Rousseau and Voltaire,
And on the mountains of Lebanon round the deceased gods   50
Of Asia, and on the deserts of Africa round the fallen angels.
The Guardian Prince of Albion burns in his nightly tent.°

America, a Prophecy (1793)

PRELUDIUM

The shadowy daughter of Urthona stood before red Orc,°
When fourteen suns had faintly journeyed o'er his dark abode.
His food she brought in iron baskets, his drink in cups of iron.
Crowned with a helmet and dark hair the nameless female stood;
A quiver with its burning stores, a bow like that of night
When pestilence is shot from heaven—no other arms she need.°
Invulnerable though naked (save where clouds roll round her loins
Their awful folds in the dark air) silent she stood as night.
For never from her iron tongue could voice or sound arise,
But dumb till that dread day when Orc essayed his fierce embrace.   10

'Dark virgin,' said the hairy youth, 'thy father stern abhorred
Rivets my tenfold chains, while still on high my spirit soars.
Sometimes an eagle screaming in the sky, sometimes a lion
Stalking upon the mountains, and sometimes a whale I lash
The raging fathomless abyss; anon a serpent folding
Around the pillars of Urthona, and round thy dark limbs,
On the Canadian wilds I fold. Feeble my spirit folds,
For chained beneath I rend these caverns. When thou bringest food
I howl my joy, and my red eyes seek to behold thy face.
In vain! These clouds roll to and fro, and hide thee from my sight.'   20

Silent as despairing love, and strong as jealousy,
The hairy shoulders rend the links; free are the wrists of fire.
Round the terrific loins he seized the panting struggling womb;
It joyed. She put aside her clouds and smiled her first-born smile,
As when a black cloud shows its lightnings to the silent deep.

Soon as she saw the terrible boy, then burst the virgin cry:

'I know thee, I have found thee, and I will not let thee go.
Thou art the image of God who dwells in darkness of Africa,
And thou art fall'n to give me life in regions of dark death.
On my American plains I feel the struggling afflictions   30
Endured by roots that writhe their arms into the nether deep:
I see a serpent in Canada, who courts me to his love;

In Mexico an eagle, and a lion in Peru;
I see a whale in the South Sea, drinking my soul away.
Oh, what limb-rending pains I feel! Thy fire and my frost
Mingle in howling pains, in furrows by thy lightnings rent.
This is eternal death, and this the torment long foretold.'

A PROPHECY

The Guardian Prince of Albion burns in his nightly tent;
Sullen fires across the Atlantic glow to America's shore,
Piercing the souls of warlike men, who rise in silent night.   40
Washington, Franklin, Paine and Warren, Gates, Hancock and
    Greene
Meet on the coast glowing with blood from Albion's fiery prince.

Washington spoke: 'Friends of America, look over the Atlantic sea:
A bended bow is lifted in heaven, and a heavy iron chain
Descends link by link from Albion's cliffs across the sea to bind
Brothers and sons of America, till our faces pale and yellow,
Heads depressed, voices weak, eyes downcast, hands work-
    bruised,
Feet bleeding on the sultry sands, and the furrows of the whip,
Descend to generations that in future times forget.'

The strong voice ceased; for a terrible blast swept over the heaving
    sea;   50
The eastern cloud rent; on his cliffs stood Albion's wrathful
    prince,
A dragon form clashing his scales. At midnight he arose,
And flamed red meteors round the land of Albion beneath.
His voice, his locks, his awful shoulders, and his glowing eyes
Appear to the Americans upon the cloudy night.

Solemn heave the Atlantic waves between the gloomy nations,
Swelling, belching from its deeps red clouds and raging fires.

Albion is sick; America faints! Enraged the zenith grew.
As human blood shooting its veins all round the orbed heaven
Red rose the clouds from the Atlantic in vast wheels of blood,   60
And in the red clouds rose a wonder o'er the Atlantic sea:
Intense! naked! a human fire fierce glowing, as the wedge
Of iron heated in the furnace. His terrible limbs were fire,

With myriads of cloudy terrors, banners dark and towers
Surrounded. Heat but not light went through the murky
  atmosphere.

The King of England looking westward trembles at the vision.

Albion's Angel stood beside the stone of night, and saw
The terror like a comet, or more like the planet red°
That once enclosed the terrible wandering comets in its sphere
(Then, Mars, thou wast our centre, and the planets three flew   70
  round
Thy crimson disc; so ere the sun was rent from thy red sphere.)
The spectre glowed, his horrid length staining the temple long
With beams of blood, and thus a voice came forth and shook the
  temple:

'The morning comes, the night decays, the watchmen leave their
  stations;
The grave is burst, the spices shed, the linen wrapped up;°
The bones of death, the cov'ring clay, the sinews shrunk and
  dried°
Reviving shake, inspiring move, breathing, awakening°—
Spring like redeemed captives when their bonds and bars are
  burst.
Let the slave grinding at the mill run out into the field;°
Let him look up into the heavens and laugh in the bright air;   80
Let the enchained soul shut up in darkness and in sighing,
Whose face has never seen a smile in thirty weary years,
Rise and look out—his chains are loose, his dungeon doors are
  open.
And let his wife and children return from the oppressor's scourge;
They look behind at every step and believe it is a dream,
Singing: "The sun has left his blackness, and has found a fresher
  morning,
And the fair moon rejoices in the clear and cloudless night.
For empire is no more, and now the lion and wolf shall cease."'

In thunders ends the voice. Then Albion's Angel wrathful burnt
Beside the Stone of Night, and, like the eternal lion's howl   90
In famine and war, replied: 'Art thou not Orc, who, serpent-
  formed,

Stands at the gate of Enitharmon to devour her children?
Blasphemous demon, Antichrist, hater of dignities,
Lover of wild rebellion, and transgressor of God's law,
Why dost thou come to angels' eyes in this terrific form?'

The terror answered: 'I am Orc, wreathed round the accursed tree.
The times are ended, shadows pass, the morning 'gins to break.
The fiery joy, that Urizen perverted to ten commands
What night he led the starry hosts through the wide wilderness—
That stony law I stamp to dust, and scatter religion abroad   100
To the four winds as a torn book, and none shall gather the leaves.
But they shall rot on desert sands and consume in bottomless
    deeps:
To make the deserts blossom and the deeps shrink to their
    fountains,°
And to renew the fiery joy and burst the stony roof—
That pale religious lechery, seeking virginity,
May find it in a harlot, and in coarse-clad honesty
The undefiled, though ravished in her cradle night and morn.
For every thing that lives is holy, life delights in life—
Because the soul of sweet delight can never be defiled.
Fires enwrap the earthly globe, yet Man is not consumed;   110
Amidst the lustful fires he walks; his feet become like brass,°
His knees and thighs like silver, and his breast and head like gold.'°

'Sound! Sound! my loud war-trumpets, and alarm my thirteen
    Angels!
Loud howls the eternal wolf! The eternal lion lashes his tail!
America is darkened, and my punishing demons terrified
Crouch howling before their caverns deep like skins dried in the
    wind.
They cannot smite the wheat, nor quench the fatness of the earth.
They cannot smite with sorrows, nor subdue the plough and
    spade.
They cannot wall the city, nor moat round the castle of princes.
They cannot bring the stubbed oak to overgrow the hills.   120
For terrible men stand on the shores, and in their robes I see
Children take shelter from the lightnings; there stands Wash-
    ington
And Paine and Warren, with their foreheads reared toward the
    east.

But clouds obscure my aged sight. A vision from afar!
Sound! Sound! my loud war-trumpets, and alarm my thirteen
    Angels.
Ah, vision from afar! Ah, rebel form that rent the ancient
Heavens, eternal viper self-renewed, rolling in clouds!
I see thee in thick clouds and darkness on America's shore,
Writhing in pangs of abhorred birth. Red flames the crest
    rebellious
And eyes of death. The harlot womb oft opened in vain   130
Heaves in enormous circles, now the times are returned upon thee,
Devourer of thy parent, now thy unutterable torment renews.
Sound! Sound! my loud war-trumpets, and alarm my thirteen
    Angels.
Ah, terrible birth! A young one bursting! Where is the weeping
    mouth?
And where the mother's milk? Instead those ever-hissing jaws
And parched lips drop with fresh gore. Now roll thou in the
    clouds;
Thy mother lays her length outstretched upon the shore beneath.
Sound! Sound! my loud war-trumpets, and alarm my thirteen
    Angels!
Loud howls the eternal wolf! The eternal lion lashes his tail!'

Thus wept the angel voice, and as he wept the terrible blasts   140
Of trumpets blew a loud alarm across the Atlantic deep.
No trumpets answer, no reply of clarion or of fifes;
Silent the Colonies remain and refuse the loud alarm.

On those vast shady hills between America and Albion's shore,
Now barred out by the Atlantic sea (called Atlantean hills,°
Because from their bright summits you may pass to the golden
    world),
An ancient palace, archetype of mighty emperies,
Rears its immortal pinnacles, built in the forest of God
By Ariston, the king of beauty, for his stolen bride.°

Here on their magic seats the thirteen Angels sat perturbed,   150
For clouds from the Atlantic hover o'er the solemn roof.

Fiery the Angels rose, and as they rose deep thunder rolled
Around their shores, indignant burning with the fires of Orc.
And Boston's Angel cried aloud as they flew through the dark
   night.

He cried: 'Why trembles honesty, and like a murderer
Why seeks he refuge from the frowns of his immortal station?
Must the generous tremble and leave his joy to the idle, to the
   pestilence,
That mock him? Who commanded this? What God? What Angel?
To keep the gen'rous from experience, till the ungenerous
Are unrestrained performers of the energies of nature;   160
Till pity is become a trade, and generosity a science
That men get rich by, and the sandy desert is given to the strong.
What God is he, writes laws of peace and clothes him in the
   tempest?
What pitying Angel lusts for tears, and fans himself with sighs?
What crawling villain preaches abstinence and wraps himself
In fat of lambs? No more I follow, no more obedience pay.'°

So cried he, rending off his robe and throwing down his sceptre
In sight of Albion's Guardian. And all the thirteen Angels
Rent off their robes to the hungry wind and threw their golden
   sceptres
Down on the land of America; indignant they descended   170
Headlong from out their heav'nly heights, descending swift as fires
Over the land. Naked and flaming are their lineaments seen
In the deep gloom. By Washington and Paine and Warren they
   stood,
And the flame folded roaring fierce within the pitchy night
Before the demon red, who burnt towards America
In black smoke, thunders and loud winds, rejoicing in its terror,
Breaking in smoky wreaths from the wild deep, and gath'ring thick
In flames as of a furnace on the land from north to south.
What time the thirteen governors that England sent convene
In Bernard's house, the flames covered the land. They rouse, they   180
   cry;°
Shaking their mental chains they rush in fury to the sea
To quench their anguish. At the feet of Washington down fall'n,
They grovel on the sand and writhing lie, while all
The British soldiers through the thirteen states sent up a howl

Of anguish, threw their swords and muskets to the earth and ran
From their encampments and dark castles, seeking where to hide
From the grim flames and from the visions of Orc, in sight
Of Albion's Angel; who enraged his secret clouds opened
From north to south, and burnt outstretched on wings of wrath
 cov'ring
The eastern sky, spreading his awful wings across the heavens. 190
Beneath him rolled his num'rous hosts; all Albion's Angels
 camped
Darkened the Atlantic mountains, and their trumpets shook the
 valleys—
Armed with diseases of the earth to cast upon the abyss,
Their numbers forty millions, must'ring in the eastern sky.

In the flames stood and viewed the armies drawn out in the sky
Washington, Franklin, Paine and Warren, Allen, Gates and Lee,
And heard the voice of Albion's Angel give the thunderous
 command.
His plagues, obedient to his voice, flew forth out of their clouds,°
Falling upon America, as a storm to cut them off,
As a blight cuts the tender corn when it begins to appear. 200
Dark is the heaven above, and cold and hard the earth beneath,
And as a plague-wind filled with insects cuts off man and beast,
And as a sea o'erwhelms a land in the day of an earthquake.

Fury, rage, madness in a wind swept through America,
And the red flames of Orc that folded roaring fierce around
The angry shores, and the fierce rushing of th' inhabitants
 together.
The citizens of New York close their books and lock their chests;
The mariners of Boston drop their anchors and unlade;
The scribe of Pennsylvania casts his pen upon the earth;
The builder of Virginia throws his hammer down in fear. 210

Then had America been lost, o'erwhelmed by the Atlantic,
And Earth had lost another portion of the infinite.
But all rush together in the night, in wrath and raging fire;
The red fires raged, the plagues recoiled, then rolled they back
 with fury
On Albion's Angels. Then the pestilence began in streaks of red
Across the limbs of Albion's Guardian; the spotted plague smote
 Bristol's

And the leprosy London's spirit, sickening all their bands.

The millions sent up a howl of anguish and threw off their
     hammered mail,

And cast their swords and spears to earth, and stood a naked
     multitude.

Albion's Guardian writhed in torment on the eastern sky,   220

Pale quiv'ring toward the brain his glimmering eyes, teeth
     chattering,

Howling and shuddering, his legs quivering, convulsed each
     muscle and sinew;

Sick'ning lay London's Guardian and the ancient mitred York,°

Their heads on snowy hills, their ensigns sick'ning in the sky.

The plagues creep on the burning winds, driven by flames of Orc,

And by the fierce Americans rushing together in the night,

Driven o'er the Guardians of Ireland and Scotland and Wales.

They, spotted with plagues, forsook the frontiers, and their
     banners seared

With fires of hell deform their ancient heavens with shame and
     woe.

Hid in his caves the Bard of Albion felt the enormous plagues,°   230

And a cowl of flesh grew o'er his head, and scales on his back and
     ribs;

And rough with black scales all his Angels fright their ancient
     heavens.

The doors of marriage are open, and the priests in rustling scales°

Rush into reptile coverts, hiding from the fires of Orc

That play around the golden roofs in wreaths of fierce desire,

Leaving the females naked and glowing with the lusts of youth.

For the female spirits of the dead, pining in bonds of religion,

Run from their fetters reddening, and in long drawn arches sitting;

They feel the nerves of youth renew, and desires of ancient times.

Over their pale limbs as a vine when the tender grape appears,   240

Over the hills, the vales, the cities, raged the red flames fierce;

The heavens melted from north to south, and Urizen, who sat

Above all heavens in thunders wrapped, emerged his leprous head

From out his holy shrine, his tears in deluge piteous

Falling into the deep sublime. Flagged with grey-browed snows

And thunderous visages, his jealous wings waved over the deep;

Weeping in dismal howling woe he dark descended, howling

Around the smitten bands, clothed in tears and trembling,
    shudd'ring cold.
His stored snows he poured forth, and his icy magazines
He opened on the deep, and on the Atlantic sea, white shiv'ring.   250
Leprous his limbs, all over white, and hoary was his visage,
Weeping in dismal howlings before the stern Americans,
Hiding the demon red with clouds and cold mists from the earth:
Till angels and weak men twelve years should govern o'er the
    strong,°
And then their end should come, when France received the
    demon's light.

Stiff shudderings shook the heav'nly thrones! France, Spain and
    Italy
In terror viewed the bands of Albion and the ancient Guardians
Fainting upon the elements, smitten with their own plagues.
They slow advance to shut the five gates of their law-built heaven,
Filled with blasting fancies and with mildews of despair,   260
With fierce disease and lust, unable to stem the fires of Orc;
But the five gates were consumed, and their bolts and hinges
    melted,°
And the fierce flames burnt round the heavens and round the
    abodes of men.

FINIS

Europe, a Prophecy (1794)

'Five windows light the caverned man: through one he breathes the air;°
Through one, hears music of the spheres; through one the eternal vine
Flourishes, that he may receive the grapes; through one, can look
And see small portions of the eternal world that ever groweth;
Through one, himself pass out what time he please but he will not,
For stolen joys are sweet, and bread eaten in secret pleasant.'°

So sang a fairy mocking as he sat on a streaked tulip,
Thinking none saw him; when he ceased I started from the trees
And caught him in my hat as boys knock down a butterfly.
'How know you this,' said I, 'small sir? Where did you learn this song?'   10
Seeing himself in my possession, thus he answered me:
'My master, I am yours; command me, for I must obey.'

'Then tell me what is the material world, and is it dead?'
He laughing answered: 'I will write a book on leaves of flowers,
If you will feed me on love-thoughts, and give me now and then
A cup of sparkling poetic fancies. So when I am tipsy,
I'll sing to you to this soft lute, and show you all alive
The world, when every particle of dust breathes forth its joy.'

I took him home in my warm bosom. As we went along
Wild flowers I gathered, and he showed me each eternal flower;   20
He laughed aloud to see them whimper because they were plucked.
They hov'red round me like a cloud of incense when I came
Into my parlour and sat down, and took my pen to write;
My fairy sat upon the table, and dictated EUROPE.

PRELUDIUM

The nameless shadowy female rose from out of the breast of Orc,
Her snaky hair brandishing in the winds of Enitharmon;
And thus her voice arose:

'O mother Enitharmon, wilt thou bring forth other sons
To cause my name to vanish, that my place may not be found?
For I am faint with travail!—   30
Like the dark cloud disburdened in the day of dismal thunder.

'My roots are brandished in the heavens, my fruits in earth
    beneath
Surge, foam, and labour into life, first born and first consumed,
Consumed and consuming!
Then why shouldst thou, accursed mother, bring me into life?

'I wrap my turban of thick clouds around my lab'ring head,
And fold the sheety waters as a mantle round my limbs—
Yet the red sun and moon
And all the overflowing stars rain down prolific pains.

'Unwilling I look up to heaven! unwilling count the stars!   40
Sitting in fathomless abyss of my immortal shrine,
I seize their burning power
And bring forth howling terrors, all-devouring fiery kings,

'Devouring and devouréd, roaming on dark and desolate mountains
In forests of eternal death, shrieking in hollow trees.
Ah, mother Enitharmon,
Stamp not with solid form this vig'rous progeny of fires!

'I bring forth from my teeming bosom myriads of flames,
And thou dost stamp them with a signet; then they roam abroad
And leave me void as death.   50
Ah! I am drowned in shady woe, and visionary joy.

'And who shall bind the infinite with an eternal band,
To compass it with swaddling bands? And who shall cherish it
With milk and honey?
I see it smile and I roll inward and my voice is past.'

She ceased, and rolled her shady clouds
Into the secret place.

A PROPHECY

The deep of winter came,
What time the secret child°
Descended through the orient gates of the eternal day.   60
War ceased, and all the troops like shadows fled to their abodes.

Then Enitharmon saw her sons and daughters rise around.
Like pearly clouds they meet together in the crystal house,
And Los, possessor of the moon, joyed in the peaceful night,
Thus speaking while his num'rous sons shook their bright fiery
wings:

'Again the night is come
That strong Urthona takes his rest,
And Urizen unloosed from chains
Glows like a meteor in the distant north.
Stretch forth your hands and strike the elemental strings!   70
Awake the thunders of the deep,
The shrill winds wake!—
Till all the sons of Urizen look out and envy Los.
Seize all the spirits of life and bind
Their warbling joys to our loud strings;
Bind all the nourishing sweets of earth
To give us bliss, that we may drink the sparkling wine of Los;
And let us laugh at war,
Despising toil and care,
Because the days and nights of joy in lucky hours renew.'   80

'Arise, O Orc, from thy deep den;
First-born of Enitharmon, rise!
And we will crown thy head with garlands of the ruddy vine—
For now thou art bound—
And I may see thee in the hour of bliss, my eldest born.'

The horrent demon rose, surrounded with red stars of fire,°
Whirling about in furious circles round the immortal fiend.
Then Enitharmon down descended into his red light,
And thus her voice rose to her children; the distant heavens reply.
'Now comes the night of Enitharmon's joy.   90
Who shall I call? Who shall I send?
That Woman, lovely Woman, may have dominion?°
Arise, O Rintrah, thee I call! And Palamabron, thee!
Go! Tell the human race that Woman's love is sin,
That an eternal life awaits the worms of sixty winters
In an allegorical abode where existence hath never come.
Forbid all joy, and from her childhood shall the little female
Spread nets in every secret path.

'My weary eyelids draw towards the evening; my bliss is yet but
   new.

'Arise, O Rintrah, eldest born, second to none but Orc.   100
O lion Rintrah, raise thy fury from thy forests black;
Bring Palamabron, horned priest, skipping upon the mountains,°
And silent Elynittria, the silver-bowed queen.
Rintrah, where hast thou hid thy bride?
Weeps she in desert shades?
Alas, my Rintrah, bring the lovely jealous Ocalythron.

'Arise, my son! Bring all thy brethren, O thou king of fire.
Prince of the sun, I see thee with thy innumerable race
Thick as the summer stars,
But each ramping his golden mane shakes;
And thine eyes rejoice because of strength, O Rintrah, furious   110
     king.'

Enitharmon slept
Eighteen hundred years. Man was a dream!°—
The night of Nature, and their harps unstrung.
She slept in middle of her nightly song:
Eighteen hundred years, a female dream.

Shadows of men in fleeting bands upon the winds
Divide the heavens of Europe;
Till Albion's Angel, smitten with his own plagues, fled with his
     bands.
The cloud bears hard on Albion's shore,   120
Filled with immortal demons of futurity.
In council gather the smitten Angels of Albion;°
The cloud bears hard upon the council house, down rushing
On the heads of Albion's Angels.

One hour they lay buried beneath the ruins of that hall;
But as the stars rise from the salt lake they arise in pain,
In troubled mists o'erclouded by the terrors of struggling times.

In thoughts perturbed, they rose from the bright ruins, silent
     following
The fiery king, who sought his ancient temple serpent-formed,°
That stretches out its shady length along the island white.   130
Round him rolled his clouds of war; silent the Angel went,
Along the infinite shores of Thames to golden Verulam.°

There stand the venerable porches that high-towering rear
Their oak-surrounded pillars, formed of massy stones, uncut
With tool, stones precious. Such eternal in the heavens,
Of colours twelve, few known on earth, give light in the opaque,°
Placed in the order of the stars when the five senses whelmed
In deluge o'er the earth-born man. Then turned the fluxile eyes°
Into two stationary orbs, concentrating all things;
The ever-varying spiral ascents to the heavens of heavens°   140
Were bended downward, and the nostrils' golden gates shut,
Turned outward, barred and petrified against the infinite.

Thought changed the infinite to a serpent, that which pitieth
To a devouring flame, and Man fled from its face and hid
In forests of night. Then all the eternal forests were divided
Into earths rolling in circles of space, that like an ocean rushed
And overwhelmed all except this finite wall of flesh.
Then was the serpent temple formed, image of infinite
Shut up in finite revolutions, and Man became an Angel,
Heaven a mighty circle turning, God a tyrant crowned.   150

Now arrived the ancient Guardian at the southern porch
That, planted thick with trees of blackest leaf, and in a vale
Obscure, enclosed the Stone of Night. Oblique it stood, o'erhung
With purple flowers and berries red, image of that sweet south
Once open to the heavens and elevated on the human neck,
Now overgrown with hair and covered with a stony roof;
Downward 'tis sunk beneath th' attractive north, that round the
    feet,
A raging whirlpool, draws the dizzy enquirer to his grave.

Albion's Angel rose upon the Stone of Night.
He saw Urizen on the Atlantic,   160
And his brazen book
That kings and priests had copied on earth
Expanded from north to south.

And the clouds and fires pale rolled round in the night of
    Enitharmon,
Round Albion's cliffs and London's walls; still Enitharmon slept.
Rolling volumes of grey mist involve churches, palaces, towers;
For Urizen unclasped his book, feeding his soul with pity.

The youth of England hid in gloom curse the pained heavens, compelled
Into the deadly night to see the form of Albion's Angel.
Their parents brought them forth and aged ignorance preaches, 170
    canting,
On a vast rock, perceived by those senses that are closed from
    thought;
Bleak, dark, abrupt it stands and overshadows London city.
They saw his bony feet on the rock, the flesh consumed in flames;
They saw the serpent temple lifted above, shadowing the island
    white;
They heard the voice of Albion's Angel howling in flames of Orc,
Seeking the trump of the last doom.

Above the rest the howl was heard from Westminster louder and
    louder.
The guardian of the secret codes forsook his ancient mansion,°
Driven out by the flames of Orc; his furred robes and false locks
Adhered and grew one with his flesh, and nerves and veins shot 180
    through them
With dismal torment sick, hanging upon the wind. He fled
Grovelling along Great George Street through the Park gate; all
    the soldiers
Fled from his sight; he dragged his torments to the wilderness.

Thus was the howl through Europe!
For Orc rejoiced to hear the howling shadows,
But Palamabron shot his lightnings trenching down his wide back,
And Rintrah hung with all his legions in the nether deep.

Enitharmon laughed in her sleep to see (Oh, woman's triumph!)
Every house a den, every man bound; the shadows are filled
With spectres, and the windows wove over with curses of iron; 190
Over the doors 'Thou shalt not', and over the chimneys 'Fear' is
    written.
With bands of iron round their necks, fastened into the walls,
The citizens; in leaden gyves the inhabitants of suburbs
Walk heavy; soft and bent are the bones of villagers.

Between the clouds of Urizen the flames of Orc roll heavy
Around the limbs of Albion's Guardian, his flesh consuming.
Howlings and hissings, shrieks and groans, and voices of despair
Arise around him in the cloudy heavens of Albion. Furious
The red-limbed angel seized, in horror and torment,
The trump of the last doom; but he could not blow the iron tube!   200
Thrice he essayed presumptuous to awake the dead to judgement.
A mighty spirit leaped from the land of Albion,
Named Newton; he seized the trump and blowed the enormous
    blast!
Yellow as leaves of autumn, the myriads of angelic hosts°
Fell through the wintry skies seeking their graves,
Rattling their hollow bones in howling and lamentation.

Then Enitharmon woke, nor knew that she had slept,
And eighteen hundred years were fled
As if they had not been.
She called her sons and daughters   210
To the sports of night
Within her crystal house,
And thus her song proceeds:

'Arise Ethinthus! Though the earth-worm call,
Let him call in vain
Till the night of holy shadows
And human solitude is past!

'Ethinthus, queen of waters, how thou shinest in the sky!
My daughter, how do I rejoice! For thy children flock around
Like the gay fishes on the wave, when the cold moon drinks the   220
    dew.
Ethinthus! thou art sweet as comforts to my fainting soul,
For now thy waters warble round the feet of Enitharmon.

'Manathu-Varcyon! I behold thee flaming in my halls,
Light of thy mother's soul! I see thy lovely eagles round;
Thy golden wings are my delight, and thy flames of soft delusion.

'Where is my luring bird of Eden? Leutha, silent love!
Leutha, the many-coloured bow delights upon thy wings:
Soft soul of flowers, Leutha!

Sweet smiling pestilence! I see thy blushing light;
Thy daughters many-changing   230
Revolve like sweet perfumes ascending, O Leutha, silken queen!

'Where is the youthful Antamon, prince of the pearly dew?
O Antamon, why wilt thou leave thy mother Enitharmon?
Alone I see thee, crystal form,
Floating upon the bosomed air
With lineaments of gratified desire,
My Antamon, the seven churches of Leutha seek thy love.°

'I hear the soft Oothoon in Enitharmon's tents.
Why wilt thou give up woman's secrecy, my melancholy child?
Between two moments bliss is ripe.   240
O Theotormon, robbed of joy, I see thy salt tears flow
Down the steps of my crystal house.

'Sotha and Thiralatha, secret dwellers of dreamful caves,
Arise and please the horrent fiend with your melodious songs.
Still all your thunders golden-hoofed, and bind your horses black.
Orc! Smile upon my children!
Smile, son of my afflictions.
Arise, O Orc, and give our mountains joy of thy red light.'

She ceased, for all were forth at sport beneath the solemn moon,
Waking the stars of Urizen with their immortal songs: 250
That Nature felt through all her pores the enormous revelry,
Till morning oped the eastern gate.
Then every one fled to his station, and Enitharmon wept.

But terrible Orc, when he beheld the morning in the east,
Shot from the heights of Enitharmon.
And in the vineyards of red France appeared the light of his fury.

The sun glowed fiery red!
The furious terrors flew around!—
On golden chariots raging, with red wheels dropping with blood.
The lions lash their wrathful tails;   260
The tigers couch upon the prey and suck the ruddy tide;
And Enitharmon groans and cries in anguish and dismay.

Then Los arose; his head he reared in snaky thunders clad,
And with a cry that shook all Nature to the utmost pole
Called all his sons to the strife of blood.

FINIS

Asia (1795)

The kings of Asia heard
The howl rise up from Europe
And each ran out from his web,
From his ancient woven den;
For the darkness of Asia was startled
At the thick-flaming, thought-creating fires of Orc.
And the kings of Asia stood
And cried in bitterness of soul:

'Shall not the king call for famine from the heath,
Nor the priest for pestilence from the fen?   10
To restrain, to dismay, to thin
The inhabitants of mountain and plain!
In the day of full-feeding prosperity
And the night of delicious songs.

'Shall not the counsellor throw his curb
Of poverty on the laborious?
To fix the price of labour,
To invent allegoric riches.

'And the privy admonishers of men
Call for fires in the city,   20
For heaps of smoking ruins,
In the night of prosperity and wantonness?

'To turn man from his path,
To restrain the child from the womb,
To cut off the bread from the city;
That the remnant may learn to obey;

'That the pride of the heart may fail;
That the lust of the eyes may be quenched;
That the delicate ear in its infancy

May be dulled, and the nostrils closed up.   30
To teach mortal worms the path
That leads from the gates of the grave.'

Urizen heard them cry,
And his shudd'ring, waving wings
Went enormous above the red flames,
Drawing clouds of despair through the heavens
Of Europe as he went.
And his books of brass, iron and gold
Melted over the land as he flew
Heavy-waving, howling, weeping.   40

And he stood over Judaea,
And stayed in his ancient place,
And stretched his clouds over Jerusalem.

For Adam, a mouldering skeleton,
Lay bleached on the garden of Eden;
And Noah as white as snow
On the mountains of Ararat.

Then the thunders of Urizen bellowed aloud
From his woven darkness above.

Orc, raging in European darkness,   50
Arose like a pillar of fire above the Alps,°
Like a serpent of fiery flame!
    The sullen earth°
    Shrunk!
Forth from the dead dust rattling, bones to bones°
Join; shaking convulsed, the shiv'ring clay breathes,
And all flesh naked stands: fathers and friends,
Mothers and infants, kings and warriors.

The grave shrieks with delight and shakes
Her hollow womb, and clasps the solid stem.   60
Her bosom swells with wild desire,
And milk and blood and glandous wine°
In rivers rush and shout and dance
On mountain, dale and plain.

            The SONG OF LOS is ended.
                Urizen wept.

THE LYRICS

Song (*before 1783*)

How sweet I roamed from field to field,
 And tasted all the summer's pride,
Till I the prince of love beheld,
 Who in the sunny beams did glide!

He showed me lilies for my hair,
 And blushing roses for my brow;
He led me through his gardens fair,
 Where all his golden pleasures grow.

With sweet May dews my wings were wet,
 And Phoebus fired my vocal rage;   10
He caught me in his silken net,
 And shut me in his golden cage.

He loves to sit and hear me sing;
 Then, laughing, sports and plays with me;
Then stretches out my golden wing,
 And mocks my loss of liberty.

Song (*before 1783*)

My silks and fine array,
 My smiles and languished air,
By love are driv'n away,
 And mournful lean Despair
Brings me yew to deck my grave;
Such end true lovers have.

His face is fair as heav'n
 When springing buds unfold;
Oh why to him was't giv'n,
 Whose heart is wintry cold?   10
His breast is love's all worshipped tomb,
Where all love's pilgrims come.

Bring me an axe and spade;
   Bring me a winding sheet.
When I my grave have made,
   Let winds and tempests beat;
Then down I'll lie, as cold as clay.
True love doth pass away!

Song (before 1783)

Love and harmony combine,
And around our souls intwine,
While thy branches mix with mine,
And our roots together join.

Joys upon our branches sit,
Chirping loud, and singing sweet;
Like gentle streams beneath our feet
Innocence and virtue meet.

Thou the golden fruit dost bear;
I am clad in flowers fair;   10
Thy sweet boughs perfume the air,
And the turtle buildeth there.

There she sits and feeds her young;
Sweet I hear her mournful song;
And thy lovely leaves among,
There is love: I hear her tongue.

There his charming nest doth lay;
There he sleeps the night away;
There he sports along the day,
And doth among our branches play.   20

Song (before 1783)

I love the jocund dance,
   The softly-breathing song,
Where innocent eyes do glance,
   And where lisps the maiden's tongue.

I love the laughing vale,
  I love the echoing hill,
Where mirth does never fail,
  And the jolly swain laughs his fill.

I love the pleasant cot,
  I love the innocent bow'r,   10
Where white and brown is our lot,°
  Or fruit in the midday hour.

I love the oaken seat,
  Beneath the oaken tree,
Where all the old villagers meet,
  And laugh our sports to see.

I love our neighbours all,
  But, Kitty, I better love thee;
And love them I ever shall,
  But thou art all to me.   20

Song (*before 1783*)

Memory, hither come,
  And tune your merry notes;
And, while upon the wind
  Your music floats,
  I'll pore upon the stream,
  Where sighing lovers dream,
  And fish for fancies as they pass
  Within the watery glass.

I'll drink of the clear stream,
  And hear the linnet's song;   10
And there I'll lie and dream
  The day along.
And, when night comes, I'll go
  To places fit for woe,
Walking along the darkened valley,
  With silent Melancholy.

Mad Song (*before 1783*)

The wild winds weep,
   And the night is a-cold;
Come hither, Sleep,
   And my griefs infold.
But lo! the morning peeps,
   Over the eastern steeps,
And the rustling birds of dawn
The earth do scorn.

Lo! to the vault
   Of paved heaven,
With sorrow fraught,
   My notes are driven.
They strike the ear of night,
   Make weep the eyes of day;
They make mad the roaring winds,
   And with tempests play.

Like a fiend in a cloud,
   With howling woe
After night I do crowd,
   And with night will go.
I turn my back to the east,
From whence comforts have increased;
For light doth seize my brain
With frantic pain.

Song (*before 1783*)

Fresh from the dewy hill, the merry year
Smiles on my head, and mounts his flaming car;
Round my young brows the laurel wreathes a shade,
And rising glories beam around my head.

My feet are winged, while o'er the dewy lawn
I meet my maiden, risen like the morn.
Oh, bless those holy feet, like angels' feet!
Oh, bless those limbs, beaming with heav'nly light!

Like as an angel glitt'ring in the sky,
In times of innocence and holy joy.   10
The joyful shepherd stops his grateful song,
To hear the music of an angel's tongue.

So when she speaks, the voice of Heaven I hear;
So when we walk, nothing impure comes near.
Each field seems Eden, and each calm retreat;
Each village seems the haunt of holy feet.

But that sweet village where my black-eyed maid
Closes her eyes in sleep beneath night's shade,
Whene'er I enter, more than mortal fire
Burns in my soul, and does my song inspire.   20

Song (*before 1783*)

When early morn walks forth in sober grey,
Then to my black-eyed maid I haste away.
When evening sits beneath her dusky bow'r,
And gently sighs away the silent hour,
The village bell alarms, away I go—
And the vale darkens at my pensive woe.

To that sweet village, where my black-eyed maid
Doth drop a tear beneath the silent shade,
I turn my eyes; and, pensive as I go,
Curse my black stars, and bless my pleasing woe.   10

Oft when the summer sleeps among the trees,
Whisp'ring faint murmurs to the scanty breeze,
I walk the village round; if at her side
A youth doth walk in stolen joy and pride,
I curse my stars in bitter grief and woe,
That made my love so high, and me so low.

O should she e'er prove false, his limbs I'd tear,
And throw all pity on the burning air;
I'd curse bright fortune for my mixed lot,
And then I'd die in peace, and be forgot.   20

Song First by a Shepherd (1784)

Welcome, stranger, to this place,
Where Joy doth sit on every bough;
Paleness flies from every face;
We reap not what we do not sow.

Innocence doth like a rose
Bloom on every maiden's cheek;
Honour twines around her brows;
The jewel Health adorns her neck

Song Third by an Old Shepherd (1784)

When silver snow decks Sylvio's clothes,
And jewel hangs at shepherd's nose,
We can abide life's pelting storm,
That makes our limbs quake, if our hearts be warm.

Whilst Virtue is our walking-staff,
And Truth a lantern to our path,
We can abide life's pelting storm,
That makes our limbs quake, if our hearts be warm.

Blow, boisterous wind; stern winter, frown.
Innocence is a winter's gown;   10
So clad, we'll abide life's pelting storm,
That makes our limbs quake, if our hearts be warm.

[The Cynic's First Song] (c.1784)

I

When old Corruption first begun,
Adorned in yellow vest,°
He committed on Flesh a whoredom:
Oh what a wicked beast!

2

From them a callow babe did spring,
And old Corruption smiles,
To think his race should never end,
For now he had a child.

3

He called him Surgery, and fed
The babe with his own milk—   10
For Flesh and he could ne'er agree:
She would not let him suck.

4

And this he always kept in mind,
And formed a crooked knife,
And ran about with bloody hands
To seek his mother's life.

5

And as he ran to seek his mother
He met with a dead woman;°
He fell in love and married her,
A deed which is not common.   20

6

She soon grew pregnant, and brought forth
Scurvy and Spotted Fever.°
The father grinned and skipped about,
And said, 'I'm made for ever.

7

'For now I have procured these imps,
I'll try experiments.'
With that he tied poor Scurvy down
And stopped up all its vents;

8

And when the child began to swell
He shouted out aloud: 30
'I've found the dropsy out, and soon
Shall do the world more good.'

9
He took up Fever by the neck
And cut out all its spots,
And through the holes which he had made
He first discovered guts.

[*Miss Gittipin's First Song*] (c.*1784*)

Phoebe dressed like beauty's queen,
Jellicoe in faint pea-green—
Sitting all beneath a grot,
Where the little lambkins trot.

Maidens dancing, loves a-sporting,
All the country folks a-courting,
Susan, Johnny, Bet, and Joe,
Lightly tripping on a row.

Happy people, who can be
In happiness compared with ye?   10
The pilgrim with his crook and hat
Sees your happiness complete.

[*The Cynic's Second Song*] (c.*1784*)

Hail Matrimony, made of Love!
To thy wide gates how great a drove
On purpose to be yoked do come,
Widows and maids, and youths also,
That lightly trip on beauty's toe,
Or sit on beauty's bum.

Hail, fingerfooted lovely creatures!°
The females of our human natures,
Formed to suckle all mankind.
'Tis you that come in time of need;   10
Without you we should never breed,
Or any comfort find.

For if a damsel's blind or lame,
Or Nature's hand has crooked her frame,
Or if she's deaf or is wall-eyed—
Yet if her heart is well inclined,
Some tender lover she shall find
That panteth for a bride.

The universal poultice this,
To cure whatever is amiss   20
In damsel or in widow gay.
It makes them smile, it makes them skip;
Like birds just cured of the pip,°
They chirp and hop away.

Then come ye maidens, come ye swains,
Come and be eased of all your pains,
In Matrimony's golden cage.

[*Obtuse Angle's Song*] (c.*1784*)

To be or not to be
Of great capacity,
Like Sir Isaac Newton,
Or Locke, or Doctor South,°
Or Sherlock upon death.
I'd rather be Sutton.°

For he did build a house
For aged men and youth,
With walls of brick and stone,
He furnished it within   10
With whatever he could win,
And all his own.

He drew out of the stocks
His money in a box,
And sent his servant
To Green the bricklayer,
And to the carpenter—
He was so fervent.

The chimneys were three score,
The windows many more;   20
And for convenience
He sinks and gutters made,
And all the way he paved
To hinder pestilence.

Was not this a good man,
Whose life was but a span,
Whose name was Sutton,
As Locke, or Doctor South,
Or Sherlock upon Death,
Or Sir Isaac Newton?   30

[*The Lawgiver's Song*] (c.*1784*)

This city and this country has brought forth many mayors,
To sit in state and give forth laws, out of their old oak chairs,
With face as brown as any nut with drinking of strong ale.
Good English hospitality, oh, then it did not fail!

With scarlet gowns and broad gold lace would make a yeoman
     sweat,
With stockings rolled above their knees, and shoes as black as jet—
With eating beef and drinking beer, oh, they were stout and hale.
Good English hospitality, oh, then it did not fail!

Thus sitting at the table wide, the Mayor and Aldermen
Were fit to give law to the City: each ate as much as ten.   10
The hungry poor entered the hall to eat good beef and ale.
Good English hospitality, oh, then it did not fail!

[*Miss Gittipin's Second Song*] (c.*1784*)

Leave, oh, leave me to my sorrows;
Here I'll sit and fade away,
Till I'm nothing but a spirit,
And I lose this form of clay.

Then if chance along this forest
Any walk in pathless ways,
Through the gloom he'll see my shadow,
Hear my voice upon the breeze.

SONGS OF INNOCENCE *(1789)*

Introduction

Piping down the valleys wild,
Piping songs of pleasant glee,
On a cloud I saw a child,
And he laughing said to me:

'Pipe a song about a lamb.'
So I piped with merry cheer.
'Piper, pipe that song again.'
So I piped; he wept to hear.

'Drop thy pipe, thy happy pipe;
Sing thy songs of happy cheer.'   10
So I sung the same again,
While he wept with joy to hear.

'Piper sit thee down and write
In a book that all may read—'
So he vanished from my sight.
And I plucked a hollow reed,

And I made a rural pen,
And I stained the water clear,°
And I wrote my happy songs
Every child may joy to hear.   20

A Dream

Once a dream did weave a shade
O'er my angel-guarded bed,
That an emmet lost its way
Where on grass methought I lay.

Troubled, wildered and forlorn,
Dark, benighted, travel-worn,
Over many a tangled spray
All heart-broke I heard her say:

'Oh my children! Do they cry?
Do they hear their father sigh?   10
Now they look abroad to see,
Now return and weep for me.'

Pitying I dropped a tear;
But I saw a glow-worm near,
Who replied: 'What wailing wight
Calls the watchman of the night?

'I am set to light the ground,
While the beetle goes his round.
Follow now the beetle's hum.
Little wanderer hie thee home.'   20

The Little Girl Lost

In futurity
I prophetic see
That the earth from sleep
(Grave the sentence deep)

Shall arise and seek
For her maker meek,
And the desert wild
Become a garden mild.

In the southern clime,
Where the summer's prime   10
Never fades away,
Lovely Lyca lay.

Seven summers old
Lovely Lyca told;
She had wandered long,
Hearing wild birds' song.

'Sweet sleep come to me
Underneath this tree;
Do father, mother weep?
Where can Lyca sleep?   20

'Lost in desert wild
Is your little child.
How can Lyca sleep,
If her mother weep?

'If her heart does ache,
Then let Lyca wake;
If my mother sleep,
Lyca shall not weep.

'Frowning, frowning night,
O'er this desert bright,   30
Let thy moon arise
While I close my eyes.'

Sleeping Lyca lay,
While the beasts of prey,
Come from caverns deep,
Viewed the maid asleep.

The kingly lion stood
And the virgin viewed,
Then he gambolled round
O'er the hallowed ground.   40

Leopards, tigers play,
Round her as she lay,
While the lion old
Bowed his mane of gold,

And her bosom lick,
And upon her neck;
From his eyes of flame
Ruby tears there came;

While the lioness
Loosed her slender dress   50
And naked they conveyed
To caves the sleeping maid.

The Little Girl Found

All the night in woe
Lyca's parents go,
Over valleys deep
While the deserts weep.

Tired and woe-begone,
Hoarse with making moan,
Arm in arm seven days
They traced the desert ways.

Seven nights they sleep
Among shadows deep,   10
And dream they see their child
Starved in desert wild.

Pale through pathless ways
The fancied image strays,
Famished, weeping, weak,
With hollow piteous shriek.

Rising from unrest,
The trembling woman pressed°
With feet of weary woe;
She could no further go.   20

In his arms he bore
Her, armed with sorrow sore,
Till before their way
A couching lion lay.

Turning back was vain.
Soon his heavy mane
Bore them to the ground;
Then he stalked around,

Smelling to his prey;
But their fears allay°   30
When he licks their hands,
And silent by them stands.

They look upon his eyes,
Filled with deep surprise,
And wondering behold
A spirit armed in gold.

On his head a crown,
On his shoulders down
Flowed his golden hair.
Gone was all their care.   40

'Follow me', he said,
'Weep not for the maid;
In my palace deep,
Lyca lies asleep.'

Then they followed
Where the vision led,
And saw their sleeping child,
Among tigers wild.

To this day they dwell
In a lonely dell,   50
Nor fear the wolvish howl,
Nor the lion's growl.

The Blossom

Merry, merry sparrow,
Under leaves so green,
A happy blossom
Sees you swift as arrow
Seek your cradle narrow
Near my bosom.

Pretty, pretty robin,
Under leaves so green,
A happy blossom
Hears you sobbing, sobbing,   10
Pretty, pretty robin
Near my bosom.

The Lamb

Little Lamb who made thee?
  Dost thou know who made thee?
Gave thee life and bid thee feed
By the stream and o'er the mead;
Gave thee clothing of delight,
Softest clothing woolly bright;
Gave thee such a tender voice,
Making all the vales rejoice.
  Little Lamb who made thee?
  Dost thou know who made thee?   10

Little Lamb I'll tell thee,
  Little Lamb I'll tell thee:
He is called by thy name,
For he calls himself a lamb.
He is meek and he is mild;
He became a little child.
I a child and thou a lamb,
We are called by his name.
  Little Lamb God bless thee.
  Little Lamb God bless thee.   20

The Shepherd

How sweet is the shepherd's sweet lot!
From the morn to the evening he strays;
He shall follow his sheep all the day,
And his tongue shall be filled with praise.°

For he hears the lamb's innocent call,
And he hears the ewe's tender reply.
He is watchful, while they are in peace,
For they know when their shepherd is nigh.

Infant Joy

'I have no name;
I am but two days old.'
What shall I call thee?
'I happy am;
Joy is my name.'
Sweet joy befall thee!

Pretty joy!
Sweet joy but two days old,
Sweet joy I call thee.
Thou dost smile;   10
I sing the while.
Sweet joy befall thee.

On Another's Sorrow

Can I see another's woe,
And not be in sorrow too?
Can I see another's grief
And not seek for kind relief?

Can I see a falling tear,
And not feel my sorrow's share?
Can a father see his child
Weep, nor be with sorrow filled?

Can a mother sit, and hear
An infant groan, an infant fear?   10
No, no, never can it be.
Never, never can it be.

And can he who smiles on all
Hear the wren with sorrows small,
Hear the small bird's grief and care,
Hear the woes that infants bear—

And not sit beside the nest,
Pouring pity in their breast;
And not sit the cradle near
Weeping tear on infant's tear;   20

And not sit both night and day,
Wiping all our tears away?°
Oh no! never can it be.
Never, never can it be.

He doth give his joy to all.
He becomes an infant small.
He becomes a man of woe.°
He doth feel the sorrow too.

Think not thou canst sigh a sigh,
And thy maker is not by.   30
Think not thou canst weep a tear,
And thy maker is not near.

Oh! he gives to us his joy,
That our grief he may destroy,
Till our grief is fled and gone
He doth sit by us and moan.

Spring

Sound the flute!
Now it's mute.
Birds delight
Day and night.
Nightingale
In the dale,
Lark in sky,
Merrily,
Merrily, merrily to welcome in the year.

Little boy
Full of joy,
Little girl
Sweet and small.
Cock does crow,
So do you.
Merry voice,
Infant noise,
Merrily, merrily to welcome in the year.

Little lamb   20
Here I am.
Come and lick
My white neck.
Let me pull
Your soft wool.
Let me kiss
Your soft face,
Merrily, merrily we welcome in the year.

The Schoolboy

I love to rise in a summer morn,
When the birds sing on every tree;
The distant huntsman winds his horn,
And the skylark sings with me.
Oh! what sweet company.

But to go to school in a summer morn,
Oh! it drives all joy away;
Under a cruel eye outworn,
The little ones spend the day
In sighing and dismay.   10

Ah! then at times I drooping sit,
And spend many an anxious hour,
Nor in my book can I take delight,
Nor sit in learning's bower,
Worn through with the dreary shower.

How can the bird that is born for joy
Sit in a cage and sing?
How can a child when fears annoy
But droop his tender wing,
And forget his youthful spring? 20

O father and mother! if buds are nipped,
And blossoms blown away,
And if the tender plants are stripped
Of their joy in the springing day,
By sorrow and care's dismay,

How shall the summer arise in joy,
Or the summer fruits appear?
Or how shall we gather what griefs destroy,
Or bless the mellowing year,
When the blasts of winter appear? 30

Laughing Song

When the green woods laugh with the voice of joy
And the dimpling stream runs laughing by,
When the air does laugh with our merry wit
And the green hill laughs with the noise of it,

When the meadows laugh with lively green
And the grasshopper laughs in the merry scene,
When Mary and Susan and Emily
With their sweet round mouths sing 'Ha, Ha, He,'

When the painted birds laugh in the shade
Where our table with cherries and nuts is spread— 10
Come live and be merry and join with me,
To sing the sweet chorus of 'Ha, Ha, He.'

The Little Black Boy

My mother bore me in the southern wild,
And I am black, but oh! my soul is white.
White as an angel is the English child;
But I am black as if bereaved of light.

My mother taught me underneath a tree,
And sitting down before the heat of day
She took me on her lap and kissed me,
And pointing to the east began to say:

'Look on the rising sun! There God does live,
And gives his light and gives his heat away;   10
And flowers and trees and beasts and men receive
Comfort in morning, joy in the noon day.

'And we are put on earth a little space,
That we may learn to bear the beams of love;
And these black bodies and this sun-burnt face
Is but a cloud, and like a shady grove.

'For when our souls have learned the heat to bear
The cloud will vanish; we shall hear his voice,
Saying: "Come out from the grove my love and care,
And round my golden tent like lambs rejoice." '   20

Thus did my mother say and kissed me,
And thus I say to little English boy.
When I from black and he from white cloud free,
And round the tent of God like lambs we joy,

I'll shade him from the heat till he can bear
To lean in joy upon our father's knee,
And then I'll stand and stroke his silver hair
And be like him, and he will then love me.°

The Voice of the Ancient Bard

Youth of delight come hither
And see the opening morn,
Image of truth new-born.
Doubt is fled and clouds of reason,
Dark disputes and artful teasing.
Folly is an endless maze;
Tangled roots perplex her ways.
How many have fallen there!
They stumble all night over bones of the dead,
And feel they know not what but care,   10
And wish to lead others when they should be led.

The Echoing Green

The sun does arise,
And make happy the skies.
The merry bells ring
To welcome the spring.
The skylark and thrush,
The birds of the bush,
Sing louder around,
To the bells' cheerful sound,
While our sports shall be seen
On the echoing green.   10

Old John with white hair
Does laugh away care,
Sitting under the oak,
Among the old folk.
They laugh at our play,
And soon they all say:
'Such, such were the joys
When we all, girls and boys,
In our youth-time were seen
On the echoing green.'   20

Till the little ones weary
No more can be merry;
The sun does descend,
And our sports have an end.
Round the laps of their mother
Many sisters and brothers,
Like birds in their nest,
Are ready for rest;
And sport no more seen
On the darkening green.   30

Nurse's Song

When the voices of children are heard on the green
And laughing is heard on the hill,
My heart is at rest within my breast
And everything else is still.

'Then come home my children: the sun is gone down
And the dews of night arise.
Come, come leave off play and let us away,
Till the morning appears in the skies.'

'No, no let us play, for it is yet day
And we cannot go to sleep.   10
Besides, in the sky the little birds fly,
And the hills are all covered with sheep.'

'Well, well go and play till the light fades away,
And then go home to bed.'
The little ones leaped and shouted and laughed
And all the hills echoed.

Holy Thursday

'Twas on a Holy Thursday, their innocent faces clean,
The children walking two and two in red and blue and green,
Grey headed beadles walked before with wands as white as snow;
Till into the high dome of Paul's they like Thames waters flow.

Oh what a multitude they seemed, these flowers of London town.°
Seated in companies they sit, with radiance all their own.°
The hum of multitudes was there, but multitudes of lambs:
Thousands of little boys and girls raising their innocent hands.

Now like a mighty wind they raise to Heaven the voice of song,°
Or like harmonious thunderings the seats of Heaven among.   10
Beneath them sit the aged men, wise guardians of the poor.
Then cherish pity, lest you drive an angel from your door.

The Divine Image

To Mercy, Pity, Peace and Love
All pray in their distress,
And to these virtues of delight
Return their thankfulness.

For Mercy, Pity, Peace and Love
Is God our father dear,
And Mercy, Pity, Peace and Love
Is Man his child and care.

For Mercy has a human heart,
Pity a human face,   10
And Love the human form divine,°
And Peace the human dress.

Then every man of every clime
That prays in his distress,
Prays to the human form divine:
Love, Mercy, Pity, Peace.

And all must love the human form,
In heathen, Turk or Jew.
Where Mercy, Love and Pity dwell,
There God is dwelling too.   20

The Chimney-Sweeper

When my mother died I was very young,
And my father sold me while yet my tongue
Could scarcely cry, 'weep weep weep weep'.
So your chimneys I sweep and in soot I sleep.

There's little Tom Dacre, who cried when his head,
That curled like a lamb's back, was shaved, so I said:
'Hush Tom, never mind it, for when your head's bare,
You know that the soot cannot spoil your white hair.'

And so he was quiet, and that very night,
As Tom was a-sleeping, he had such a sight: 10
That thousands of sweepers, Dick, Joe, Ned and Jack,
Were all of them locked up in coffins of black,

And by came an angel who had a bright key,
And he opened the coffins and set them all free.
Then down a green plain leaping, laughing they run,
And wash in a river and shine in the sun.

Then naked and white, all their bags left behind,
They rise upon clouds, and sport in the wind.
And the angel told Tom if he'd be a good boy,
He'd have God for his father and never want joy.   20

And so Tom awoke, and we rose in the dark,
And got with our bags and our brushes to work.
Though the morning was cold, Tom was happy and warm.
So if all do their duty, they need not fear harm.

A Cradle Song

Sweet dreams, form a shade
O'er my lovely infant's head,
Sweet dreams of pleasant streams,
By happy, silent, moony beams.

Sweet sleep, with soft down
Weave thy brows an infant crown.
Sweet sleep, angel mild,
Hover o'er my happy child.

Sweet smiles in the night,
Hover over my delight.
Sweet smiles, mother's smiles,
All the livelong night beguiles

10

Sweet moans, dovelike sighs,
Chase not slumber from thy eyes.
Sweet moans, sweeter smiles,
All the dovelike moans beguiles.

Sleep, sleep happy child.
All creation slept and smiled.
Sleep, sleep, happy sleep,
While o'er thee thy mother weep.

20

Sweet babe, in thy face,
Holy image I can trace.
Sweet babe, once like thee
Thy maker lay, and wept for me,

Wept for me, for thee, for all,
When he was an infant small.
Thou his image ever see,
Heavenly face that smiles on thee,

Smiles on thee, on me, on all,
Who became an infant small.
Infant smiles are his own smiles;
Heaven and earth to peace beguiles.

30

The Little Boy Lost

'Father, father where are you going?
Oh do not walk so fast.
Speak father, speak to your little boy,
Or else I shall be lost.'

The night was dark, no father was there,
The child was wet with dew.
The mire was deep, and the child did weep,
And away the vapour flew.°

The Little Boy Found

The little boy lost in the lonely fen,
Led by the wand'ring light,
Began to cry, but God ever nigh
Appeared like his father in white.

He kissed the child and by the hand led,
And to his mother brought,
Who in sorrow pale through the lonely dale
Her little boy weeping sought.

Night

The sun descending in the west,
The evening star does shine.
The birds are silent in their nest,
And I must seek for mine.
The moon, like a flower
In heaven's high bower,
With silent delight
Sits and smiles on the night.

Farewell green fields and happy groves,
Where flocks have took delight;   10
Where lambs have nibbled, silent moves
The feet of angels bright.
Unseen they pour blessing,
And joy without ceasing,
On each bud and blossom,
And each sleeping bosom.

They look in every thoughtless nest,
Where birds are covered warm;
They visit caves of every beast,
To keep them all from harm.   20
If they see any weeping
That should have been sleeping,
They pour sleep on their head
And sit down by their bed.

When wolves and tigers howl for prey
They pitying stand and weep,
Seeking to drive their thirst away,
And keep them from the sheep.
But if they rush dreadful,
The angels most heedful,   30
Receive each mild spirit,
New worlds to inherit.

And there the lion's ruddy eyes
Shall flow with tears of gold,
And pitying the tender cries
And walking round the fold,
Saying: 'Wrath by his meekness,
And by his health sickness,
Is driven away
From our immortal day.   40

'And now beside thee, bleating lamb,
I can lie down and sleep,°
Or think on him who bore thy name,
Graze after thee and weep.
For washed in life's river,°
My bright mane for ever
Shall shine like the gold,
As I guard o'er the fold.'

MANUSCRIPT LYRICS BETWEEN *INNOCENCE* AND *EXPERIENCE*

'I told my love'

I told my love, I told my love,
I told her all my heart,
Trembling, cold, in ghastly fears.
Ah, she doth depart.

Soon as she was gone from me,
A traveller came by
Silently, invisibly.
Oh, was no deny.

'I laid me down'

I laid me down upon a bank
Where love lay sleeping.
I heard among the rushes dank
Weeping, weeping.

Then I went to the heath and the wild,
To the thistles and thorns of the waste,
And they told me how they were beguiled,
Driven out, and compelled to be chaste.

'I saw a chapel'

I saw a chapel all of gold
That none did dare to enter in,
And many weeping stood without,
Weeping, mourning, worshipping.

I saw a serpent rise between
The white pillars of the door,
And he forced and forced and forced;
Down the gold hinges tore.

And along the pavement sweet,
Set with pearls and rubies bright,   10
All his slimy length he drew
Till upon his altar white,

Vomiting his poison out
On the bread and on the wine.
So I turned into a sty
And laid me down among the swine.

'I asked a thief'

I asked a thief to steal me a peach;
He turned up his eyes.
I asked a lithe lady to lie her down;
Holy and meek, she cries.

As soon as I went
An angel came.
He winked at the thief
And smiled at the dame,

And, without one word said,
Had a peach from the tree   10
And, still as a maid,
Enjoyed the lady.

'I heard an angel'

I heard an angel singing,
When the day was springing,
'Mercy, Pity, Peace,
Is the world's release.'

Thus he sung all day
Over the new-mown hay,
Till the sun went down
And haycocks looked brown.

I heard a devil curse
Over the heath and the furze,   10
'Mercy could be no more
If there was nobody poor,

'And pity no more could be
If all were as happy as we.'
At his curse the sun went down,
And the heavens gave a frown.

A Cradle Song

Sleep, sleep, beauty bright,
Dreaming o'er the joys of night.
Sleep, sleep; in thy sleep
Little sorrows sit and weep.

Sweet babe, in thy face
Soft desires I can trace,
Secret joys and secret smiles,
Little pretty infant wiles.

As thy softest limbs I feel,
Smiles as of the morning steal   10
O'er thy cheek and o'er thy breast,
Where thy little heart does rest.

Oh, the cunning wiles that creep
In thy little heart asleep.
When thy little heart does wake,
Then the dreadful lightnings break

From thy cheek and from thy eye,
O'er the youthful harvests nigh.
Infant wiles and infant smiles
Heaven and Earth of peace beguiles.   20

'I feared the fury'

I feared the fury of my wind
Would blight all blossoms fair and true;
And my sun it shined and shined,
And my wind it never blew.

But a blossom fair or true
Was not found on any tree,
For all blossoms grew and grew
Fruitless, false—though fair to see.

'Silent, silent night'

Silent, silent night,
Quench the holy light
Of thy torches bright;

For, possessed of day,
Thousand spirits stray
That sweet joys betray.

Why should joys be sweet
Used with deceit,
Nor with sorrows meet?

But an honest joy   10
Does itself destroy
For a harlot coy.

'Why should I care'

Why should I care for the men of Thames,
Or the cheating waves of chartered streams,°
Or shrink at the little blasts of fear
That the hireling blows into my ear?

Though born on the cheating banks of Thames,
Though his waters bathed my infant limbs,
The Ohio shall wash his stains from me.
I was born a slave, but I go to be free.

'O lapwing'

O lapwing, thou flyest around the heath,
Nor seest the net that is spread beneath.
Why dost thou not fly among the cornfields?
They cannot spread nets where a harvest yields.

'Thou hast a lap full of seed'

'Thou hast a lap full of seed
And this is a fine country;°
Why dost thou not cast thy seed
And live in it merrily?'

'Shall I cast it on the sand
And turn it into fruitful land?
For on no other ground
Can I sow my seed,
Without tearing up
Some stinking weed.'   10

In a Myrtle Shade

Why should I be bound to thee,
O my lovely myrtle tree?
Love, free love, cannot be bound°
To any tree that grows on ground.

Oh how sick and weary I
Underneath my myrtle lie,
Like to dung upon the ground,
Underneath my myrtle bound.

Oft my myrtle signed in vain,
To behold my heavy chain.   10
Oft my father saw us sigh,
And laughed at our simplicity.

So I smote him, and his gore
Stained the roots my myrtle bore;
But the time of youth is fled,
And grey hairs are on my head.

'As I wandered'

As I wandered the forest,
The green leaves among,
I heard a wild flower
Singing a song:

'I slept in the earth
In the silent night,
I murmured my fears
And I felt delight.

'In the morning I went,
As rosy as morn,
To seek for new joy,
But I met with scorn.'

'Are not the joys'

Are not the joys of morning sweeter
Than the joys of night,
And are the vig'rous joys of youth
Ashamed of the light?

Let age and sickness silent rob
The vineyards in the night,
But those who burn with vig'rous youth
Pluck fruits before the light.

To Nobodaddy

Why art thou silent and invisible,
Father of Jealousy?
Why dost thou hide thyself in clouds
From every searching eye?

Why darkness and obscurity
In all thy words and laws,
That none dare eat the fruit but from
The wily serpent's jaws?

Or is it because secrecy gains females' loud applause?

How to Know Love from Deceit

Love to faults is always blind,
Always is to joy inclined,
Lawless, winged and unconfined,
And breaks all chains from every mind.

Deceit to secrecy confined,
Lawful, cautious and refined,
To everything but interest blind,
And forges fetters for the mind.

Soft Snow

I walked abroad in a snowy day;
I asked the soft snow with me to play.
She played and she melted in all her prime,
And the winter called it a dreadful crime.

An Ancient Proverb

Remove away that black'ning church,
Remove away that marriage hearse,
Remove away that man of blood—
You'll quite remove the ancient curse.

To my Myrtle

To a lovely myrtle bound,
Blossoms show'ring all around,
Oh, how sick and weary I
Underneath my myrtle lie.
Why should I be bound to thee,
O my lovely myrtle tree?

Merlin's Prophecy

The harvest shall flourish in wintry weather,
When two virginities meet together.

The king and the priest must be tied in a tether,
Before two virgins can meet together.

Day

The sun arises in the east,
Clothed in robes of blood and gold;
Swords and spears and wrath increased,
All around his bosom rolled,
Crowned with warlike fires and raging desires.

The Fairy

'Come hither my sparrows,
My little arrows.
If a tear or a smile
Will a man beguile;
If an amorous delay
Clouds a sunshiny day;
If the step of a foot
Smites the heart to its root—
'Tis the marriage ring
Makes each fairy a king.'

10

So a fairy sung—
From the leaves I sprung.
He leaped from the spray
To flee away,
But, in my hat caught,
He soon shall be taught.
Let him laugh, let him cry,
He's my butterfly.
For I've pulled out the sting
Of the marriage ring   20

'The sword sung on the barren heath'

The sword sung on the barren heath,
The sickle in the fruitful field;
The sword he sung a song of death,
But could not make the sickle yield.

'Abstinence sows sand'

Abstinence sows sand all over
The ruddy limbs and flaming hair;
But desire gratified
Plants fruits of life and beauty there.

'In a wife I would desire'

In a wife I would desire
(What in whores is always found),
The lineaments of gratified desire.

'If you trap the moment'

If you trap the moment before it's ripe,
The tears of repentance you'll certainly wipe;
Buf if once you let the ripe moment go,
You can never wipe off the tears of woe.

Eternity

He who binds to himself a joy
Does the winged life destroy;
But he who kisses the joy as it flies
Lives in eternity's sunrise.

The Question Answered

What is it men in women do require?
The lineaments of gratified desire.
What is it women do in men require?
The lineaments of gratified desire.

Lacedaemonian Instruction

'Come hither my boy; tell me what thou seest there.'
'A fool tangled in a religious snare'.

Riches

The countless gold of a merry heart,
The rubies and pearls of a loving eye,
The indolent never can bring to the mart,
Nor the secret hoard up in his treasury.

An Answer to the Parson

'Why of the sheep do you not learn peace?'
'Because I don't want you to shear my fleece'.

'The look of love alarms'

The look of love alarms
Because 'tis filled with fire;
But the look of soft deceit
Shall win the lover's hire.

'Her whole life is an epigram'

Her whole life is an epigram, smack-smooth and neatly penned,°
Platted quite neat to catch applause—with a sliding noose at the
    end.

'An old maid early'

An old maid early, ere I knew
Ought but the love that on me grew;
And now I'm covered o'er and o'er,
And wish that I had been a whore.

Oh, I cannot, cannot find
The undaunted courage of a virgin mind,
For early I in love was crossed,
Before my flower of love was lost.

Motto to the Songs of Innocence and Experience

The good are attracted by men's perceptions,
    And think not for themselves,
Till experience teaches them to catch
    And to cage the fairies and elves.

And then the knave begins to snarl;
    And the hypocrite to howl;
And all his good friends show their private ends,
    And the eagle is known from the owl.

SONGS OF EXPERIENCE (*1793*)

Introduction

Hear the voice of the bard!
Who present, past, and future sees;
Whose ears have heard
The Holy Word,
That walked among the ancient trees°

Calling the lapsed soul,
And weeping in the evening dew;
That might control
The starry pole,°
And fallen, fallen light renew!   10

'O Earth, O Earth return!
Arise from out the dewy grass.
Night is worn,
And the morn
Rises from the slumberous mass.

'Turn away no more;
Why wilt thou turn away?
The starry floor,
The wat'ry shore,
Is giv'n thee till the break of day.'   20

Earth's Answer

Earth raised up her head
From the darkness dread and drear.
Her light fled—
Stony dread!—
And her locks covered with grey despair.

'Prisoned on wat'ry shore
Starry jealousy does keep my den.
Cold and hoar,
Weeping o'er,
I hear the father of the ancient men.   10

'Selfish father of men,
Cruel, jealous, selfish fear!
Can delight
Chained in night
The virgins of youth and morning bear?

'Does spring hide its joy
When buds and blossoms grow?
Does the sower
Sow by night?
Or the ploughman in darkness plough?   20

'Break this heavy chain
That does freeze my bones around.
Selfish! Vain!
Eternal bane!
That free love with bondage bound.'°

The Clod and the Pebble

'Love seeketh not itself to please,
Nor for itself hath any care,
But for another gives its ease,
And builds a Heaven in Hell's despair.'

So sang a little Clod of Clay,
Trodden with the cattle's feet,
But a Pebble of the brook
Warbled out these metres meet:

'Love seeketh only self to please,
To bind another to its delight,   10
Joys in another's loss of ease,
And builds a Hell in Heaven's despite.'

Holy Thursday

Is this a holy thing to see,
In a rich and fruitful land:
Babes reduced to misery,
Fed with cold and usurous hand?°

Is that trembling cry a song?
Can it be a song of joy?
And so many children poor?
It is a land of poverty!

And their sun does never shine,
And their fields are bleak and bare,
And their ways are filled with thorns;
It is eternal winter there.

For where'er the sun does shine,
And where'er the rain does fall—
Babe can never hunger there,
Nor poverty the mind appal.

The Chimney-Sweeper

A little black thing among the snow,
Crying 'weep, weep' in notes of woe!
'Where are thy father and mother? Say!'
'They are both gone up to the church to pray.

'Because I was happy upon the heath,
And smiled among the winter's snow,
They clothed me in the clothes of death,
And taught me to sing the notes of woe.

'And because I am happy and dance and sing,
They think they have done me no injury,
And are gone to praise God and his priest and king,
Who make up a heaven of our misery.'

Nurse's Song

When the voices of children are heard on the green,
And whisperings are in the dale,
The days of my youth rise fresh in my mind,
My face turns green and pale.

Then come home my children, the sun is gone down,
And the dews of night arise.
Your spring and your day are wasted in play,
And your winter and night in disguise.

The Sick Rose

O rose, thou art sick;
The invisible worm
That flies in the night,
In the howling storm,

Has found out thy bed
Of crimson joy,
And his dark secret love
Does thy life destroy.

The Fly

Little fly,
Thy summer's play
My thoughtless hand
Has brushed away.

Am not I
A fly like thee?
Or art not thou
A man like me?

For I dance
And drink and sing,
Till some blind hand
Shall brush my wing.

If thought is life
And strength and breath,
And the want
Of thought is death,

Then am I
A happy fly,
If I live,
Or if I die.

The Angel

I dreamt a dream!—what can it mean?—
And that I was a maiden queen,
Guarded by an angel mild.
Witless woe was ne'er beguiled!

And I wept both night and day,
And he wiped my tears away,
And I wept both day and night,
And hid from him my heart's delight.

So he took his wings and fled;
Then the morn blushed rosy red.   10
I dried my tears, and armed my fear
With ten thousand shields and spears.

Soon my angel came again.
I was armed; he came in vain,
For the time of youth was fled,
And grey hairs were on my head.

The Tiger

Tiger, tiger, burning bright,
In the forests of the night:
What immortal hand or eye
Could frame thy fearful symmetry?

In what distant deeps or skies,
Burnt the fire of thine eyes?
On what wings dare he aspire?
What the hand dare seize the fire?

And what shoulder, and what art,
Could twist the sinews of thy heart?   10
And when thy heart began to beat,
What dread hand? and what dread feet?

What the hammer? what the chain?
In what furnace was thy brain?
What the anvil? what dread grasp
Dare its deadly terrors clasp?

When the stars threw down their spears,
And watered Heaven with their tears,
Did he smile his work to see?
Did he who made the lamb make thee?   20

Tiger, tiger, burning bright,
In the forests of the night:
What immortal hand or eye
Dare frame thy fearful symmetry?

My Pretty Rose Tree

A flower was offered to me,
Such a flower as May never bore,
But I said, 'I've a pretty rose tree,'
And I passed the sweet flower o'er.

Then I went to my pretty rose tree,
To tend her by day and by night,
But my rose turned away with jealousy,
And her thorns were my only delight.

Ah! Sunflower

Ah! sunflower, weary of time,
Who countest the steps of the sun,
Seeking after that sweet golden clime
Where the traveller's journey is done;

Where the youth pined away with desire,°
And the pale virgin shrouded in snow,
Arise from their graves and aspire;
Where my sunflower wishes to go.

The Lily

The modest rose puts forth a thorn,
The humble sheep a threat'ning horn;
While the lily white shall in love delight,
Nor a thorn nor a threat stain her beauty bright.

The Garden of Love

I went to the Garden of Love,
And saw what I never had seen:
A chapel was built in the midst,
Where I used to play on the green.

And the gates of this chapel were shut,
And 'Thou shalt not' writ over the door;
So I turned to the Garden of Love,
That so many sweet flowers bore.

And I saw it was filled with graves,
And tomb-stones where flowers should be,   10
And priests in black gowns were walking their rounds,
And binding with briars my joys and desires.

The Little Vagabond

Dear mother, dear mother, the church is cold,
But the ale-house is healthy and pleasant and warm.
Besides, I can tell where I am used well;
Such usage in Heaven will never do well.

But if at the church they would give us some ale,
And a pleasant fire our souls to regale,
We'd sing and we'd pray all the live-long day,
Nor ever once wish from the church to stray.

Then the parson might preach and drink and sing,
And we'd be as happy as birds in the spring;   10
And modest dame Lurch, who is always at church,
Would not have bandy children, nor fasting, nor birch.°

And God, like a father rejoicing to see
His children as pleasant and happy as he,
Would have no more quarrel with the Devil or the barrel,
But kiss him and give him both drink and apparel.

London

I wander through each chartered street,°
Near where the chartered Thames does flow,
And mark in every face I meet
Marks of weakness, marks of woe.

In every cry of every man,
In every infant's cry of fear,
In every voice, in every ban,
The mind-forged manacles I hear:

How the chimney-sweeper's cry
Every black'ning church appalls,°   10
And the hapless soldier's sigh
Runs in blood down palace walls.

But most through midnight streets I hear
How the youthful harlot's curse
Blasts the new-born infant's tear,
And blights with plagues the marriage hearse.

The Human Abstract

Pity would be no more
If we did not make somebody poor,
And Mercy no more could be
If all were as happy as we.

And mutual fear brings Peace,
Till the selfish loves increase.
Then Cruelty knits a snare,
And spreads his baits with care.

He sits down with holy fears,
And waters the ground with tears;   10
Then Humility takes its root
Underneath his foot.

Soon spreads the dismal shade
Of Mystery over his head,
And the caterpillar and fly
Feed on the Mystery.

And it bears the fruit of Deceit,
Ruddy and sweet to eat,
And the raven his nest has made
In its thickest shade.   20

The gods of the earth and sea
Sought through Nature to find this tree,
But their search was all in vain.
There grows one in the human brain.

Infant Sorrow

My mother groaned, my father wept!
Into the dangerous world I leapt,
Helpless, naked, piping loud,
Like a fiend hid in a cloud.

Struggling in my father's hands,
Striving against my swaddling bands,°
Bound and weary, I thought best
To sulk upon my mother's breast.

A Poison Tree

I was angry with my friend;
I told my wrath—my wrath did end.
I was angry with my foe;
I told it not—my wrath did grow.

And I watered it in fears,
Night and morning with my tears,
And I sunned it with smiles,
And with soft deceitful wiles.

And it grew both day and night,
Till it bore an apple bright.   10
And my foe beheld it shine,
And he knew that it was mine,

And into my garden stole
When the night had veiled the pole.°
In the morning glad I see
My foe outstretched beneath the tree.°

A Little Boy Lost

'Nought loves another as itself,
Nor venerates another so,
Nor is it possible to thought
A greater than itself to know.

'And father, how can I love you,
Or any of my brothers more?
I love you like the little bird
That picks up crumbs around the door.'

The priest sat by and heard the child;
In trembling zeal he seized his hair.   10
He led him by his little coat,
And all admired the priestly care.

And, standing on the altar high,
'Lo, what a fiend is here!' said he,
'One who sets reason up for judge
Of our most holy mystery.'

The weeping child could not be heard;
The weeping parents wept in vain.
They stripped him to his little shirt,
And bound him in an iron chain,   20

And burned him in a holy place,
Where many had been burned before.
The weeping parents wept in vain.
Are such things done on Albion's shore?

A Little Girl Lost

Children of the future age,
Reading this indignant page,
Know that in a former time
Love! sweet love! was thought a crime.

In the age of gold,
Free from winter's cold,
Youth and maiden bright,
To the holy light,
Naked in the sunny beams delight.

Once a youthful pair,   10
Filled with softest care,
Met in garden bright,
Where the holy light
Had just removed the curtains of the night.

There in rising day
On the grass they play.
Parents were afar;
Strangers came not near;
And the maiden soon forgot her fear.

Tired with kisses sweet,   20
They agree to meet
When the silent sleep
Waves o'er heavens deep,
And the weary, tired wanderers weep.

To her father white
Came the maiden bright;
But his loving look,
Like the holy book,
All her tender limbs with terror shook.

'Ona! pale and weak!   30
To thy father speak.
Oh the trembling fear!
Oh the dismal care,
That shakes the blossoms of my hoary hair!'

To Tirzah

Whate'er is born of mortal birth
Must be consumed with the earth,
To rise from generation free.°
Then what have I to do with thee?

The sexes sprung from shame and pride,
Blowed in the morn, in evening died.
But mercy changed death into sleep;
The sexes rose to work and weep.

Thou, mother of my mortal part,
With cruelty didst mould my heart,   10
And with false self-deceiving tears
Didst bind my nostrils, eyes and ears;

Didst close my tongue in senseless clay,
And me to mortal life betray.
The death of Jesus set me free.°
Then what have I to do with thee?

A Divine Image

Cruelty has a human heart,
And Jealousy a human face;
Terror the human form divine,
And Secrecy the human dress.

The human dress is forged iron;
The human form, a fiery forge;
The human face, a furnace sealed;
The human heart, its hungry gorge.

MANUSCRIPT LYRICS OF THE FELPHAM YEARS
(1800–1803)

'When a man has married a wife'

When a man has married a wife he finds out whether
Her knees and elbows are only glued together.

'A woman scaly'

A woman scaly and a man all hairy
Is such a match as he who dares,
Will find the woman's scales scrape off the man's hairs.

'A fairy stepped upon my knee'

A fairy stepped upon my knee,
Singing and dancing merrily.
I said: 'Thou thing of patches, rings,
Pins, necklaces and suchlike things,
Disguiser of the female form,
Thou paltry, gilded pois'nous worm!'
Weeping he fell upon my thigh
And thus in tears did soft reply:
'Knowest thou not, O fairies' lord,
How much by us contemned, abhorred,
Whatever hides the female form,
That cannot bear the mental storm?
Therefore in pity still we give
Our lives to make the female live,
And what would turn into disease
We turn to what will joy and please.'

10

On the Virginity of the Virgin Mary and Joanna Southcott

Whate'er is done to her she cannot know,
And if you'll ask her she will swear it so.
Whether 'tis good or evil none's to blame—
No one can take the pride, no one the shame.

The Golden Net

Three virgins at the break of day:
'Whither young man, whither away?
Alas for woe! alas for woe!'
They cry, and tears for ever flow.
The one was clothed in flames of fire,
The other clothed in iron wire,
The other clothed in tears and sighs,
Dazzling bright before my eyes.
They bore a net of golden twine
To hang upon the branches fine.   10
Pitying I wept to see the woe
That love and beauty undergo.
To be consumed in burning fires
And in ungratified desires,
And in tears clothed night and day,
Melted all my soul away.
When they saw my tears, a smile
That did heaven itself beguile
Bore the golden net aloft.
As on downy pinions soft,   20
Over the morning of my day.
Underneath the net I stray,
Now entreating Burning Fire,
Now entreating Iron Wire,
Now entreating Tears and Sighs:
'Oh, when will the morning rise?'

The Birds

HE: Where thou dwellest, in what grove,
         Tell me, fair one, tell me, love;
         Where thou thy charming nest dost build,
         O thou pride of every field.

SHE: Yonder stands a lonely tree,
         There I live and mourn for thee;
         Morning drinks my silent tear,
         And evening winds my sorrows bear.

HE: O thou summer's harmony,
         I have lived and mourned for thee;   10
         Each day I mourn along the wood,
         And night hath heard my sorrows loud.

SHE: Dost thou truly long for me,
         And am I thus sweet to thee?
         Sorrow now is at an end,
         O my lover, and my friend

HE: Come, on wings of joy we'll fly
         To where my bower hangs on high;
         Come and make thy calm retreat,
         Among green leaves and blossoms sweet.   20

The Grey Monk

'I die, I die,' the mother said,
'My children die for lack of bread.
What more has the merciless tyrant said?'
The monk sat down on the stony bed.

The blood red ran from the grey monk's side;
His hands and feet were wounded wide,
His body bent, his arms and knees
Like to the roots of ancient trees.

His eye was dry, no tear could flow;
A hollow groan first spoke his woe.   10
He trembled and shuddered upon the bed;
At length with a feeble cry he said:

'When God commanded this hand to write
In the studious hours of deep midnight,
He told me the writing I wrote should prove
The bane of all that on earth I loved.

'My brother starved between two walls;
His children's cry my soul appalls.
I mocked at the rack and griding chain;
My bent body mocks their torturing pain.   20

'Thy father drew his sword in the north;
With his thousands strong he marched forth.
Thy brother has armed himself in steel,
To avenge the wrongs thy children feel.

'But vain the sword, and vain the bow;
They never can work war's overthrow.
The hermit's prayer, and the widow's tear,
Alone can free the world from fear.

'The hand of vengeance sought the bed
To which the purple tyrant fled;   30
The iron hand crushed the tyrant's head,
And became a tyrant in his stead.

'Until the tyrant himself relent,
The tyrant who first the black bow bent,
Slaughter shall heap the bloody plain;
Resistance and war is the tyrant's gain.

'But the tear of love and forgiveness sweet,
And submission to death beneath his feet—
The tear shall melt the sword of steel,
And every wound it has made shall heal.   40

'A tear is an intellectual thing,
And a sigh is the sword of an angel king,
And the bitter groan of the martyr's woe
Is an arrow from the Almighty's bow.'

Morning

To find the western path,
Right through the gates of Wrath,
I urge my way.
Sweet Mercy leads me on
With soft repentant moan.
I see the break of day.

The war of swords and spears,
Melted by dewy tears,
Exhales on high.
The sun is freed from fears,   10
And with soft grateful tears
Ascends the sky.

'Terror in the house'

Terror in the house does roar,
But Pity stands before the door.

'Mock on'

Mock on, mock on, Voltaire, Rousseau;
Mock on, mock on: 'tis all in vain!
You throw the sand against the wind,
And the wind blows it back again.

And every sand becomes a gem,°
Reflected in the beams divine.
Blown back they blind the mocking eye,
But still in Israel's paths they shine.

The atoms of Democritus,°
And Newton's particles of light,°
Are sands upon the Red Sea shore,
Where Israel's tents do shine so bright.

'My Spectre around me'

My Spectre around me night and day
Like a wild beast guards my way;
My Emanation far within
Weeps incessantly for my sin.

A fathomless and boundless deep—
There we wander, there we weep.
On the hungry craving wind
My Spectre follows thee behind.

He scents thy footsteps in the snow,
Wheresoever thou dost go,
Through the wintry hail and rain.
'When wilt thou return again?

'Dost thou not in pride and scorn
Fill with tempests all my morn,
And, with jealousies and fears,
Fill my pleasant nights with tears?

'Seven of my sweet loves thy knife
Has bereaved of their life.
Their marble tombs I built with tears,
And with cold and shuddering fears.

'Seven more loves weep night and day
Round the tombs where my loves lay;
And seven more loves attend each night
Around my couch with torches bright;

'And seven more loves in my bed
Crown with wine my mournful head,
Pitying and forgiving all
Thy transgressions great and small.

'When wilt thou return and view
My loves, and them to life renew?   30
When wilt thou return and live?
When wilt thou pity as I forgive?'

'Never, never I return.
Still for victory I burn.
Living, thee alone I'll have,
And when dead I'll be thy grave.

'Through the Heaven and Earth and Hell
Thou shalt never, never quell,
I will fly and thou pursue;
Night and morn the flight renew.'   40

'Till I turn from female love,
And root up the infernal grove,°
I shall never worthy be
To step into eternity;

'And to end thy cruel mocks
Annihilate thee on the rocks,
And another form create
To be subservient to my fate.

'Let us agree to give up love,
And root up the infernal grove;   50
Then shall we return and see
The worlds of happy eternity,

'And throughout all eternity
I forgive you, you forgive me.
As our dear redeemer said:
"This the wine, and this the bread." '

'O'er my sins'

O'er my sins thou sit and moan;
Hast thou no sins of thy own?
O'er my sins thou sit and weep,
And lull thy own sins fast asleep.

What transgressions I commit,
Are for thy transgressions fit;
They thy harlots, thou their slave,
And my bed becomes their grave.

Poor, pale, pitiable form
That I follow in a storm,   10
Iron tears and groans of lead
Bind around my aching head.

'And let us go to the highest downs
With many pleasing wiles.
The woman that does not love your frowns
Will never embrace your smiles.'

The Smile

There is a smile of love,
And there is a smile of deceit;
And there is a smile of smiles,
In which these two smiles meet.

(And there is a frown of hate,
And there is a frown of disdain;
And there is a frown of frowns
Which you strive to forget in vain,

For it sticks in the heart's deep core,
And it sticks in the deep backbone.)   10
And no smile that ever was smiled,
But only one smile alone—

That betwixt the cradle and grave
It only once smiled can be.
But when it once is smiled
There's an end to all misery.

The Mental Traveller

I travelled through a land of men,
A land of men and women too,
And heard and saw such dreadful things
As cold earth-wanderers never knew.

For there the babe is born in joy
That was begotten in dire woe,
Just as we reap in joy the fruit
Which we in bitter tears did sow.°

And if the babe is born a boy
He's given to a woman old,   10
Who nails him down upon a rock,
Catches his shrieks in cups of gold.

She binds iron thorns around his head,
She pierces both his hands and feet,
She cuts his heart out at his side
To make it feel both cold and heat.

Her fingers number every nerve,
Just as a miser counts his gold.
She lives upon his shrieks and cries,
And she grows young as he grows old,   20

Till he becomes a bleeding youth,
And she becomes a virgin bright.
Then he rends up his manacles,
And binds her down for his delight.

He plants himself in all her nerves,
Just as a husbandman his mould,
And she becomes his dwelling-place,
And garden fruitful seventyfold.

An aged shadow soon he fades,
Wand'ring round an earthly cot,   30
Full filled all with gems and gold
Which he by industry had got.

And these are the gems of the human soul,
The rubies and pearls of a lovesick eye,
The countless gold of the aching heart,
The martyr's groan, and the lover's sigh;

They are his meat, they are his drink.
He feeds the beggar and the poor,
And the wayfaring traveller;
For ever open is his door.   40

His grief is their eternal joy;
They make the roofs and walls to ring;
Till from the fire on the hearth
A little female babe does spring.

And she is all of solid fire,
And gems and gold, that none his hand
Dares stretch to touch her baby form,
Or wrap her in his swaddling-band;

But she comes to the man she loves,
If young or old, or rich or poor.   50
They soon drive out the aged host,
A beggar at another's door.

He wanders weeping far away,
Until some other take him in;
Oft blind and age-bent, sore distressed,
Until he can a maiden win.

And to allay his freezing age
The poor man takes her in his arms;
The cottage fades before his sight,
The garden and its lovely charms;   60

The guests are scattered through the land.
For the eye altering, alters all;
The senses roll themselves in fear,
And the flat earth becomes a ball;

The stars, sun, moon, all shrink away,
A desert vast without a bound—
And nothing left to eat or drink,
And a dark desert all around.

The honey of her infant lips,
The bread and wine of her sweet smile,   70
The wild game of her roving eye,
Does him to infancy beguile.

For as he eats and drinks he grows
Younger and younger every day;
And on the desert wild they both
Wander in terror and dismay.

Like the wild stag she flees away;
Her fear plants many a thicket wild.
While he pursues her night and day,
By various arts of love beguiled,   80

By various arts of love and hate;
Till the wide desert planted o'er
With labyrinths of wayward love,
Where roams the lion, wolf and boar;

Till he becomes a wayward babe,
And she a weeping woman old.
Then many a lover wanders here;
The sun and stars are nearer rolled;

The trees bring forth sweet ecstasy
To all who in the desert roam—   90
Till many a city there is built,
And many a pleasant shepherd's home.

But when they find the frowning babe
Terror strikes through the region wide.
They cry: 'The babe, the babe is born,'
And flee away on every side.

For who dare touch the frowning form
His arm is withered to its root;
Lions, boars, wolves, all howling flee,
And every tree does shed its fruit.   100

And none can touch that frowning form,
Except it be a woman old;
She nails him down upon the rock,
And all is done as I have told.

The Land of Dreams

'Awake, awake my little boy,
Thou wast thy mother's only joy.
Why dost thou weep in thy gentle sleep?
Awake, thy father does thee keep.'

'Oh, what land is the Land of Dreams?
What are its mountains and what are its streams?
O father, I saw my mother there,
Among the lilies by waters fair.

'Among the lambs clothed in white,
She walked with her Thomas in sweet delight.   10
I wept for joy; like a dove I mourn.
Oh, when shall I again return?'

'Dear child, I also by pleasant streams
Have wandered all night in the land of dreams;
But though calm and warm the waters wide,
I could not get to the other side.'

'Father, O father, what do we here,
In this land of unbelief and fear?
The Land of Dreams is better far—
Above the light of the morning star.'   20

Mary

Sweet Mary, the first time she ever was there,
Came into the ballroom among the fair.
The young men and maidens around her throng,
And these are the words upon every tongue:

'An angel is here from the heavenly climes,
Or again does return the golden times.
Her eyes outshine every brilliant ray;
She opens her lips—'tis the month of May.'

Mary moves in soft beauty and conscious delight,
To augment with sweet smiles all the joys of the night,   10
Nor once blushes to own to the rest of the fair
That sweet love and beauty are worthy our care.

In the morning the villagers rose with delight,
And repeated with pleasure the joys of the night;
And Mary arose among friends to be free—
But no friend from henceforward thou, Mary, shalt see.

Some said she was proud, some called her a whore,
And some when she passed by shut to the door.
A damp cold came o'er her, her blushes all fled,
Her lilies and roses are blighted and shed.   20

'Oh, why was I born with a different face?
Why was I not born like this envious race?
Why did Heaven adorn me with bountiful hand,
And then set me down in an envious land?

'To be weak as a lamb and smooth as a dove,
And not to raise envy, is called Christian love.
But if you raise envy your merit's to blame
For planting such spite in the weak and the tame.

'I will humble my beauty; I will not dress fine;
I will keep from the ball and my eyes shall not shine.   30
And if any girl's lover forsakes her for me,
I'll refuse him my hand, and from envy be free.'

She went out in morning attired plain and neat;
'Proud Mary's gone mad,' said the child in the street.
She went out in morning in plain neat attire,
And came home in evening bespattered with mire.

She trembled and wept, sitting on the bed-side;
She forgot it was night and she trembled and cried.
She forgot it was night, she forgot it was morn,
Her soft memory imprinted with faces of scorn—   40

With faces of scorn and with eyes of disdain,
Like foul fiends inhabiting Mary's mild brain.
She remembers no face like the human divine;°
All faces have envy, sweet Mary, but thine.

And thine is a face of sweet love in despair,
And thine is a face of mild sorrow and care,
And thine is a face of wild terror and fear,
That shall never be quiet till laid on its bier.

The Crystal Cabinet

The maiden caught me in the wild,
Where I was dancing merrily.
She put me into her cabinet,
And locked me up with a golden key.

This cabinet is formed of gold
And pearl, and crystal shining bright,
And within it opens into a world,
And a little, lovely, moony night.

Another England there I saw,
Another London with its Tower,   10
Another Thames and other hills,
And another pleasant Surrey bower,

Another maiden like herself,
Translucent, lovely, shining clear—
Threefold, each in the other closed.
Oh, what a pleasant trembling fear!

Oh, what a smile, a threefold smile,
Filled me, that like a flame I burned!
I bent to kiss the lovely maid,
And found a threefold kiss returned.   20

I strove to seize the inmost form
With ardour fierce and hands of flame,
But burst the crystal cabinet,
And like a weeping babe became:

A weeping babe upon the wild,
And weeping woman, pale, reclined.
And in the outward air again
I filled with woes the passing wind.

Long John Brown and Little Mary Bell

Little Mary Bell had a fairy in a nut;
Long John Brown had the Devil in his gut.
Long John Brown loved little Mary Bell,
And the fairy drew the Devil into the nut-shell.

Her fairy skipped out and her fairy skipped in;
He laughed at the Devil, saying, 'Love is a sin'.
The Devil he raged and the Devil he was wroth,
And the Devil entered into the young man's broth.

He was soon in the gut of the loving young swain,
For John ate and drank to drive away love's pain.   10
But, all he could do, he grew thinner and thinner,
Though he ate and drank as much as ten men for his dinner.

Some said he had a wolf in his stomach day and night;
Some said he had the Devil, and they guessed right.
The fairy skipped about in his glory, joy and pride,
And he laughed at the Devil till poor John Brown died.

Then the fairy skipped out of the old nutshell,
And woe and alack for pretty Mary Bell—
For the Devil crept in when the fairy skipped out,
And there goes Miss Bell with her fusty old nut.   20

William Bond

I wonder whether the girls are mad,
And I wonder whether they mean to kill,
And I wonder if William Bond will die,
For assuredly he is very ill.

He went to church in a May morning
Attended by fairies, one, two and three;
But the angels of providence drove them away,
And he returned home in misery.

He went not out to the field nor fold;
He went not out to the village nor town;   10
But he came home in a black, black cloud,
And took to his bed and there lay down.

And an angel of providence at his feet,
And an angel of providence at his head,
And in the midst a black, black cloud,
And in the midst the sick man on his bed.

And on his right hand was Mary Green,
And on his left hand was his sister Jane,
And their tears fell through the black, black cloud
To drive away the sick man's pain.   20

'O William, if thou dost another love—
Dost another love better than poor Mary—
Go and take that other to be thy wife,
And Mary Green shall her servant be.'

'Yes, Mary, I do another love;
Another I love far better than thee,
And another I will have for my wife.
Then what have I to do with thee?

'For thou art melancholy pale,
And on thy head is the cold moon's shine;   30
But she is ruddy and bright as day,
And the sunbeams dazzle from her eyne.'

Mary trembled and Mary chilled,
And Mary fell down on the right-hand floor—
That William Bond and his sister Jane
Scarce could recover Mary more.

When Mary woke and found her laid
On the right hand of her William dear,
On the right hand of his loved bed,
And saw her William Bond so near,   40

The fairies that fled from William Bond
Danced around her shining head;
They danced over the pillow white,
And the angels of providence left the bed.

'I thought Love lived in the hot sunshine,
But oh, he lives in the moony light;
I thought to find Love in the heat of day,
But sweet Love is the comforter of night.

Seek Love in the pity of others' woe,
In the gentle relief of another's care,   50
In the darkness of night and the winter's snow,
In the naked and outcast, seek Love there.'

LYRICS FROM THE EPIC POEMS

From the Preface to Milton

And did those feet in ancient time
Walk upon England's mountains green?
And was the holy Lamb of God°
On England's pleasant pastures seen?

And did the countenance divine
Shine forth upon our clouded hills?
And was Jerusalem builded here,
Among these dark Satanic mills?

Bring me my bow of burning gold;
Bring me my arrows of desire;   10
Bring me my spear; O clouds, unfold!
Bring me my chariot of fire!°

I will not cease from mental fight,
Nor shall my sword sleep in my hand,
Till we have built Jerusalem,
In England's green and pleasant land.

From Jerusalem (*plate 27*)

The fields from Islington to Marybone,
To Primrose Hill and Saint John's Wood,°
    Were builded over with pillars of gold,
And there Jerusalem's pillars stood.

    Her little ones ran on the fields,
The Lamb of God among them seen,
    And fair Jerusalem his bride,
Among the little meadows green.

    Pancras and Kentish Town repose
Among her golden pillars high,   10
    Among her golden arches which
Shine upon the starry sky.

The Jews' Harp House and the Green Man,°
The ponds where boys to bathe delight,
    The fields of cows by Willan's farm,°
Shine in Jerusalem's pleasant sight.

    She walks upon our meadows green,
The Lamb of God walks by her side,
    And every English child is seen
Children of Jesus and his bride,   20

    Forgiving trespasses and sins,
Lest Babylon with cruel Og,
    With moral and self-righteous law
Should crucify in Satan's synagogue!

    What are those golden builders doing
Near mournful ever-weeping Paddington?—
    Standing above that mighty ruin
Where Satan the first victory won,

    Where Albion slept beneath the fatal tree,
And the Druid's golden knife   30
    Rioted in human gore,
In offerings of human life.

    They groaned aloud on London Stone,
They groaned aloud on Tyburn's brook;°
    Albion gave his deadly groan,
And all the Atlantic mountains shook.

    Albion's Spectre from his loins
Tore forth in all the pomp of war.
    Satan his name; in flames of fire
He stretched his Druid pillars far.   40

    Jerusalem fell from Lambeth's vale,
Down through Poplar and Old Bow,
    Through Maldon and across the sea,
In war and howling, death and woe.

The Rhine was red with human blood,
The Danube rolled a purple tide;
   On the Euphrates Satan stood,
And over Asia stretched his pride.

   He withered up sweet Zion's hill
From every nation of the earth.
   He withered up Jerusalem's gates,
And in a dark land gave her birth.

   He withered up the human form
By laws of sacrifice for sin,
   Till it became a mortal worm,
But, oh, translucent all within!

   The Divine Vision still was seen,
Still was the human form divine,
   Weeping in weak and mortal clay;
O Jesus, still the form was thine.

   And thine the human face, and thine
The human hands and feet and breath,
   Entering through the gates of birth
And passing through the gates of death.

   And, O thou Lamb of God, whom I
Slew in my dark self-righteous pride!
   Art thou returned to Albion's land,
And is Jerusalem thy bride?

   Come to my arms and never more
Depart, but dwell for ever here.
   Create my spirit to thy love;
Subdue my Spectre to thy fear.

   Spectre of Albion! Warlike fiend!
In clouds of blood and ruin rolled.
   I here reclaim thee as my own,
My selfhood! Satan! armed in gold.

Is this thy soft family love,
Thy cruel patriarchal pride:
  Planting thy family alone,
Destroying all the world beside?   80

A man's worst enemies are those
Of his own house and family;
  And he who makes his law a curse,
By his own law shall surely die.

In my exchanges every land
Shall walk, and mine in every land;
  Mutual shall build Jerusalem,
Both heart in heart, and hand in hand.

From Jerusalem (*plate 41*)

Each man is in his Spectre's power
Until the arrival of that hour,
When his humanity awake,
And cast his Spectre into the lake.

From Jerusalem (*plate 52*)

I saw a monk of Charlemagne°
Arise before my sight.
  I talked with the grey monk as we stood
In beams of infernal light.

Gibbon arose with a lash of steel,
And Voltaire with a racking wheel;
  The schools, in clouds of learning rolled,°
Arose with war in iron and gold.

'Thou lazy monk,' they sound afar,
'In vain condemning glorious War,   10
  And in your cell you shall ever dwell.
Rise War, and bind him in his cell.'

The blood red ran from the grey monk's side;
His hands and feet were wounded wide,
  His body bent, his arms and knees
Like to the roots of ancient trees.

When Satan first the black bow bent,
And the moral law from the Gospel rent,
  He forged the law into a sword,
And spilled the blood of mercy's lord.   20

Titus! Constantine! Charlemagne!°
O Voltaire! Rousseau! Gibbon! Vain
  Your Grecian mocks and Roman sword
Against this image of his lord.

For a tear is an intellectual thing!
And a sigh is the sword of an angel king,
  And the bitter groan of a martyr's woe
Is an arrow from the Almighty's bow!

From Jerusalem (*plate 77*)

England! awake! awake! awake!
  Jerusalem thy sister calls!
Why wilt thou sleep the sleep of death,
  And close her from thy ancient walls?

Thy hills and valleys felt her feet
  Gently upon their bosoms move;
Thy gates beheld sweet Zion's ways;
  Then was a time of joy and love.

And now the time returns again.
  Our souls exult, and London's towers   10
Receive the Lamb of God, to dwell
  In England's green and pleasant bowers.

LATE LYRICS

'Grown old in love'

Grown old in love from seven till seven times seven,
I oft have wished for Hell, for ease from Heaven.

'Madman I have been called'

Madman I have been called; fool they call thee.
I wonder which they envy, thee or me?

'He's a blockhead'

He's a blockhead who wants a proof of what he can't perceive,
And he's a fool who tries to make such a blockhead believe.

'I am no Homer's hero'

I am no Homer's hero you all know;
I profess not generosity to a foe.
My generosity is to my friends,
That for their friendship I may make amends.
The generous to enemies promotes their ends,
And becomes the enemy and betrayer of his friends.

'The angel that presided o'er my birth'

The angel that presided o'er my birth
Said: 'Little creature formed of joy and mirth,
Go love without the help of any king on earth.'

'Some men created for destruction'

Some men created for destruction come
Into the world, and make the world their home.
Be they as vile and base as e'er they can,
They'll still be called 'The World's' honest man.

Imitation of Pope: A Compliment to the Ladies

Wondrous the gods; more wondrous are the men;
More wondrous, wondrous still the cock and hen;
More wondrous still the table, stool, and chair—
But ah! more wondrous still the Charming Fair!

'If I e'er grow'

If I e'er grow to man's estate
Oh, give to me a woman's fate:
May I govern all, both great and small,
Have the last word, and take the wall.°

'You don't believe'

You don't believe—I won't attempt to make ye.
You are asleep—I won't attempt to wake ye.
Sleep on, sleep on, while in your pleasant dreams
Of reason you may drink of life's clear streams.
Reason and Newton, they are quite two things,
For so the swallow and the sparrow sings.
Reason says 'Miracle', Newton says 'Doubt'.°
Aye, that's the way to make all Nature out:
Doubt, doubt, and don't believe without experiment.
That is the very thing that Jesus meant   10
When he said: 'Only believe.' Believe and try,°
Try, try, and never mind the reason why.

'Great things are done'

Great things are done when men and mountains meet;
This is not done by jostling in the street.

'If you play a game of chance'

If you play a game of chance, know before you begin;
If you are benevolent you will never win.

'I rose up'

I rose up at the dawn of day:
'Get thee away, get thee away!
Pray'st thou for riches? Away, away!
This is the throne of Mammon grey.'

Said I: 'This sure is very odd;
I took it to be the throne of God.
For everything besides I have;
It is only for riches that I can crave.

'I have mental joy and mental health,
And mental friends and mental wealth;   10
I've a wife I love and that loves me;
I've all but riches bodily.

'I am in God's presence night and day,
And he never turns his face away.
The accuser of sins by my side does stand,°
And he holds my money-bag in his hand.

'For my worldly things God makes him pay,
And he'd pay for more if to him I would pray.
And so you may do the worst you can do;
Be assured, Mr Devil, I won't pray to you.   20

'Then if for riches I must not pray,
God knows I little of prayers need say.
So as a church is known by its steeple,
If I pray it must be for other people.

'He says if I do not worship him for a God
I shall eat coarser food and go worse shod;
So as I don't value such things as these,
You must do, Mr Devil, just as God please.'

'Why was Cupid a boy'

Why was Cupid a boy,
And why a boy was he?
He should have been a girl,
For aught that I can see.

For he shoots with his bow
And the girl shoots with her eye,
And they both are merry and glad
And laugh when we do cry.

And to make Cupid a boy
Was the Cupid-girl's mocking plan;   10
For a boy can't interpret the thing
Till he is become a man.

And then he's so pierced with care,
And wounded with arrowy smarts,
That the whole business of his life
Is to pick out the heads of the darts.

'Twas the Greeks' love of war
Turned Love into a boy,°
And woman into a statue of stone;
And away fled every joy.   20

'Great men and fools'

Great men and fools do often me inspire;
But the greater fool the greater liar.

To God

If you have formed a circle to go into,
Go into it yourself, and see how you would do.

'Since all the riches'

Since all the riches of this world
May be gifts from the Devil and earthly kings,
I should suspect that I worshipped the Devil
If I thanked my God for worldly things.

'To Chloe's breast'

To Chloe's breast young Cupid slyly stole,
But he crept in at Myra's pocket-hole.°

'Anger and wrath'

Anger and wrath my bosom rends;
I thought them the errors of friends.
But all my limbs with warmth glow;
I find them the errors of the foe.

THE LOS POEMS

The Book of Urizen (1794)

PRELUDIUM

Of the primeval priest's assumed power,°
When Eternals spurned back his religion,
And gave him a place in the north,°
Obscure, shadowy, void, solitary.

Eternals, I hear your call gladly.
Dictate swift-winged words, and fear not°
To unfold your dark visions of torment.

CHAPTER I

1. Lo, a shadow of horror is risen
In Eternity! Unknown, unprolific,
Self-closed, all-repelling: what demon
Hath formed this abominable void,
This soul-shudd'ring vacuum? Some said,
'It is Urizen'. But unknown, abstracted,
Brooding secret, the dark power hid.°

2. Times on times he divided, and measured
Space by space in his ninefold darkness,°
Unseen, unknown. Changes appeared
In his desolate mountains, rifted furious
By the black winds of perturbation.

3. For he strove in battles dire,
In unseen conflictions with shapes
Bred from his forsaken wilderness
Of beast, bird, fish, serpent, and element,
Combustion, blast, vapour and cloud:

4. Dark revolving in silent activity,
Unseen in tormenting passions,
An activity unknown and horrible,
A self-contemplating shadow,
In enormous labours occupied.

5. But Eternals beheld his vast forests.
Age on ages he lay, closed, unknown,
Brooding, shut in the deep; all avoid
The petrific abominable chaos.°

6. His cold horrors silent, dark Urizen
Prepared; his ten thousands of thunders°
Ranged in gloomed array stretch out across
The dread world, and the rolling of wheels,
As of swelling seas, sound in his clouds,
In his hills of stored snows, in his mountains
Of hail and ice. Voices of terror   40
Are heard, like thunders of autumn,
When the cloud blazes over the harvests.

CHAPTER II

1. Earth was not, nor globes of attraction.°
The will of the Immortal expanded
Or contracted his all-flexible senses.
Death was not, but eternal life sprung.

2. The sound of a trumpet the heavens
Awoke, and vast clouds of blood rolled
Round the dim rocks of Urizen, so named,
That solitary one in immensity.   50

3. Shrill the trumpet; and myriads of Eternity
Muster around the bleak deserts,
Now filled with clouds, darkness and waters
That rolled perplexed, lab'ring, and uttered
Words articulate, bursting in thunders
That rolled on the tops of his mountains.

4. 'From the depths of dark solitude, from
The eternal abode in my holiness,
Hidden, set apart in my stern counsels
Reserved for the days of futurity,   60
I have sought for a joy without pain,
For a solid without fluctuation.
Why will you die, O Eternals,
Why live in unquenchable burnings?

5. 'First, I fought with the fire, consumed
Inwards, into a deep world within,
A void immense, wild, dark and deep,°
Where nothing was, Nature's wide womb.
And self-balanced, stretched o'er the void,
I alone, even I, the winds merciless   70
Bound. But condensing, in torrents
They fall and fall; strong I repelled
The vast waves and arose on the waters,
A wide world of solid obstruction.

6. 'Here alone I, in books formed of metals,
Have written the secrets of wisdom,
The secrets of dark contemplation,
By fightings and conflicts dire
With terrible monsters sin-bred,
Which the bosoms of all inhabit,   80
Seven deadly sins of the soul.

7. 'Lo! I unfold my darkness, and on
This rock place with strong hand the book
Of eternal brass, written in my solitude:

8. 'Laws of peace, of love, of unity,
Of pity, compassion, forgiveness.
Let each choose one habitation,
His ancient infinite mansion,
One command, one joy, one desire,
One curse, one weight, one measure,   90
One King, one God, one Law.'

CHAPTER III

1. The voice ended; they saw his pale visage
Emerge from the darkness, his hand
On the rock of Eternity unclasping
The book of brass. Rage seized the strong,

2. Rage, fury, intense indignation
In cataracts of fire, blood and gall,
In whirlwinds of sulphurous smoke,
And enormous forms of energy.
All the seven deadly sins of the soul   100
In living creations appeared
In the flames of eternal fury.

3. Sund'ring, dark'ning, thund'ring!
Rent away with a terrible crash,
Eternity rolled wide apart,
Wide asunder rolling
Mountainous, all around
Departing, departing, departing,
Leaving ruinous fragments of life
Hanging, frowning cliffs, and all between   110
An ocean of voidness unfathomable.

4. The roaring fires ran o'er the heav'ns
In whirlwinds and cataracts of blood;
And o'er the dark deserts of Urizen
Fires pour through the void on all sides,
On Urizen's self-begotten armies.

5. But no light from the fires; all was darkness°
In the flames of eternal fury.

6. In fierce anguish and quenchless flames
To the deserts and rocks he ran raging   120
To hide, but he could not; combining,
He dug mountains and hills in vast strength.
He piled them in incessant labour,
In howlings and pangs and fierce madness,
Long periods in burning fires labouring,
Till hoary, and age-broke, and aged,
In despair and the shadows of death.

7. And a roof, vast, petrific, around
On all sides he framed, like a womb
Where thousands of rivers in veins   130
Of blood pour down the mountains to cool

The eternal fires beating without
From Eternals; and like a black globe
Viewed by sons of Eternity, standing
On the shore of the infinite ocean,
Like a human heart struggling and beating,
The vast world of Urizen appeared.

8. And Los round the dark glow of Urizen
Kept watch for Eternals, to confine
The obscure separation alone;   140
For Eternity stood wide apart,
As the stars are apart from the earth.

9. Los wept, howling around the dark demon,
And cursing his lot; for in anguish
Urizen was rent from his side,
And a fathomless void for his feet,
And intense fires for his dwelling.

10. But Urizen laid in a stony sleep
Unorganized, rent from Eternity.

11. The Eternals said: 'What is this? Death.   150
Urizen is a clod of clay.'

12. Los howled in a dismal stupor,
Groaning! gnashing! groaning!
Till the wrenching apart was healed.

13. But the wrenching of Urizen healed not;
Cold, featureless, flesh or clay,
Rifted with direful changes,
He lay in a dreamless night,

14. Till Los roused his fires, affrighted
At the formless unmeasurable death.   160

CHAPTER IVa

1. Los, smitten with astonishment,
Frightened at the hurtling bones°

2. And at the surging, sulphureous
Perturbed Immortal, mad-raging

3. In whirlwinds and pitch and nitre
Round the furious limbs of Los.

4. And Los formed nets and gins,
And threw the nets round about.

5. He watched in shudd'ring fear
The dark changes, and bound every change   170
With rivets of iron and brass.

6. And these were the changes of Urizen.

CHAPTER IVb

1. Ages on ages rolled over him!
In stony sleep ages rolled over him!
Like a dark waste stretching, changeable,
By earthquakes riven, belching sullen fires,
On ages rolled ages in ghastly
Sick torment. Around him in whirlwinds
Of darkness the Eternal Prophet howled,
Beating still on his rivets of iron,   180
Pouring solder of iron, dividing
The horrible night into watches.

2. And Urizen (so his eternal name)
His prolific delight obscured more and more
In dark secrecy, hiding in surging
Sulphureous fluid his fantasies.
The Eternal Prophet heaved the dark bellows,
And turned restless the tongs, and the hammer
Incessant beat, forging chains new and new,
Numb'ring with links hours, days and years.   190

3. The eternal mind, bounded, began to roll
Eddies of wrath ceaseless round and round,
And the sulphureous foam surging thick
Settled, a lake bright and shining clear,
White as the snow on the mountains cold.

4. Forgetfulness, dumbness, necessity!
In chains of the mind locked up,
Like fetters of ice shrinking together,
Disorganized, rent from Eternity.
Los beat on his fetters of iron,   200
And heated his furnaces, and poured
Iron solder and solder of brass.

5. Restless turned the Immortal, enchained,
Heaving dolorous, anguished! unbearable
Till a roof, shaggy wild, enclosed
In an orb his fountain of thought.

6. In a horrible dreamful slumber,
Like the linked infernal chain,
A vast spine writhed in torment
Upon the winds, shooting pained   210
Ribs, like a bending cavern,
And bones of solidness froze
Over all his nerves of joy.
And a first age passed over,
And a state of dismal woe.

7. From the caverns of his jointed spine,
Down sunk with fright a red
Round globe hot-burning, deep,
Deep down into the abyss,
Panting, conglobing, trembling,   220
Shooting out ten thousand branches
Around his solid bones.
And a second age passed over,
And a state of dismal woe.

8. In harrowing fear rolling round,
His nervous brain shot branches
Round the branches of his heart
On high into two little orbs;
And fixed in two little caves,
Hiding carefully from the wind,   230
His eyes beheld the deep.
And a third age passed over,
And a state of dismal woe.

9. The pangs of hope began.
In heavy pain striving, struggling,
Two ears in close volutions
From beneath his orbs of vision
Shot spiring out and petrified
As they grew. And a fourth age passed,
And a state of dismal woe.   240

10. In ghastly torment sick,
Hanging upon the wind,
Two nostrils bent down to the deep.
And a fifth age passed over,
And a state of dismal woe.

11. In ghastly torment sick,
Within his ribs bloated round,
A craving hungry cavern:
Thence arose his channelled throat,
And like a red flame a tongue   250
Of thirst and of hunger appeared.
And a sixth age passed over,
And a state of dismal woe.

12. Enraged and stifled with torment,
He threw his right arm to the north,
His left arm to the south
Shooting out in anguish deep;
And his feet stamped the nether abyss
In trembling and howling and dismay.
And a seventh age passed over,°   260
And a state of dismal woe.

CHAPTER V

1. In terrors Los shrunk from his task;
His great hammer fell from his hand;
His fires beheld and, sickening,
Hid their strong limbs in smoke.
For with noises ruinous loud,
With hurtlings and clashings and groans,
The Immortal endured his chains
Though bound in a deadly sleep.

2. All the myriads of Eternity,   270
All the wisdom and joy of life,
Roll like a sea around him,
Except what his little orbs
Of sight by degrees unfold.

3. And now his eternal life
Like a dream was obliterated.

4. Shudd'ring, the Eternal Prophet smote
With a stroke from his north to south region.
The bellows and hammer are silent now,
A nerveless silence; his prophetic voice   280
Seized; a cold solitude and dark void
The Eternal Prophet and Urizen closed.

5. Ages on ages rolled over them,
Cut off from life and light, frozen
Into horrible forms of deformity.
Los suffered his fires to decay,
Then he looked back with an anxious desire,
But the space undivided by existence
Struck horror into his soul.

6. Los wept, obscured with mourning;   290
His bosom earthquaked with sighs;
He saw Urizen, deadly black,
In his chains bound, and pity began,

7. In anguish dividing and dividing
(For pity divides the soul),
In pangs, eternity on eternity,
Life in cataracts poured down his cliffs.
The void shrunk the lymph into nerves,
Wand'ring wide on the bosom of night,
And left a round globe of blood   300
Trembling upon the void.
Thus the Eternal Prophet was divided
Before the death-image of Urizen;
For in changeable clouds and darkness,
In a winterly night beneath,

The abyss of Los stretched immense,
And now seen, now obscured to the eyes
Of Eternals, the visions remote
Of the dark separation appeared.
As glasses discover worlds   310
In the endless abyss of space,
So the expanding eyes of Immortals
Beheld the dark visions of Los,
And the globe of life-blood trembling.

8. The globe of life-blood trembled,
Branching out into roots,
Fibrous, writing upon the winds:
Fibres of blood, milk and tears,
In pangs, eternity on eternity.
At length in tears and cries, embodied,   320
A female form trembling and pale
Waves before his deathy face.

9. All Eternity shuddered at sight
Of the first female now separate,
Pale as a cloud of snow,
Waving before the face of Los.

10. Wonder, awe, fear, astonishment,
Petrify the eternal myriads
At the first female form now separate;
They called her Pity, and fled: 330

11. 'Spread a tent, with strong curtains around them;°
Let cords and stakes bind in the void,
That Eternals may no more behold them.'

12. They began to weave curtains of darkness;
They erected large pillars round the void,
With golden hooks fastened in the pillars.
With infinite labour the Eternals
A woof wove, and called it Science.

CHAPTER VI

1. But Los saw the female and pitied.
He embraced her; she wept; she refused;   340
In perverse and cruel delight
She fled from his arms, yet he followed.

2. Eternity shuddered when they saw
Man begetting his likeness
On his own divided image.

3. A time passed over. The Eternals
Began to erect the tent—
When Enitharmon, sick,
Felt a worm within her womb.

4. Yet helpless it lay like a worm   350
In the trembling womb,
To be moulded into existence.

5. All day the worm lay on her bosom;
All night within her womb
The worm lay, till it grew to a serpent,
With dolorous hissings and poisons
Round Enitharmon's loins folding.

6. Coiled within Enitharmon's womb
The serpent grew, casting its scales.
With sharp pangs the hissings began   360
To change to a grating cry.
Many sorrows and dismal throes,
Many forms of fish, bird and beast,
Brought forth an infant form
Where was a worm before.

7. The Eternals their tent finished,
Alarmed with these gloomy visions—
When Enitharmon, groaning,
Produced a man-child to the light.

8. A shriek ran through Eternity   370
And a paralytic stroke,
At the birth of the human shadow.

9. Delving earth in his resistless way,
Howling, the child with fierce flames
Issued from Enitharmon.

10. The Eternals closed the tent;
They beat down the stakes; the cords
Stretched for a work of Eternity.
No more Los beheld Eternity.

11. In his hands he seized the infant;   380
He bathed him in springs of sorrow;
He gave him to Enitharmon.

CHAPTER VII

1. They named the child Orc; he grew,
Fed with milk of Enitharmon.

2. Los awoke her. Oh sorrow and pain!
A tight'ning girdle grew
Around his bosom. In sobbings
He burst the girdle in twain,
But still another girdle
Oppressed his bosom. In sobbings   390
Again he burst it. Again
Another girdle succeeds.
The girdle was formed by day;
By night was burst in twain.

3. These, falling down on the rock
Into an iron chain,
In each other link by link locked.

4. They took Orc to the top of a mountain;
Oh, how Enitharmon wept!
They chained his young limbs to the rock   400
With the chain of jealousy,
Beneath Urizen's deathful shadow.

5. The dead heard the voice of the child,
And began to awake from sleep.
All things heard the voice of the child,
And began to awake to life.

6. And Urizen, craving with hunger,
Stung with the odours of Nature,
Explored his dens around.

7. He formed a line and a plummet   410
To divide the abyss beneath;
He formed a dividing rule;

8. He formed scales to weigh;
He formed massy weights;
He formed a brazen quadrant;
He formed golden compasses,
And began to explore the abyss;
And he planted a garden of fruits.

9. But Los encircled Enitharmon
With fires of prophecy   420
From the sight of Urizen and Orc.

10. And she bore an enormous race.

CHAPTER VIII

1. Urizen explored his dens—
Mountain, moor and wilderness—
With a globe of fire lighting his journey,
A fearful journey, annoyed
By cruel enormities, forms
Of life on his forsaken mountains.

2. And his world teemed vast enormities:
Fright'ning, faithless, fawning   430
Portions of life, similitudes
Of a foot, or a hand, or a head,
Or a heart, or an eye; they swam, mischievous
Dread terrors! delighting in blood.

3. Most Urizen sickened to see
His eternal creations appear,
Sons and daughters of sorrow on mountains,
Weeping! wailing! First Thiriel appeared,°

Astonished at his own existence,
Like a man from a cloud born; and Utha,   440
From the waters emerging, laments!
Grodna rent the deep earth howling,
Amazed! His heavens immense cracks
Like the ground parched with heat. Then Fuzon
Flamed out! first begotten, last born.
All his eternal sons in like manner;
His daughters from green herbs and cattle,
From monsters, and worms of the pit.

4. He, in darkness closed, viewed all his race,°
And his soul sickened. He cursed   450
Both sons and daughters, for he saw
That no flesh nor spirit could keep
His iron laws one moment.

5. For he saw that life lived upon death;
The ox in the slaughterhouse moans,
The dog at the wintry door.
And he wept, and he called it pity,
And his tears flowed down on the winds.

6. Cold he wandered on high, over their cities,°
In weeping and pain and woe!   460
And wherever he wandered in sorrows
Upon the aged heavens,
A cold shadow followed behind him
Like a spider's web, moist, cold, and dim,
Drawing out from his sorrowing soul,
The dungeon-like heaven dividing,
Wherever the footsteps of Urizen
Walked over the cities in sorrow;

7. Till a web dark and cold throughout all
The tormented element stretched   470
From the sorrows of Urizen's soul.
And the web is a female in embryo;
None could break the web, no wings of fire,

8. So twisted the cords, and so knotted
The meshes, twisted like to the human brain.

9. And all called it The Net of Religion.

CHAPTER IX

1. Then the inhabitants of those cities°
Felt their nerves change into marrow,
And hardening bones began
In swift diseases and torments,   480
In throbbings and shootings and grindings
Through all the coasts; till, weakened,
The senses inward rushed, shrinking,
Beneath the dark net of infection;

2. Till the shrunken eyes, clouded over,
Discerned not the woven hypocrisy.
But the streaky slime in their heavens,
Brought together by narrowing perceptions,
Appeared transparent air; for their eyes
Grew small like the eyes of a man,   490
And in reptile forms shrinking together
Of seven feet stature they remained.

3. Six days they shrunk up from existence,
And on the seventh day they rested;°
And they blessed the seventh day, in sick hope,
And forgot their eternal life.

4. And their thirty cities divided
In form of a human heart.
No more could they rise at will
In the infinite void, but, bound down   500
To earth by their narrowing perceptions,
They lived a period of years,
Then left a noisome body
To the jaws of devouring darkness.

5. And their children wept, and built
Tombs in the desolate places,
And formed laws of prudence, and called them
The eternal laws of God.

6. And the thirty cities remained°
Surrounded by salt floods, now called
Africa (its name was then Egypt).

    510

7. The remaining sons of Urizen
Beheld their brethren shrink together
Beneath the net of Urizen.
Persuasion was in vain,
For the ears of the inhabitants
Were withered, and deafened, and cold,
And their eyes could not discern
Their brethren of other cities.

8. So Fuzon called all together,
The remaining children of Urizen,
And they left the pendulous earth;
They called it Egypt, and left it.

    520

9. And the salt ocean rolled englobed.

The Book of Los (1795)

1. Eno, aged mother,
Who the chariot of Leutha guides,
Since the day of thunders in old time,

2. Sitting beneath the eternal oak,
Trembled and shook the steadfast earth,
And thus her speech broke forth:

3. 'Oh times remote!
When love and joy were adoration,
And none impure were deemed,

Not eyeless Covet,   10
Nor thin-lipped Envy,
Nor bristled Wrath,
Nor curled Wantonness.°

4. 'But Covet was poured full,
Envy fed with fat of lambs,
Wrath with lions' gore,
Wantonness lulled to sleep
With the virgin's lute,
Or sated with her love.

5. 'Till Covet broke his locks and bars,   20
And slept with open doors;
Envy sung at the rich man's feast;
Wrath was followed up and down
By a little ewe lamb;
And Wantonness on his own true love
Begot a giant race'.

6. Raging furious, the flames of desire
Ran through heaven and earth, living flames,
Intelligent, organized, armed
With destruction and plagues. In the midst   30
The Eternal Prophet, bound in a chain,
Compelled to watch Urizen's shadow,

7. Raged with curses and sparkles of fury.
Round the flames roll as Los hurls his chains,
Mounting up from his fury, condensed,
Rolling round and round, mounting on high
Into vacuum, into non-entity,
Where nothing was! Dashed wide apart,
His feet stamp the eternal fierce-raging
Rivers of wide flame; they roll round   40
And round on all sides making their way
Into darkness and shadowy obscurity.

8. Wide apart stood the fires. Los remained
In the void between fire and fire.
In trembling and horror they beheld him;
They stood wide apart, driven by his hands
And his feet, which the nether abyss
Stamped in fury and hot indignation.

9. But no light from the fires. All was
Darkness round Los. Heat was not; for bound up   50
Into fiery spheres from his fury
The gigantic flames trembled and hid.

10. Coldness, darkness, obstruction, a solid
Without fluctuation, hard as adamant,
Black as marble of Egypt, impenetrable,
Bound in the fierce raging Immortal.
And the separated fires froze in
A vast solid without fluctuation,
Bound in his expanding clear senses.

CHAPTER II°

1. The Immortal stood frozen amidst   60
The vast Rock of Eternity times
And times, a night of vast durance:
Impatient, stifled, stiffened, hardened,

2. Till impatience no longer could bear
The hard bondage. Rent, rent, the vast solid
With a crash from immense to immense

3. Cracked across into numberless fragments.
The prophetic wrath, struggling for vent,
Hurls apart, stamping furious to dust
And crumbling with bursting sobs, heaves   70
The black marble on high into fragments

4. Hurled apart on all sides, as a falling
Rock. The Innumerable fragments away
Fell asunder, and horrible vacuum
Beneath him, and on all sides round,

5. Falling, falling! Los fell and fell,
Sunk precipitant, heavy, down, down,
Times on times, night on night, day on day.
Truth has bounds, error none; falling, falling,
Years on years, and ages on ages,   80
Still he fell through the void, still a void
Found for falling day and night without end.
For though day or night was not, their spaces
Were measured by his incessant whirls
In the horrid vacuity bottomless.

6. The Immortal, revolving, indignant,
First in wrath threw his limbs, like the babe
New-born into our world. Wrath subsided,
And contemplative thoughts first arose.
Then aloft his head reared in the abyss,   90
And his downward-borne fall changed oblique

7. Many ages of groans, till there grew
Branchy forms, organizing the human
Into finite inflexible organs,

8: Till in process from falling he bore
Sidelong on the purple air, wafting
The weak breeze in efforts o'erwearied.

9. Incessant the falling mind laboured,
Organizing itself, till the vacuum
Became element, pliant to rise,°   100
Or to fall, or to swim, or to fly—
With ease searching the dire vacuity.

CHAPTER III

1. The lungs heave incessant, dull and heavy;
For as yet were all other parts formless,
Shiv'ring, clinging around like a cloud,
Dim and glutinous as the white polypus°
Driven by waves and englobed on the tide.

2. And the unformed part craved repose.
Sleep began; the lungs heave on the wave,
Weary, overweighed, sinking beneath   110
In a stifling black fluid. He woke;

3. He arose on the waters but, soon
Heavy falling, his organs like roots
Shooting out from the seed shot beneath,
And a vast world of waters around him
In furious torrents began.

4. Then he sunk, and around his spent lungs
Began intricate pipes that drew in
The spawn of the waters, outbranching
An immense fibrous form, stretching out,   120
Through the bottoms of immensity raging.

5. He rose on the floods; then he smote
The wild deep with his terrible wrath,
Separating the heavy and thin.

6. Down the heavy sunk, cleaving around
To the fragments of solid; up rose
The thin, flowing round the fierce fires
That glowed furious in the expanse.

CHAPTER IV

1. Then light first began; from the fires
Beams, conducted by fluid so pure,   130
Flowed around the immense. Los beheld
Forthwith, writhing upon the dark void,
The backbone of Urizen appear,
Hurtling upon the wind
Like a serpent! like an iron chain
Whirling about in the deep.

2. Upfolding his fibres together
To a form of impregnable strength,
Los, astonished and terrified, built
Furnaces; he formed an anvil,   140
A hammer of adamant. Then began
The binding of Urizen day and night.

3. Circling round the dark demon, with howlings,
Dismay and sharp blightings, the prophet
Of Eternity beat on his iron links.

4. And, first, from those infinite fires
The light that flowed down on the winds
He seized: beating incessant, condensing
The subtle particles in an orb.°

5. Roaring, indignant the bright sparks   150
Endured the vast hammer; but unwearied
Los beat on the anvil, till glorious
An immense orb of fire he framed.

6. Oft he quenched it beneath in the deeps,°
Then surveyed the all-bright mass. Again
Seizing fires from the terrific orbs
He heated the round globe, then beat,
While, roaring, his furnaces endured
The chained orb in their infinite wombs.

7. Nine ages completed their circles   160
When Los heated the glowing mass, casting
It down into the deeps. The deeps fled
Away in redounding smoke; the sun°
Stood self-balanced, and Los smiled with joy.
He the vast spine of Urizen seized
And bound down to the glowing illusion.

8. But no light, for the deep fled away
On all sides, and left an unformed
Dark vacuity. Here Urizen lay
In fierce torments on his glowing bed;   170

9. Till his brain in a rock, and his heart
In a fleshy slough formed four rivers,°
Obscuring the immense orb of fire
Flowing down into night; till a form
Was completed, a human illusion,
In darkness and deep clouds involved.

The end of the Book of Los

Milton
A Poem in Two Books (*1804*)

To justify the ways of God to men°

PREFACE

[*see above, p. 71*]

BOOK THE FIRST

Daughters of Beulah, muses who inspire the poet's song!°
Record the journey of immortal Milton through your realms
Of terror and mild moony lustre, in soft sexual delusions
Of varied beauty, to delight the wanderer and repose
His burning thirst and freezing hunger. Come into my hand
By your mild power, descending down the nerves of my right arm
From out the portals of my brain, where by your ministry
The eternal great Humanity Divine planted his paradise,
And in it caused the Spectres of the dead to take sweet forms
In likeness of himself. Tell also of the false tongue! vegetated°   10
Beneath your land of shadows; of its sacrifices and
Its offerings, even till Jesus, the image of the invisible God,
Became its prey—a curse, an offering and an atonement
For death eternal, in the heavens of Albion, and before the gates
Of Jerusalem his Emanation, in the heavens beneath Beulah.

Say first, what moved Milton—who walked about in Eternity
One hundred years, pond'ring the intricate mazes of providence°
(Unhappy though in Heaven, he obeyed, he murmured not, he was
    silent),
Viewing his sixfold Emanation scattered through the deep°
In torment—to go into the deep, her to redeem and himself   20
    perish?
That cause at length moved Milton to this unexampled deed:
A bard's prophetic song! For, sitting at eternal tables,
Terrific among the sons of Albion, in chorus solemn and loud
A bard broke forth! All sat attentive to the awful man.

'Mark well my words! They are of your eternal salvation.

'Three classes are created by the hammer of Los, and woven
By Enitharmon's looms, and spun beneath the spindle of Tirzah.
The first, the Elect from before the foundation of the world;
The second, the Redeemed; the third, the Reprobate and formed
To destruction from the mother's womb. Follow with me my   30
    plough.

'Of the first class was Satan. With incomparable mildness
His primitive tyrannical attempts on Los; with most endearing
    love
He soft entreated Los to give to him Palamabron's station—
For Palamabron returned with labour wearied every evening.
Palamabron oft refused, and as often Satan offered
His service, till by repeated offers and repeated entreaties
Los gave to him the harrow of the Almighty; alas, blamable,
Palamabron feared to be angry lest Satan should accuse him of
Ingratitude, and Los believe the accusation through Satan's
    extreme
Mildness. Satan laboured all day (it was a thousand years);°   40
In the evening returning terrified, overlaboured and astonished,
Embraced soft with a brother's tears Palamabron, who also wept.

'Mark well my words! They are of your eternal salvation.

'Next morning Palamabron rose. The horses of the harrow
Were maddened with tormenting fury, and the servants of the
    harrow,
The gnomes, accused Satan with indignation, fury and fire.
Then Palamabron, reddening like the moon in an eclipse,
Spoke, saying: "You know Satan's mildness and his self-
    imposition,
Seeming a brother, being a tyrant, even thinking himself a brother
While he is murdering the just. Prophetic I behold   50
His future course through darkness and despair to eternal death.
But we must not be tyrants also; he hath assumed my place
For one whole day, under pretence of pity and love to me.
My horses hath he maddened! and my fellow servants injured.
How should he—*he*—know the duties of another? Oh foolish
    forbearance,
Would I had told Los all my heart! But patience, O my friends;
All may be well. Silent remain, while I call Los and Satan."

'Loud as the wind of Beulah that unroots the rocks and hills
Palamabron called, and Los and Satan came before him;
And Palamabron showed the horses and the servants. Satan wept   60
And, mildly cursing Palamabron, him accused of crimes
Himself had wrought. Los trembled; Satan's blandishment almost
Persuaded the Prophet of Eternity that Palamabron
Was Satan's enemy, and that the gnomes, being Palamabron's
    friends,
Were leagued together against Satan through ancient enmity.
What could Los do? How could he judge, when Satan's self
    believed
That he had not oppressed the horses of the harrow, nor the
    servants?

'So Los said: "Henceforth, Palamabron, let each his own station
Keep; nor in pity false, nor in officious brotherhood, where
None needs, be active." Meantime Palamabron's horses   70
Raged with thick flames redundant, and the harrow maddened
    with fury.
Trembling Palamabron stood: the strongest of demons trembled,
Curbing his living creatures. Many of the strongest gnomes
They bit in their wild fury, who also maddened like wildest beasts.

'Mark well my words! They are of your eternal salvation.

'Meanwhile wept Satan before Los, accusing Palamabron,
Himself exculpating with mildest speech; for himself believed
That he had not oppressed nor injured the refractory servants.

'But Satan returning to his mills (for Palamabron had served
The mills of Satan as the easier task) found all confusion,   80
And back returned to Los, not filled with vengeance but with tears,
Himself convinced of Palamabron's turpitude. Los beheld
The servants of the mills drunken with wine, and dancing wild
With shouts and Palamabron's songs, rending the forests green
With echoing confusion, though the sun was risen on high.

'Then Los took off his left sandal, placing it on his head,
Signal of solemn mourning. When the servants of the mills
Beheld the signal they in silence stood, though drunk with wine.
Los wept! But Rintrah also came, and Enitharmon on
His arm leaned tremblingly, observing all these things.   90

'And Los said: "Ye genii of the mills, the sun is on high;
Your labours call you. Palamabron is also in sad dilemma:
His horses are mad, his harrow confounded, his companions
    enraged.
Mine is the fault! I should have remembered that pity divides the
    soul,
And Man unmans. Follow with me my plough; this mournful day
Must be a blank in Nature. Follow with me, and tomorrow again
Resume your labours, and this day shall be a mournful day."

'Wildly they followed Los and Rintrah, and the mills were silent.
They mourned all day, this mournful day of Satan and Palama-
    bron;
And all the Elect and all the Redeemed mourned one toward   100
    another
Upon the mountains of Albion among the cliffs of the dead.

'They ploughed in tears! Incessant poured Jehovah's rain, and
    Molech's
Thick fires, contending with the rain, thundered above, rolling
Terrible over their heads. Satan wept over Palamabron.
Theotormon and Bromion contended on the side of Satan,
Pitying his youth and beauty, trembling at eternal death.
Michael contended against Satan in the rolling thunder.
Thulloh, the friend of Satan, also reproved him; faint their
    reproof.

'But Rintrah, who is of the Reprobate, of those formed to
    destruction,
In indignation for Satan's soft dissimulation of friendship   110
Flamed above all the ploughed furrows, angry, red and furious,
Till Michael sat down in the furrow, weary, dissolved in tears.
Satan, who drove the team beside him, stood angry and red;
He smote Thulloh and slew him, and he stood terrible over
    Michael
Urging him to arise. He wept. Enitharmon saw his tears,
But Los hid Thulloh from her sight, lest she should die of grief.
She wept; she trembled! She kissed Satan; she wept over Michael;
She formed a space for Satan and Michael and for the poor
    infected.
Trembling she wept over the space, and closed it with a tender
    moon.

'Los secret buried Thulloh, weeping disconsolate over the moony   120
    space.

'But Palamabron called down a great solemn assembly,
That he who will not defend truth may be compelled to
Defend a lie, that he may be snared and caught and taken.

'And all Eden descended into Palamabron's tent,
Among Albion's Druids and bards, in the caves beneath Albion's
Death-couch, in the caverns of death, in the corner of the Atlantic.
And in the midst of the great assembly Palamabron prayed:
"O God, protect me from my friends, that they have not power
    over me;
Thou hast given me power to protect myself from my bitterest
    enemies."

'Mark well my words! They are of your eternal salvation.   130

'Then rose the two witnesses, Rintrah and Palamabron,
And Palamabron appealed to all Eden and received
Judgement. And, lo, it fell on Rintrah and his rage,
Which now flamed high and furious in Satan against Palamabron,
Till it became a proverb in Eden: "Satan is among the Reprobate".

'Los in his wrath cursed heaven and earth; he rent up nations,
Standing on Albion's rocks among high-reared Druid temples,
Which reach the stars of heaven and stretch from pole to pole.
He displaced continents; the oceans fled before his face.
He altered the poles of the world, east, west and north and south,   140
But he closed up Enitharmon from the sight of all these things.

'For Satan, flaming with Rintrah's fury hidden beneath his own
    mildness,
Accused Palamabron before the assembly of ingratitude, of malice.
He created seven deadly sins, drawing out his infernal scroll
Of moral laws and cruel punishments upon the clouds of
    Jehovah—
To pervert the divine voice in its entrance to the earth
With thunder of war and trumpets' sound, with armies of disease,
Punishments and deaths mustered and numbered, saying: "I am
    God alone.

There is no other. Let all obey my principles of moral individua-
   lity.°
I have brought them from the uppermost, innermost recesses   150
Of my eternal mind; transgressors I will rend off for ever,
As now I rend this accursed family from my covering."

'Thus Satan raged amidst the assembly, and his bosom grew
Opaque against the Divine Vision. The paved terraces of
His bosom inwards shone with fires, but the stones, becoming
   opaque,
Hid him from sight in an extreme blackness and darkness.
And there a world of deeper Ulro was opened, in the midst
Of the assembly; in Satan's bosom, a vast unfathomable abyss.

'Astonishment held the assembly in an awful silence; and tears
Fell down as dews of night; and a loud, solemn, universal groan   160
Was uttered from the east and from the west and from the south
And from the north. And Satan stood opaque, immeasurable,
Covering the east with solid blackness round his hidden heart,
With thunders uttered from his hidden wheels, accusing loud
The divine mercy for protecting Palamabron in his tent.

'Rintrah reared up walls of rocks, and poured rivers and moats
Of fire round the walls. Columns of fire guard around
Between Satan and Palamabron in the terrible darkness.

'And Satan, not having the science of wrath but only of pity,
Rent them asunder, and wrath was left to wrath, and pity to pity.   170
He sunk down, a dreadful death, unlike the slumbers of Beulah.

'The separation was terrible. The dead was reposed on his couch,
Beneath the couch of Albion, on the seven mountains of Rome,
In the whole space of the covering cherub, Rome, Babylon and
   Tyre.°
His Spectre raging furious descended into its space.
He set his face against Jerusalem, to destroy the eon of Albion.°

'But Los hid Enitharmon from the sight of all these things
Upon the Thames, whose lulling harmony reposed her soul,
Where Beulah lovely terminates in rocky Albion,
Terminating in Hyde Park, on Tyburn's awful brook.   180

'And the mills of Satan were separated into a moony space
Among the rocks of Albion's temple, and Satan's Druid sons
Offer the human victims throughout all the earth, and Albion's
Dread tomb, immortal on his rock, overshadowed the whole earth;
Where Satan, making to himself laws from his own identity,
Compelled others to serve him in moral gratitude and submission,
Being called God, setting himself above all that is called God.
And all the spectres of the dead, calling themselves sons of God,
In his synagogues worship Satan under the unutterable name.

'And it was enquired why, in a great solemn assembly,   190
The innocent should be condemned for the guilty? Then an
    Eternal rose,
Saying: "If the guilty should be condemned, he must be an eternal
    death;
And one must die for another throughout all eternity.
Satan is fallen from his station and never can be redeemed,
But must be new-created continually, moment by moment.
And therefore the class of Satan shall be called the Elect, and those
Of Rintrah the Reprobate, and those of Palamabron the
    Redeemed;
For he is redeemed from Satan's law, the wrath falling on Rintrah.
And therefore Palamabron dared not to call a solemn assembly
Till Satan had assumed Rintrah's wrath in the day of mourning, 200
In a feminine delusion of false pride, self-deceived."

'So spake the Eternal, and confirmed it with a thunderous oath.

'But when Leutha, a daughter of Beulah, beheld Satan's condem-
    nation
She down descended into the midst of the great solemn assembly,
Offering herself a ransom for Satan, taking on her his sin.

'Mark well my words! They are of your eternal salvation.

'And Leutha stood glowing with varying colours, immortal, heart-
    piercing
And lovely, and her moth-like elegance shone over the assembly.

'At length, standing upon the golden floor of Palamabron,
She spake: "I am the author of this sin; by my suggestion   210
My parent-power Satan has committed this transgression.

I loved Palamabron, and I sought to approach his tent;
But beautiful Elynittria with her silver arrows repelled me.
For her light is terrible to me; I fade before her immortal beauty.
Oh wherefore doth a dragon-form forth issue from my limbs
To seize her new-born son? Ah me! the wretched Leutha!
This to prevent, entering the doors of Satan's brain night after
    night
Like sweet perfumes, I stupefied the masculine perceptions
And kept only the feminine awake. Hence rose his soft
Delusory love to Palamabron, admiration joined with envy,   220
Cupidity unconquerable! My fault, when at noon of day
The horses of Palamabron called for rest and pleasant death,
I sprang out of the breast of Satan, over the harrow beaming
In all my beauty, that I might unloose the flaming steeds
As Elynittria used to do. But too well those living creatures
Knew that I was not Elynittria, and they brake the traces.
But me the servants of the harrow saw not, but as a bow
Of varying colours on the hills. Terribly raged the horses.
Satan, astonished, and with power above his own control
Compelled the gnomes to curb the horses, and to throw banks of 230
    sand
Around the fiery flaming harrow in labyrinthine forms,
And brooks between to intersect the meadows in their course.
The harrow cast thick flames; Jehovah thundered above;
Chaos and ancient night fled from beneath the fiery harrow.
The harrow cast thick flames and orbed us round in concave fires,
A hell of our own making. See, its flames still gird me round.
Jehovah thundered above. Satan in pride of heart
Drove the fierce harrow among the constellations of Jehovah,
Drawing a third part in the fires as stubble north and south,
To devour Albion and Jerusalem the Emanation of Albion,   240
Driving the harrow in pity's paths. 'Twas then, with our dark fires
Which now gird round us (oh eternal torment), I formed the
    serpent
Of precious stones and gold, turned poisons on the sultry wastes.
The gnomes in all that day spared not; they cursed Satan bitterly.
To do unkind things in kindness! With power armed, to say
The most irritating things in the midst of tears and love!
These are the stings of the serpent. Thus did we by them, till thus
They in return retaliated, and the living creatures maddened.
The gnomes laboured. I weeping hid in Satan's inmost brain;

But when the gnomes refused to labour more, with blandishments   250
I came forth from the head of Satan. Back the gnomes recoiled
And called me Sin, and for a sign portentous held me. Soon°
Day sunk and Palamabron returned. Trembling I hid myself
In Satan's inmost palace of his nervous fine-wrought brain.
For Elynittria met Satan with all her singing women,
Terrific in their joy, and pouring wine of wildest power.
They gave Satan their wine; indignant at the burning wrath,
Wild with prophetic fury, his former life became like a dream.
Clothed in the serpent's folds, in selfish holiness demanding purity
(Being most impure), self-condemned to eternal tears, he drove   260
Me from his inmost brain, and the doors closed with thunder's
    sound.
O Divine Vision, who didst create the female to repose
The sleepers of Beulah, pity the repentant Leutha. My
Sick couch bears the dark shades of eternal death, enfolding
The Spectre of Satan. He furious refuses to repose in sleep.
I humbly bow in all my sin before the throne divine;
Not so the sick one. Alas, what shall be done him to restore—
Who calls the individual law holy, and despises the Saviour,
Glorying to involve Albion's body in fires of eternal war?"

'Now Leutha ceased; tears flowed; but the divine pity supported   270
    her.

' "All is my fault! We are the Spectre of Luvah, the murderer
Of Albion. Oh Vala! Oh Luvah! Oh Albion! Oh lovely Jerusalem!
The sin was begun in eternity, and will not rest to eternity,
Till two eternities meet together. Ah! lost! lost! lost! for ever!"

'So Leutha spoke. But when she saw that Enitharmon had
Created a new space to protect Satan from punishment,
She fled to Enitharmon's tent and hid herself. Loud raging
Thundered the assembly dark and clouded, and they ratified°
The kind decision of Enitharmon, and gave a time to the space,
Even six thousand years, and sent Lucifer for its guard.   280
But Lucifer refused to die, and in pride he forsook his charge.
And they elected Molech, and when Molech was impatient
The divine hand found the two limits: first of opacity, then of
    contraction.
Opacity was named Satan; contraction was named Adam.

Triple Elohim came; Elohim wearied, fainted; they elected
    Shaddai.°
Shaddai angry, Pahad descended; Pahad terrified, they sent
    Jehovah.°
And Jehovah was leprous; loud he called, stretching his hand to
    Eternity°
(For then the body of death was perfected in hypocritic holiness
Around the Lamb, a female tabernacle woven in Cathedron's
    looms).
He died as a Reprobate; he was punished as a transgressor.   290
Glory! glory! glory to the holy Lamb of God!
I touch the heavens as an instrument to glorify the Lord!

'The Elect shall meet the Redeemed; on Albion's rocks they shall
    meet,
Astonished at the transgressor, in him beholding the Saviour.
And the Elect shall say to the Redeemed: "We behold it is of divine
Mercy alone, of free gift and election that we live.
Our virtues and cruel goodnesses have deserved eternal death."
Thus they weep upon the fatal brook of Albion's river.

'But Elynittria met Leutha in the place where she was hidden,
And threw aside her arrows, and laid down her sounding bow.   300
She soothed her with soft words, and brought her to Palamabron's
    bed,
In moments new-created for delusion interwoven round about.
In dreams she bore the shadowy Spectre of sleep, and named him
    Death.
In dreams she bore Rahab, the mother of Tirzah, and her sisters,
In Lambeth's vales, in Cambridge and in Oxford, places of
    thought;
Intricate labyrinths of times and spaces unknown that Leutha
    lived
In Palamabron's tent, and Oothoon was her charming guard.'

The bard ceased. All considered, and a loud resounding murmur
Continued round the halls, and much they questioned the
    immortal
Loud-voiced bard. And many condemned the high-toned song,   310
Saying: 'Pity and love are too venerable for the imputation
Of guilt.' Others said: 'If it is true, if the acts have been performed,
Let the bard himself witness. Where hadst thou this terrible song?'

The bard replied: 'I am inspired! I know it is truth, for I sing
According to the inspiration of the Poetic Genius,
Who is the eternal, all-protecting Divine Humanity,
To whom be glory and power and dominion evermore. Amen.'°

Then there was great murmuring in the heavens of Albion
Concerning generation and the vegetative power, and concerning
The Lamb, the Saviour. Albion trembled to Italy, Greece and   320
   Egypt,
To Tartary and Hindustan and China, and to great America,
Shaking the roots and fast foundations of the earth in doubtful-
   ness.
The loud-voiced bard terrified took refuge in Milton's bosom.

Then Milton rose up from the heavens of Albion ardorous!
The whole assembly wept prophetic, seeing in Milton's face
And in his lineaments divine the shades of death and Ulro.
He took off the robe of the promise, and ungirded himself from the
   oath of God.

And Milton said: 'I go to eternal death! The nations still
Follow after the detestable gods of Priam, in pomp
Of warlike selfhood, contradicting and blaspheming.   330
When will the resurrection come to deliver the sleeping body
From corruptibility? Oh when, Lord Jesus, wilt thou come?
Tarry no longer, for my soul lies at the gates of death.
I will arise and look forth for the morning of the grave.
I will go down to the sepulchre to see if morning breaks!
I will go down to self-annihilation and eternal death,
Lest the Last Judgement come and find me unannihilate,
And I be seized and given into the hands of my own selfhood.
The Lamb of God is seen through mists and shadows, hov'ring
Over the sepulchres in clouds of Jehovah and winds of Elohim,   340
A disc of blood, distant, and heavens and earths roll dark between.
What do I here before the Judgement? Without my Emanation?
With the Daughters of Memory, and not with the Daughters of
   Inspiration?
I in my selfhood am that Satan. I am that evil one!
He is my Spectre! In my obedience to loose him from my hells,
To claim the hells, my furnaces, I go to eternal death.'

And Milton said; 'I go to eternal death!' Eternity shuddered,
For he took the outside course, among the graves of the dead,
A mournful shade. Eternity shuddered at the image of eternal
    death.

Then on the verge of Beulah he beheld his own shadow: 350
A mournful form, double, hermaphroditic, male and female
In one wonderful body. And he entered into it
In direful pain; for the dread shadow, twenty-seven-fold,
Reached to the depths of direst Hell, and thence to Albion's land,
Which is this earth of vegetation on which now I write.

The seven angels of the presence wept over Milton's shadow.

As, when a man dreams, he reflects not that his body sleeps,
Else he would wake; so seemed he entering his shadow. But
With him the spirits of the seven angels of the presence
Entering, they gave him still perceptions of his sleeping body,   360
Which now arose and walked with them in Eden, as an eighth
Image, divine though darkened, and though walking as one walks
In sleep. And the seven comforted and supported him.

Like as a polypus that vegetates beneath the deep
They saw his shadow vegetated underneath the couch
Of death. For when he entered into his shadow himself,
His real and immortal self, was, as appeared to those
Who dwell in immortality, as one sleeping on a couch
Of gold, and those in immortality gave forth their Emanations,
Like females of sweet beauty, to guard round him and to feed   370
His lips with food of Eden in his cold and dim repose.
But to himself he seemed a wanderer lost in dreary night.

Onwards his shadow kept its course among the Spectres, called
Satan, but swift as lightning passing them; startled, the shades
Of Hell beheld him in a trail of light, as a comet
That travels into chaos. So Milton went guarded within.

The nature of infinity is this: that everything has its
Own vortex, and when once a traveller through Eternity
Has passed that vortex, he perceives it roll backward behind
His path, into a globe itself enfolding like a sun,   380

Or like a moon, or like a universe of starry majesty
(While he keeps onwards in his wondrous journey on the earth),
Or like a human form, a friend with whom he lived benevolent;
As the eye of Man views both the east and west, encompassing
Its vortex, and the north and south, with all their starry host,
Also the rising sun and setting moon he views, surrounding
His cornfields and his valleys of five hundred acres square.
Thus is the earth one infinite plane, and not as apparent
To the weak traveller confined beneath the moony shade.
Thus is the heaven a vortex passed already, and the earth   390
A vortex not yet passed by the traveller through Eternity.

First Milton saw Albion upon the Rock of Ages,°
Deadly pale, outstretched and snowy cold, storm-covered,
A giant form of perfect beauty outstretched on the rock
In solemn death. The sea of time and space thundered aloud
Against the rock, which was enwrapped with the weeds of death.
Hovering over the cold bosom, in its vortex, Milton bent down
To the bosom of death. What was underneath soon seemed above;
A cloudy heaven mingled with stormy seas in loudest ruin.
But as a wintry globe descends precipitant through Beulah,   400
    bursting
With thunders loud and terrible, so Milton's shadow fell,
Precipitant loud-thund'ring into the sea of time and space.

Then first I saw him in the zenith as a falling star,
Descending perpendicular, swift as the swallow or swift;
And on my left foot falling, on the tarsus, entered there;°
But from my left foot a black cloud redounding spread over
    Europe.

Then Milton knew that the three heavens of Beulah were beheld
By him on earth, in his bright pilgrimage of sixty years,
In those three females whom his wives, and those three whom his
    daughters,
Had represented and contained; that they might be resumed   410
By giving up of selfhood. And they distant viewed his journey
In their eternal spheres, now human, though their bodies remain
    closed
In the dark Ulro till the Judgement. Also Milton knew they and
Himself was human, though now wandering through death's vale

In conflict with those female forms, which in blood and jealousy
Surrounded him, dividing and uniting without end or number.

He saw the cruelties of Ulro, and he wrote them down
In iron tablets. And his wives' and daughters' names were these:
Rahab, and Tirzah, and Milcah, and Malah, and Noah, and
  Hoglah.°
They sat ranged round him as the rocks of Horeb round the land°   420
Of Canaan, and they wrote in thunder, smoke and fire
His dictate, and his body was the rock Sinai: that body
Which was on earth born to corruption. And the six females
Are Hor, and Peor, and Bashan, and Abarim, and Lebanon, and
  Hermon:°
Seven rocky masses terrible in the deserts of Midian.

But Milton's human shadow continued journeying above
The rocky masses of the Mundane Shell, in the lands
Of Edom, and Aram, and Moab, and Midian, and Amalek.°

The Mundane Shell is a vast concave Earth: an immense
Hardened shadow of all things upon our vegetated Earth,   430
Enlarged into dimension and deformed into indefinite space,
In twenty-seven heavens and all their hells, with chaos
And ancient night and purgatory. It is a cavernous Earth
Of labyrinthine intricacy, twenty-seven folds of opaqueness,
And finishes where the lark mounts. Here Milton journeyed,
In that region called Midian, among the rocks of Horeb;
For travellers from Eternity pass outward to Satan's seat,
But travellers to Eternity pass inward to Golgonooza.

Los, the vehicular terror, beheld him, and divine Enitharmon°
Called all her daughters, saying: 'Surely to unloose my bond   440
Is this man come! Satan shall be unloosed upon Albion.'

Los heard in terror Enitharmon's words. In fibrous strength
His limbs shot forth like roots of trees against the forward path
Of Milton's journey. Urizen beheld the immortal man,
And he also darked his brows, freezing dark rocks between
The footsteps and infixing deep the feet in marble beds—
That Milton laboured with his journey, and his feet bled sore
Upon the clay now changed to marble. Also Urizen rose
And met him on the shores of Arnon, and by the streams of the
  brooks.°

Silent they met, and silent strove among the streams of Arnon   450
Even to Mahanaim, when with cold hand Urizen stooped down°
And took up water from the river Jordan, pouring on
To Milton's brain the icy fluid from his broad cold palm.
But Milton took of the red clay of Succoth, moulding it with care°
Between his palms and filling up the furrows of many years,
Beginning at the feet of Urizen, and on the bones
Creating new flesh on the demon cold and building him,
As with new clay, a human form in the valley of Beth Peor.°

Four universes round the mundane egg remain chaotic:
One to the north, named Urthona; one to the south, named Urizen;   460
One to the east, named Luvah; one to the west, named Tharmas.
They are the four Zoas that stood around the throne divine.°
But when Luvah assumed the world of Urizen to the south,
And Albion was slain upon his mountains and in his tent,
All fell towards the centre in dire ruin, sinking down.
And in the south remains a burning fire, in the east a void,
In the west, a world of raging waters, in the north a solid,
Unfathomable, without end. But in the midst of these
Is built eternally the universe of Los and Enitharmon,
Towards which Milton went; but Urizen opposed his path.   470

The man and demon strove many periods. Rahab beheld,
Standing on Carmel; Rahab and Tirzah trembled to behold°
The enormous strife, one giving life, the other giving death
To his adversary. And they sent forth all their sons and daughters
In all their beauty to entice Milton across the river.

The twofold form hermaphroditic, and the double-sexed,
The female-male and the male-female, self-dividing, stood
Before him in their beauty and in cruelties of holiness,
Shining in darkness, glorious upon the deeps of Entuthon,

Saying: 'Come thou to Ephraim! Behold the kings of Canaan!°   480
The beautiful Amalekites behold the fires of youth
Bound with the chain of jealousy by Los and Enitharmon.
The banks of Cam, cold learning's streams, London's dark-
    frowning towers,°
Lament upon the winds of Europe in Rephaim's vale°
Because Ahania, rent apart into a desolate night,

Laments! And Enion wanders like a weeping inarticulate voice,
And Vala labours for her bread and water among the furnaces.
Therefore bright Tirzah triumphs, putting on all beauty
And all perfection in her cruel sports among the victims.
Come, bring with thee Jerusalem with songs on the Grecian lyre!   490
In natural religion, in experiments on men,
Let her be offered up to holiness! Tirzah numbers her;
She numbers with her fingers every fibre ere it grow.
Where is the Lamb of God? Where is the promise of his coming?°
Her shadowy sisters form the bones, even the bones of Horeb,
Around the marrow, and the orbed skull around the brain.
His images are born for war, for sacrifice to Tirzah,
To natural religion! to Tirzah, the daughter of Rahab the holy!
She ties the knot of nervous fibres into a white brain!
She ties the knot of bloody veins into a red-hot heart!   500
Within her bosom Albion lies embalmed, never to awake;
Hand is become a rock; Sinai and Horeb is Hyle and Coban;°
Skofield is bound in iron armour before Reuben's gate.°
She ties the knot of milky seed into two lovely heavens,
Two, yet but one, each in the other sweet reflected; these
Are our three heavens beneath the shades of Beulah, land of rest.
Come then to Ephraim and Manasseh, O beloved one!°
Come to my ivory palaces, O beloved of thy mother!
And let us bind thee in the bands of war, and be thou king
Of Canaan and reign in Hazor where the twelve tribes meet.'°   510

So spoke they as in one voice. Silent Milton stood before
The darkened Urizen, as the sculptor silent stands before
His forming image; he walks round it patient labouring.
Thus Milton stood forming bright Urizen, while his mortal part
Sat frozen in the rock of Horeb, and his redeemed portion
Thus formed the clay of Urizen; but within that portion
His real human walked above in power and majesty
Though darkened; and the seven angels of the presence attended
    him.

Oh how can I, with my gross tongue that cleaveth to the dust,
Tell of the fourfold man, in starry numbers fitly ordered,   520
Or how can I with my cold hand of clay? But thou, O Lord,°
Do with me as thou wilt, for I am nothing, and vanity.°
If thou choose to elect a worm, it shall remove the mountains.

For that portion named the Elect, the spectrous body of Milton,
Redounding from my left foot into Los's mundane space,
Brooded over his body in Horeb against the Resurrection,
Preparing it for the great consummation. Red the cherub on Sinai
Glowed, but in terrors folded round his clouds of blood.

Now Albion's sleeping humanity began to turn upon his couch,
Feeling the electric flame of Milton's awful precipitate descent.   530
Seest thou the little winged fly, smaller than a grain of sand?
It has a heart like thee, a brain open to Heaven and Hell,
Withinside wondrous and expansive. Its gates are not closed;
I hope thine are not. Hence it clothes itself in rich array;
Hence thou art clothed with human beauty, O thou mortal man.
Seek not thy heavenly father then beyond the skies;
There chaos dwells and ancient night, and Og and Anak old.°
For every human heart has gates of brass and bars of adamant,
Which few dare unbar, because dread Og and Anak guard the gates
Terrific! And each mortal brain is walled and moated round   540
Within, and Og and Anak watch here; here is the seat
Of Satan in its webs. For in brain and heart and loins
Gates open behind Satan's seat to the city of Golgonooza,
Which is the spiritual fourfold London in the loins of Albion.

Thus Milton fell through Albion's heart, travelling outside of
    humanity,
Beyond the stars in chaos, in caverns of the Mundane Shell.

But many of the Eternals rose up from eternal tables
Drunk with the spirit; burning round the couch of death they
    stood
Looking down into Beulah, wrathful, filled with rage.
They rend the heavens round the watchers in a fiery circle,   550
And round the shadowy eighth; the eight close up the couch
Into a tabernacle and flee with cries down to the deeps,
Where Los opens his three wide gates, surrounded by raging fires.
They soon find their own place and join the watchers of the Ulro.

Los saw them and a cold pale horror covered o'er his limbs.
Pondering, he knew that Rintrah and Palamabron might depart
Even as Reuben and as Gad, gave up himself to tears.°
He sat down on his anvil-stock and leaned upon the trough,°
Looking into the black water, mingling with it tears.

At last, when desperation almost tore his heart in twain,   560
He recollected an old prophecy, in Eden recorded
And often sung to the loud harp at the immortal feasts,
That Milton of the land of Albion should up ascend
Forwards from Ulro, from the Vale of Felpham, and set free
Orc from his chain of jealousy. He started at the thought
And down descended into Udan-Adan. It was night,
And Satan sat sleeping upon his couch in Udan-Adan;
His Spectre slept, his shadow woke (when one sleeps the other
    wakes).

But Milton entering my foot, I saw in the nether
Regions of the imagination; also all men on Earth   570
And all in Heaven saw in the nether regions of the imagination,
In Ulro beneath Beulah, the vast breach of Milton's descent.
But I knew not that it was Milton, for Man cannot know
What passes in his members till periods of space and time
Reveal the secrets of Eternity; for more extensive
Than any other earthly things are Man's earthly lineaments.
And all this vegetable world appeared on my left foot,
As a bright sandal formed immortal of precious stones and gold.
I stooped down and bound it on to walk forward through
    Eternity.°

There is in Eden a sweet river, of milk and liquid pearl,   580
Named Ololon, on whose mild banks dwelt those who Milton
    drove°
Down into Ulro. And they wept in long resounding song
For seven days of eternity, and the river's living banks,
The mountains, wailed, and every plant that grew in solemn sighs
    lamented.

When Luvah's bulls each morning drag the sulphur sun out of the
    deep,
Harnessed with starry harness black and shining, kept by black
    slaves
That work all night at the starry harness, strong and vigorous
They drag the unwilling orb. At this time all the family
Of Eden heard the lamentation, and providence began.
But when the clarions of day sounded they drowned the   590
    lamentations,

And when night came all was silent in Ololon, and all refused to
    lament
In the still night, fearing lest they should others molest.

Seven mornings Los heard them, as the poor bird within the shell
Hears its impatient parent bird, and Enitharmon heard them,
But saw them not, for the blue Mundane Shell enclosed them in.

And they lamented that they had in wrath and fury and fire
Driven Milton into the Ulro. For now they knew too late
That it was Milton the awakener; they had not heard the bard
Whose song called Milton to the attempt. And Los heard these
    laments.
He heard them call in prayer all the Divine Family,   600
And he beheld the cloud of Milton stretching over Europe.

But all the Family Divine collected as four suns
In the four points of heaven, east, west and north and south,
Enlarging and enlarging till their discs approached each other
And, when they touched, closed together southward in one sun
Over Ololon. And as one man, who weeps over his brother
In a dark tomb, so all the Family Divine wept over Ololon,

Saying, 'Milton goes to eternal death!' So saying they groaned in
    spirit
And were troubled; and again the Divine Family groaned in spirit!

And Ololon said: 'Let us descend also, and let us give   610
Ourselves to death in Ulro among the transgressors.
Is virtue a punisher? Oh no! How is this wondrous thing?—
This world beneath, unseen before, this refuge from the wars
Of great Eternity! unnatural refuge! unknown by us till now.
Or are these the pangs of repentance? Let us enter into them.'

Then the Divine Family said: 'Six thousand years are now
Accomplished in this world of sorrows; Milton's angel knew
The universal dictate, and you also feel this dictate.
And now you know this world of sorrow and feel pity. Obey
The dictate! Watch over this world, and with your brooding wings   620
Renew it to eternal life. Lo! I am with you always.°
But you cannot renew Milton; he goes to eternal death.'

So spake the Family Divine as one man, even Jesus,
Uniting in one with Ololon, and the appearance of one man,
Jesus the Saviour, appeared coming in the clouds of Ololon.°
Though driven away with the seven starry ones into the Ulro,
Yet the Divine Vision remains everywhere for ever. Amen.
And Ololon lamented for Milton with a great lamentation;

While Los heard indistinct in fear. What time I bound my sandals
On to walk forward through Eternity, Los descended to me;   630
And Los behind me stood, a terrible flaming sun, just close
Behind my back. I turned round in terror and, behold,
Los stood in that fierce glowing fire, and he also stooped down
And bound my sandals on in Udan-Adan. Trembling I stood
Exceedingly with fear and terror, standing in the vale
Of Lambeth, but he kissed me and wished me health,
And I became one man with him, arising in my strength.
'Twas too late now to recede. Los had entered into my soul;
His terrors now possessed me whole! I arose in fury and strength.

'I am that shadowy prophet who six thousand years ago   640
Fell from my station in the eternal bosom. Six thousand years
Are finished; I return! Both time and space obey my will.
I in six thousand years walk up and down, for not one moment
Of time is lost, nor one event of space unpermanent,
But all remain; every fabric of six thousand years
Remains permanent. Though on the earth, where Satan
Fell and was cut off, all things vanish and are seen no more,
They vanish not from me and mine; we guard them first and last.
The generations of men run on in the tide of time,
But leave their destined lineaments permanent, for ever and ever.'   650

So spoke Los as we went along to his supreme abode.

Rintrah and Palamabron met us at the gate of Golgonooza,
Clouded with discontent and brooding in their minds terrible
     things.

They said: 'O father most beloved, O merciful parent,
Pitying and permitting evil, though strong and mighty to destroy,
Whence is this shadow terrible? Wherefore dost thou refuse
To throw him into the furnaces? Knowest thou not that he

Will unchain Orc, and let loose Satan, Og, Sihon and Anak°
Upon the body of Albion? For this he is come. Behold it written
Upon his fibrous left foot black, most dismal to our eyes.   660
The shadowy female shudders through heaven in torment
    inexpressible,
And all the daughters of Los prophetic wail; yet in deceit
They weave a new religion from new jealousy of Theotormon.
Milton's religion is the cause; there is no end to destruction.
Seeing the churches at their period in terror and despair,
Rahab created Voltaire, Tirzah created Rousseau,
Asserting the self-righteousness against the universal Saviour,
Mocking the confessors and martyrs, claiming self-righteousness,
With cruel virtue making war upon the Lamb's Redeemed,
To perpetuate war and glory, to perpetuate the laws of sin.   670
They perverted Swedenborg's visions in Beulah and in Ulro,
To destroy Jerusalem as a harlot and her sons as Reprobates,
To raise up Mystery, the virgin harlot, mother of war,
Babylon the great, the abomination of desolation.
O Swedenborg! strongest of men, the Samson shorn by the
    churches,
Showing the transgressors in Hell, the proud warriors in Heaven,
Heaven as a punisher and Hell as one under punishment,
With laws from Plato and his Greeks to renew the Trojan gods
In Albion, and to deny the value of the Saviour's blood.
But then I raised up Whitefield; Palamabron raised up Wesley.°   680
And these are the cries of the churches before the two witnesses'
Faith in God, the dear Saviour, who took on the likeness of men,
Becoming obedient to death, even the death of the cross:°
"The witnesses lie dead in the street of the great city;
No faith is in all the earth; the book of God is trodden under foot.
He sent his two servants, Whitefield and Wesley. Were they
    prophets,
Or were they idiots or madmen? Show us miracles!"
Can you have greater miracles than these? Men who devote
Their life's whole comfort to entire scorn and injury, and death?
Awake, thou sleeper on the Rock of Eternity, Albion, awake!   690
The trumpet of Judgement hath twice sounded; all nations are
    awake,
But thou art still heavy and dull. Awake, Albion, awake!
Lo, Orc arises on the Atlantic. Lo, his blood and fire
Glow on America's shore. Albion turns upon his couch;

He listens to the sounds of war, astonished and confounded;
He weeps into the Atlantic deep, yet still in dismal dreams
Unawakened, and the covering cherub advances from the east.
How long shall we lay dead in the street of the great city,
How long beneath the covering cherub give our Emanations?
Milton will utterly consume us and thee, our beloved father.   700
He hath entered into the covering cherub, becoming one with
Albion's dread sons; Hand, Hyle and Coban surround him as
A girdle, Gwendolen and Conwenna as a garment woven°
Of war and religion. Let us descend and bring him chained
To Bowlahoola. O father most beloved! O mild parent!
Cruel in thy mildness, pitying and permitting evil
Though strong and mighty to destroy, O Los our beloved father!'

Like the black storm coming out of chaos, beyond the stars,
It issues through the dark and intricate caves of the Mundane
    Shell,
Passing the planetary vision and the well-adorned firmament.   710
The sun rolls into chaos and the stars into the deserts,
And then the storms become visible, audible and terrible,
Covering the light of day, and, rolling down upon the mountains,
Deluge all the country round. Such is a vision of Los
When Rintrah and Palamabron spoke, and such his stormy face
Appeared, as does the face of heaven when covered with thick
    storms,
Pitying and loving, though in frowns of terrible perturbation.

But Los dispersed the clouds, even as the strong winds of Jehovah,
And Los thus spoke: 'O noble sons, be patient yet a little.
I have embraced the falling death, he is become one with me.   720
O sons, we live not by wrath; by mercy alone we live!
I recollect an old prophecy in Eden recorded in gold, and oft
Sung to the harp: that Milton of the land of Albion
Should up ascend, forward from Felpham's vale, and break the
    chain
Of jealousy from all its roots. Be patient therefore, O my sons.
These lovely females form sweet night and silence and secret
Obscurities to hide from Satan's watch-fiends human loves
And graces: lest they write them in their books and in the scroll
Of mortal life, to condemn the accused who, at Satan's bar,
Tremble in spectrous bodies continually, day and night,   730

While on the earth they live in sorrowful vegetations.
Oh when shall we tread our winepresses in Heaven, and reap°
Our wheat with shoutings of joy, and leave the earth in peace?
Remember how Calvin and Luther in fury premature
Sowed war and stern division between papists and protestants.
Let it not be so now! Oh, go not forth in martyrdoms and wars.
We were placed here by the universal brotherhood and mercy,
With powers fitted to circumscribe this dark Satanic death,
And that the seven eyes of God may have space for redemption.°
But how this is as yet we know not, and we cannot know   740
Till Albion is arisen; then patient wait a little while.
Six thousand years are passed away; the end approaches fast.
This mighty one is come from Eden; he is of the Elect
Who died from earth and he is returned before the Judgement.
   This thing
Was never known: that one of the holy dead should willing return.
Then patient wait a little while, till the last vintage is over,
Till we have quenched the sun of Salah in the Lake of Udan-Adan.
O my dear sons! leave not your father, as your brethren left me.
Twelve sons successive fled away in that thousand years of sorrow,
Of Palamabron's harrow and of Rintrah's wrath and fury.   750
Reuben, and Manazzoth, and Gad, and Simeon, and Levi,°
And Ephraim, and Judah were generated because
They left me, wandering with Tirzah. Enitharmon wept
One thousand years, and all the earth was in a watery deluge.
We called him Menassheh because of the generations of Tirzah,
Because of Satan, and the seven eyes of God continually
Guard round them. But I, the fourth Zoa, am also set
The watchman of Eternity; the three are not, and I am preserved!
Still my four mighty ones are left to me in Golgonooza:
Still Rintrah fierce, and Palamabron mild and piteous,   760
Theotormon filled with care, Bromion loving science.
You, O my sons, still guard round Los. O wander not and leave me.
Rintrah, thou well rememberest when Amalek and Canaan°
Fled with their sister Moab into that abhorred void;
They became nations in our sight beneath the hands of Tirzah.
And Palamabron, thou rememberest when Joseph, an infant°
Stolen from his nurse's cradle wrapped in needlework
Of emblematic texture, was sold to the Amalekite,
Who carried him down into Egypt, where Ephraim and Menas-
   sheh°

Gathered my sons together in the sands of Midian.   770
And if you also flee away and leave your father's side,
Following Milton into Ulro, although your power is great,
Surely you also shall become poor mortal vegetations
Beneath the moon of Ulro. Pity then your father's tears.
When Jesus raised Lazarus from the grave I stood and saw
Lazarus (who is the vehicular body of Albion the Redeemed)
Arise into the covering cherub who is the Spectre of Albion—
By martyrdoms to suffer, to watch over the sleeping body
Upon his rock beneath his tomb. I saw the covering cherub
Divide fourfold into four churches when Lazarus arose: 780
Paul, Constantine, Charlemagne, Luther. Behold, they stand
    before us,
Stretched over Europe and Asia. Come, O sons, come, come away.
Arise, O sons, give all your strength against eternal death,
Lest we are vegetated, for Cathedron's looms weave only death,
A web of death. And were it not for Bowlahoola and Allamanda
No human form, but only a fibrous vegetation,
A polypus of soft affections without thought or vision,
Must tremble in the heavens and earths through all the Ulro space.
Throw all vegetated mortals into Bowlahoola,
But as to this Elected form who is returned again—   790
He is the signal that the last vintage now approaches,
Nor vegetation may go on till all the earth is reaped.

So Los spoke. Furious they descended to Bowlahoola and
    Allamanda,
Indignant, unconvinced by Los's arguments and thunders rolling.
They saw that wrath now swayed, and now pity absorbed him.
As it was, so it remained, and no hope of an end.

Bowlahoola is named Law by mortals. Tharmas founded it,
Because of Satan, before Luban in the city of Golgonooza;
But Golgonooza is named Art and Manufacture by mortal men.

In Bowlahoola Los's anvils stand and his furnaces rage;   800
Thundering the hammers beat, and the bellows blow loud,
Living, self-moving, mourning, lamenting and howling inces-
    santly.
Bowlahoola through all its porches feels, though too fast-founded
Its pillars and porticoes to tremble at the force

Of mortal or immortal arm. And softly lilling flutes°
Accordant with the horrid labours make sweet melody.
The bellows are the animal lungs, the hammers the animal heart,
The furnaces the stomach for digestion; terrible their fury.
Thousands and thousands labour, thousands play on instruments,
Stringed or fluted, to ameliorate the sorrows of slavery;   810
Loud sport the dancers in the dance of death, rejoicing in carnage.
The hard-dentant hammers are lulled by the flutes' lula lula;
The bellowing furnaces' blare by the long sounding clarion.
The double drum drowns howls and groans; the shrill fife shrieks
    and cries;
The crooked horn mellows the hoarse raving serpent, terrible but
    harmonious.°
Bowlahoola is the stomach in every individual man.

Los is by mortals named Time; Enitharmon is named Space.
But they depict him bald and aged who is in eternal youth,
All-powerful, and his locks flourish like the brows of morning.
He is the spirit of prophecy, the ever-apparent Elias.°   820
Time is the mercy of Eternity; without time's swiftness,
Which is the swiftest of all things, all were eternal torment.
All the gods of the kingdoms of earth labour in Los's halls;
Every one is a fallen son of the spirit of prophecy.
He is the fourth Zoa, that stood around the throne divine.

Loud shout the sons of Luvah at the winepresses, as Los
    descended
With Rintrah and Palamabron in his fires of resistless fury.

The winepress on the Rhine groans loud, but all its central beams
Act more terrific in the central cities of the nations,
Where human thought is crushed beneath the iron hand of power.   830
There Los puts all into the press, the oppressor and the oppressed
Together, ripe for the harvest and vintage, and ready for the loom.

They sang at the vintage: 'This is the last vintage! And seed
Shall no more be sown upon earth, till all the vintage is over,
And all gathered in; till the plough has passed over the nations,
And the harrow and heavy thundering roller upon the mountains.'

And loud the souls howl round the porches of Golgonooza,
Crying: 'O God, deliver us to the heavens or to the earths,
That we may preach righteousness and punish the sinner with
    death.'
But Los refused, till all the vintage of earth was gathered in.   840

And Los stood and cried to the labourers of the vintage in voice of
    awe:

'Fellow labourers! The great vintage and harvest is now upon
    earth.
The whole extent of the globe is explored; every scattered atom
Of human intellect now is flocking to the sound of the trumpet;
All the wisdom which was hidden in caves and dens, from ancient
Time, is now sought out from animal and vegetable and mineral.
The awakener is come, outstretched over Europe; the vision of
    God is fulfilled.
The ancient man upon the rock of Albion awakes;
He listens to the sounds of war, astonished and ashamed;
He sees his children mock at faith and deny providence.   850
Therefore you must bind the sheaves not by nations or families,
You shall bind them in three classes; according to their classes
So shall you bind them, separating what has been mixed
Since men began to be wove into nations by Rahab and Tirzah,
Since Albion's death and Satan's cutting-off from our awful
    fields—
When, under pretence to benevolence, the Elect subdued all
From the foundation of the world. The Elect is one class; you
Shall bind them separate; they cannot believe in eternal life
Except by miracle and a new birth. The other two classes—
The Reprobate who never cease to believe, and the Redeemed   860
Who live in doubts and fears perpetually tormented by the Elect—
These you shall bind in a twin-bundle for the consummation;
But the Elect must be saved from fires of eternal death,
To be formed into the churches of Beulah that they destroy not the
    earth.
For in every nation and every family the three classes are born,
And in every species of earth: metal, tree, fish, bird and beast.
We form the mundane egg that Spectres, coming by fury or amity,
All is the same, and every one remains in his own energy.
Go forth, reapers, with rejoicing; you sowed in tears,°

But the time of your refreshing cometh. Only a little moment   870
Still abstain from pleasure and rest in the labours of Eternity,
And you shall reap the whole earth from pole to pole, from sea to
    sea!
Beginning at Jerusalem's inner court, Lambeth ruined and given
To the detestable gods of Priam, to Apollo, and at the asylum
Given to Hercules (who labour in Tirzah's looms for bread,°
Who set pleasure against duty, who create Olympic crowns
To make learning a burden, and the work of the Holy Spirit strife,
The Thor and cruel Odin, who first reared the polar caves.)
Lambeth mourns, calling Jerusalem; she weeps and looks abroad
For the Lord's coming, that Jerusalem may overspread all nations.   880
Crave not for the mortal and perishing delights, but leave them
To the weak, and pity the weak as your infant care; break not
Forth in your wrath lest you also are vegetated by Tirzah.
Wait till the Judgement is past, till the creation is consumed,
And then rush forward with me into the glorious spiritual
Vegetation, the supper of the Lamb and his bride, and the°
Awakening of Albion, our friend and ancient companion.'

So Los spoke. But lightnings of discontent broke on all sides
    round,
And murmurs of thunder rolling, heavy, long and loud over the
    mountains,
While Los called his sons around him to the harvest and the   890
    vintage.

Thou seest the constellations in the deep and wondrous night;
They rise in order and continue their immortal courses
Upon the mountains and in vales, with harp and heavenly song,
With flute and clarion, with cups and measures filled with foaming
    wine.
Glittering the streams reflect the vision of beatitude,
And the calm ocean joys beneath and smoothes his awful waves.
These are the sons of Los, and these the labourers of the vintage.
Thou seest the gorgeous clothed flies that dance and sport in
    summer
Upon the sunny brooks and meadows. Every one the dance
Knows in its intricate mazes of delight artful to weave,   900
Each one to sound his instruments of music in the dance,
To touch each other and recede, to cross and change and return.

These are the children of Los. Thou seest the trees on mountains;
The wind blows heavy, loud they thunder through the darksome
    sky,
Uttering prophecies and speaking instructive words to the sons
Of men. These are the sons of Los! these the visions of Eternity,
But we see only as it were the hem of their garments°
When with our vegetable eyes we view these wondrous visions.

There are two gates through which all souls descend: one
    southward°
From Dover Cliff to Lizard Point; the other toward the north,   910
Caithness and rocky Durness, Pentland and John Groat's House.

The souls descending to the body wail on the right hand
Of Los, and those delivered from the body on the left hand.
For Los against the east his force continually bends,
Along the valleys of Middlesex from Hounslow to Blackheath,
Lest those three heavens of Beulah should the creation destroy,
And lest they should descend before the north and south gates;
Groaning with pity, he among the wailing souls laments.

And these the labours of the sons of Los in Allamanda,
And in the city of Golgonooza, and in Luban, and around   920
The Lake of Udan-Adan, in the forests of Entuthon Benython,
Where souls incessant wail, being piteous passions and desires
With neither lineament nor form, but like to watery clouds.
The passions and desires descend upon the hungry winds;
For such alone sleepers remain, mere passion and appetite.
The sons of Los clothe them and feed, and provide houses and
    fields.

And every generated body in its inward form
Is a garden of delight, and a building of magnificence
Built by the sons of Los in Bowlahoola and Allamanda—
And the herbs and flowers and furniture and beds and chambers   930
Continually woven in the looms of Enitharmon's daughters,
In bright Cathedron's golden dome, with care and love and tears.
For the various classes of men are all marked out determinate
In Bowlahoola, and as the Spectres choose their affinities
So they are born on earth, and every class is determinate;
But not by natural, but by spiritual power alone, because

The natural power continually seeks and tends to destruction
Ending in death, which would of itself be eternal death.
And all are classed by spiritual and not by natural power.

And every natural effect has a spiritual cause, and not   940
A natural. For a natural cause only seems; it is a delusion
Of Ulro, and a ratio of the perishing vegetable memory.

But the winepress of Los is eastward of Golgonooza, before the
    seat
Of Satan; Luvah laid the foundation, and Urizen finished it in
    howling woe.
How red the sons and daughters of Luvah! Here they tread the
    grapes;
Laughing and shouting, drunk with odours, many fall o'erwearied;
Drowned in the wine is many a youth and maiden. Those around
Lay them on skins of tigers and of the spotted leopard and the wild
    ass
Till they revive, or bury them in cool grots, making lamentation.

This winepress is called War on earth; it is the printing-press   950
Of Los, and here he lays his words in order above the mortal brain,
As cogs are formed in a wheel to turn the cogs of the adverse wheel.

Timbrels and violins sport round the winepresses; the little seed,
The sportive root, the earth-worm, the gold beetle, the wise
    emmet
Dance round the winepresses of Luvah; the centipede is there,
The ground-spider with many eyes, the mole clothed in velvet,
The ambitious spider in his sullen web, the lucky golden spinner,°
The earwig armed, the tender maggot, emblem of immortality,
The flea, louse, bug, the tape-worm, all the armies of disease,
Visible or invisible to the slothful vegetating man,   960
The slow slug, the grasshopper that sings and laughs and drinks
(Winter comes, he folds his slender bones without a murmur).
The cruel scorpion is there, the gnat, wasp, hornet and the honey
    bee,
The toad and venomous newt, the serpent clothed in gems and
    gold.
They throw off their gorgeous raiment; they rejoice with loud
    jubilee°
Around the winepresses of Luvah, naked and drunk with wine.

There is a nettle that stings with soft down, and there
The indignant thistle, whose bitterness is bred in his milk,
Who feeds on contempt of his neighbour; there all the idle weeds
That creep around the obscure places show their various limbs,   970
Naked in all their beauty dancing round the winepresses.

But in the winepresses the human grapes sing not nor dance.
They howl and writhe in shoals of torment, in fierce flames
    consuming,
In chains of iron and in dungeons circled with ceaseless fires,
In pits and dens and shades of death, in shapes of torment and woe:
The plates and screws and racks and saws and cords and fires and
    cisterns,
The cruel joys of Luvah's daughters lacerating with knives
And whips their victims, and the deadly sport of Luvah's sons.

They dance around the dying, and they drink the howl and groan;
They catch the shrieks in cups of gold; they hand them to one   980
    another.
These are the sports of love, and these the sweet delights of
    amorous play:
Tears of the grape, the death-sweat of the cluster, the last sigh
Of the mild youth who listens to the luring songs of Luvah.

But Allamanda (called on earth, Commerce) is the cultivated land
Around the city of Golgonooza in the forests of Entuthon.
Here the sons of Los labour against death eternal. Through all
The twenty-seven heavens of Beulah, in Ulro, seat of Satan
(Which is the false tongue beneath Beulah: it is the sense of touch),
The plough goes forth in tempests and lightnings, and the harrow
    cruel
In blights of the east; the heavy roller follows in howlings of woe.   990

Urizen's sons here labour also, and here are seen the mills
Of Theotormon, on the verge of the Lake of Udan-Adan.
These are the starry voids of night, and the depths and caverns of
    earth;
These mills are oceans, clouds and waters ungovernable in their
    fury.
Here are the stars created and the seeds of all things planted,
And here the sun and moon receive their fixed destinations.

But in Eternity the four arts—poetry, painting, music,
And architecture (which is science)—are the four faces of Man.
Not so in time and space; there three are shut out, and only
Science remains, through mercy. And by means of science the 1000
    three
Become apparent in time and space in the three professions:
Poetry in religion; Music, law; Painting in physic and surgery—
That Man may live upon earth till the time of his awakening.
And from these three science derives every occupation of men,
And science is divided into Bowlahoola and Allamanda.

Some sons of Los surround the passions with porches of iron and
    silver,
Creating form and beauty around the dark regions of sorrow,
Giving to airy nothing a name and a habitation°
Delightful, with bounds to the infinite putting off the indefinite
Into most holy forms of thought (such is the power of inspiration). 1010
They labour incessant, with many tears and afflictions,
Creating the beautiful house for the piteous sufferer.

Others cabinets richly fabricate of gold and ivory,
For doubts and fears unformed and wretched and melancholy.
The little weeping Spectre stands on the threshold of death
Eternal, and sometimes two Spectres like lamps quivering;
And often malignant they combat (heart-breaking, sorrowful and
    piteous).
Antamon takes them into his beautiful flexible hands,
As the sower takes the seed, or as the artist his clay
Or fine wax, to mould artful a model for golden ornaments.   1020
The soft hands of Antamon draw the indelible line,
Form immortal, with golden pen, such as the Spectre admiring
Puts on the sweet form; then smiles Antamon bright through his
    windows.
The daughters of beauty look up from their loom and prepare
The integument soft for its clothing with joy and delight.

But Theotormon and Sotha stand in the gate of Luban, anxious.
Their numbers are seven million and seven thousand and seven
    hundred.
They contend with the weak Spectres; they fabricate soothing
    forms.

The Spectre refuses; he seeks cruelty. They create the crested
    cock;
Terrified, the Spectre screams and rushes in fear into their net   1030
Of kindness and compassion, and is born a weeping terror.
Or they create the lion and tiger in compassionate thunderings.
Howling the Spectres flee; they take refuge in human lineaments.

The sons of Ozoth within the optic nerve stand fiery glowing;
And the number of his sons is eight million and eight.
They give delights to the man unknown; artificial riches
They give to scorn, and their possessors to trouble and sorrow and
    care—
Shutting the sun, and moon, and stars, and trees, and clouds, and
    waters,
And hills out from the optic nerve, and hardening it into a bone
Opaque, and like the black pebble on the enraged beach.   1040
While the poor indigent is like the diamond which, though clothed
In rugged covering in the mine, is open all within,
And in his hallowed centre holds the heavens of bright Eternity.
Ozoth here builds walls of rocks against the surging sea,
And timbers cramped with iron cramps bar in the joys of life
From fell destruction in the spectrous cunning or rage. He creates
The speckled newt, the spider and beetle, the rat and mouse,
The badger and fox; they worship before his feet in trembling fear.

But others of the sons of Los build moments, and minutes, and
    hours,
And days, and months, and years, and ages, and periods: wondrous   1050
    buildings.
And every moment has a couch of gold for soft repose
(A moment equals a pulsation of the artery);
And between every two moments stands a daughter of Beulah
To feed the sleepers on their couches with maternal care.
And every minute has an azure tent with silken veils;
And every hour has a bright golden gate carved with skill;
And every day and night has walls of brass and gates of adamant,
Shining like precious stones and ornamented with appropriate
    signs;
And every month a silver-paved terrace builded high;
And every year invulnerable barriers with high towers;   1060
And every age is moated deep with bridges of silver and gold;

And every seven ages is encircled with a flaming fire.
Now seven ages is amounting to two hundred years;
Each has its guard, each moment, minute, hour, day, month and
    year.
All are the work of fairy hands of the four elements;
The guard are angels of providence on duty evermore.
Every time less than a pulsation of the artery
Is equal in its period and value to six thousand years;
For in this period the poet's work is done, and all the great
Events of time start forth and are conceived in such a period,   1070
Within a moment, a pulsation of the artery.

The sky is an immortal tent built by the sons of Los,
And every space that a man views around his dwelling-place,
Standing on his own roof, or in his garden on a mount
Of twenty-five cubits in height—such space is his universe.°
And on its verge the sun rises and sets; the clouds bow
To meet the flat earth and the sea in such an ordered space;
The starry heavens reach no further, but here bend and set
On all sides; and the two poles turn on their valves of gold.
And if he move his dwelling-place, his heavens also move   1080
Where'er he goes, and all his neighbourhood bewail his loss.
Such are the spaces called Earth, and such its dimension.
As to the false appearance which appears to the reasoner
(As of a globe rolling through voidness) it is a delusion of Ulro.
The microscope knows not of this, nor the telescope; they alter
The ratio of the spectator's organs, but leave objects untouched.
For every space larger than a red globule of Man's blood
Is visionary and is created by the hammer of Los;
And every space smaller than a globule of Man's blood opens
Into Eternity, of which this vegetable earth is but a shadow.   1090
The red globule is the unwearied sun, by Los created
To measure time and space to mortal men every morning.
Bowlahoola and Allamanda are placed on each side
Of that pulsation and that globule; terrible their power.

But Rintrah and Palamabron govern over day and night
In Allamanda and Entuthon Benython; where souls wail;
Where Orc incessant howls, burning in fires of eternal youth
Within the vegetated mortal nerves. For every man born is joined
Within into one mighty polypus, and this polypus is Orc.

But in the optic vegetative nerves sleep was transformed   1100
To death in old time by Satan, the father of Sin and Death,
And Satan is the spectre of Orc, and Orc is the generate Luvah.

But in the nerves of the nostrils, accident being formed
Into substance and principle, by the cruelties of demonstration
It became opaque and indefinite; but the divine Saviour
Formed it into a solid by Los's mathematic power.
He named the opaque Satan; he named the solid Adam.

And in the nerves of the ear (for the nerves of the tongue are
    closed)
On Albion's rock Los stands, creating the glorious sun each
    morning,
And when, unwearied, in the evening he creates the moon   1110
(Death to delude, who all in terror at their splendour leaves
His prey while Los appoints), and Rintrah and Palamabron guide
The souls clear from the rock of Death, that Death himself may
    wake
In his appointed season when the ends of heaven meet.

Then Los conducts the spirits to be vegetated into
Great Golgonooza, free from the four iron pillars of Satan's throne
(Temperance, prudence, justice, fortitude, the four pillars of
    tyranny),
That Satan's watch-fiends touch them not before they vegetate.

But Enitharmon and her daughters take the pleasant charge,
To give them to their lovely heavens till the great judgement day. 1120
Such is their lovely charge, but Rahab and Tirzah pervert
Their mild influences. Therefore the seven eyes of God walk
    round
The three heavens of Ulro, where Tirzah and her sisters
Weave the black woof of death upon Entuthon Benython,
In the vale of Surrey where Horeb terminates in Rephaim.
The stamping feet of Zelophehad's daughters are covered with
    human gore°
Upon the treadles of the loom; they sing to the winged shuttle.
The river rises above its banks to wash the woof;
He takes it in his arms; he passes it in strength through his current.
The veil of human miseries is woven over the ocean   1130
From the Atlantic to the great South Sea, the Erythrean.°

Such is the world of Los, the labour of six thousand years.
Thus Nature is a vision of the science of the Elohim.

End of the First Book

BOOK THE SECOND

There is a place where contrarieties are equally true;
This place is called Beulah. It is a pleasant lovely shadow,
Where no dispute can come, because of those who sleep.
Into the place the sons and daughters of Ololon descended
With solemn mourning, into Beulah's moony shades and hills,
Weeping for Milton. Mute wonder held the daughters of Beulah,
Enraptured with affection sweet and mild benevolence.

Beulah is evermore created around Eternity, appearing
To the inhabitants of Eden around them on all sides.
But Beulah to its inhabitants appears, within each district,   10
As the beloved infant in his mother's bosom, round encircled
With arms of love and pity and sweet compassion. But to
The sons of Eden the moony habitations of Beulah
Are from great Eternity a mild and pleasant rest.

And it is thus created. Lo, the eternal great humanity—
To whom be glory and dominion evermore, Amen—
Walks among all his awful family, seen in every face
As the breath of the Almighty; such are the words of man to man
In the great wars of Eternity, in fury of poetic inspiration
To build the universe stupendous, mental forms creating.   20

But the Emanations trembled exceedingly, nor could they
Live, because the life of Man was too exceeding unbounded.
His joy became terrible to them; they trembled and wept,
Crying with one voice: 'Give us a habitation and a place
In which we may be hidden under the shadow of wings.
For if we, who are but for a time, and who pass away in winter,
Behold these wonders of Eternity, we shall consume;
But you, O our fathers and brothers, remain in Eternity.
But grant us a temporal habitation; do you speak
To us. We will obey your words, as you obey Jesus   30
The Eternal, who is blessed for ever and ever, Amen.'

So spake the lovely Emanations; and there appeared a pleasant
Mild shadow above, beneath, and on all sides round.
Into this pleasant shadow all the weak and weary,
Like women and children, were taken away as on wings
Of dovelike softness, and shadowy habitations prepared for them.
But every man returned and went, still going forward through
The bosom of the Father in Eternity on Eternity;
Neither did any lack or fall into error, without
A shadow to repose in, all the days of happy Eternity.   40

Into this pleasant shadow, Beulah, all Ololon descended.
And when the daughters of Beulah heard the lamentation
All Beulah wept, for they saw the Lord coming in the clouds,
And the shadows of Beulah terminate in rocky Albion.

And all nations wept in affliction, family by family.
Germany wept towards France and Italy; England wept and
    trembled
Towards America; India rose up from his golden bed,
As one awakened in the night. They saw the Lord coming°
In the clouds of Ololon with power and great glory.

And all the living creatures of the four elements wailed   50
With bitter wailing. These in the aggregate are named Satan
And Rahab; they know not of regeneration, but only of generation.
The fairies, nymphs, gnomes and genii of the four elements,
Unforgiving and unalterable—these cannot be regenerated
But must be created, for they know only of generation.
These are the gods of the kingdoms of the earth: in contrarious
And cruel opposition, element against element, opposed in war
Not mental, as the wars of Eternity, but a corporeal strife,
In Los's halls continual labouring, in the furnaces of Golgonooza.
Orc howls on the Atlantic; Enitharmon trembles; all Beulah   60
    weeps.

Thou hearest the nightingale begin the song of spring.
The lark sitting upon his earthy bed, just as the morn
Appears, listens silent; then, springing from the waving cornfield,
    loud
He leads the choir of day. Trill, trill, trill, trill:
Mounting upon the wings of light into the great expanse,

Re-echoing against the lovely blue and shining heavenly shell.
His little throat labours with inspiration; every feather
On throat and breast and wings vibrates with the effluence divine.
All nature listens silent to him, and the awful sun
Stands still upon the mountain looking at this little bird   70
With eyes of soft humility and wonder, love and awe.
Then loud from their green covert all the birds begin their song;
The thrush, the linnet and the goldfinch, robin and the wren,
Awake the sun from his sweet reverie upon the mountain;
The nightingale again essays his song, and through the day
And through the night warbles luxuriant, every bird of song
Attending his loud harmony with admiration and love.°
This is a vision of the lamentation of Beulah over Ololon.

Thou perceivest the flowers put forth their precious odours,
And none can tell how from so small a centre comes such sweets: 80
Forgetting that within that centre Eternity expands
Its ever-during doors, that Og and Anak fiercely guard.°
First, ere the morning breaks, joy opens in the flowery bosoms,
Joy even to tears, which the sun rising dries. First the wild thyme,
And meadowsweet downy and soft, waving among the reeds,
Light springing on the air, lead the sweet dance. They wake
The honeysuckle sleeping on the oak. The flaunting beauty
Revels along upon the wind. The white-thorn, lovely may,
Opens her many lovely eyes. Listening, the rose still sleeps;
None dare to wake her. Soon she burst her crimson-curtained bed,   90
And comes forth in the majesty of beauty. Every flower—
The pink, the jessamine, the wall-flower, the carnation,
The jonquil, the mild lily—opes her heavens. Every tree
And flower and herb soon fill the air with an innumerable dance,
Yet all in order sweet and lovely. Men are sick with love.
Such is a vision of the lamentation of Beulah over Ololon.

And the divine voice was heard in the songs of Beulah, saying:
'When I first married you I gave you all my whole soul.
I thought that you would love my loves and joy in my delights,
Seeking for pleasures in my pleasures, O daughter of Babylon.°   100
Then thou wast lovely, mild and gentle; now thou art terrible
In jealousy, and unlovely in my sight, because thou hast cruelly
Cut off my loves in fury, till I have no love left for thee.
Thy love depends on him thou lovest, and on his dear loves

Depend thy pleasures, which thou hast cut off by jealousy.
Therefore I show my jealousy, and set before you death.
Behold Milton descended to redeem the female shade
From death eternal. Such your lot: to be continually redeemed
By death and misery of those you love, and by annihilation.
When the sixfold female perceives that Milton annihilates   110
Himself, that seeing all his loves by her cut off he leaves
Her also, entirely abstracting himself from female loves,
She shall relent in fear of death; she shall begin to give
Her maidens to her husband, delighting in his delight.
And then, and then alone, begins the happy female joy,
As it is done in Beulah, and thou, O virgin Babylon, mother of
      whoredoms,
Shalt bring Jerusalem in thine arms in the night watches and,
No longer turning her a wandering harlot in the streets,
Shalt give her into the arms of God, your lord and husband.'
Such are the songs of Beulah in the lamentations of Ololon.   120

And all the songs of Beulah sounded comfortable notes
To comfort Ololon's lamentation, for they said:
'Are you the fiery circle that late drove, in fury and fire,
The eight immortal starry ones down into Ulro dark,
  Rending the heavens of Beulah with your thunders and lightnings?
And can you thus lament, and can you pity and forgive?
Is terror changed to pity? Oh wonder of Eternity!'

And the four states of humanity in its repose
Were showed them. First of Beulah, a most pleasant sleep
On couches soft, with mild music, tended by flowers of Beulah,   130
Sweet female forms winged or floating in the air spontaneous.
The second state is Alla, and the third state Al-Ulro;
But the fourth state is dreadful; it is named Or-Ulro.
The first state is in the head, the second is in the heart,
The third in the loins and seminal vessels, and the fourth
In the stomach and intestines, terrible, deadly, unutterable.
And he whose gates are opened in those regions of his body
Can from those gates view all these wondrous imaginations.

But Ololon sought the Or-Ulro and its fiery gates,
And the couches of the martyrs; and many daughters of Beulah   140

Accompany them down to the Ulro with soft melodious tears—
A long journey and dark through chaos, in the track of Milton's
    course,
To where the contraries of Beulah war beneath negation's banner.

Then viewed from Milton's track they see the Ulro: a vast polypus
Of living fibres down into the sea of time and space growing,
A self-devouring monstrous human death, twenty-seven fold.
Within it sit five females and the nameless shadowy mother,
Spinning it from their bowels with songs of amorous delight
And melting cadences, that lure the sleepers of Beulah down
The River Storgé (which is Arnon) into the Dead Sea.°   150

Around this polypus Los continual builds the Mundane Shell.

Four universes round the universe of Los remain chaotic,
Four intersecting globes—and the egg-formed world of Los
In midst, stretching from zenith to nadir in midst of chaos.
One of these ruined universes is to the north, named Urthona;
One to the south—this was the glorious world of Urizen;
One to the east, of Luvah: one to the west, of Tharmas.
But when Luvah assumed the world of Urizen in the south,
All fell towards the centre, sinking downward in dire ruin.

Here in these chaoses the sons of Ololon took their abode,   160
In chasms of the Mundane Shell which open on all sides round,
Southward, and by the east within the breach of Milton's
    descent—
To watch the time, pitying and gentle, to awaken Urizen.
They stood in a dark land of death, of fiery corroding waters,
Where lie in evil death the four Immortals, pale and cold,
And the eternal man, even Albion, upon the Rock of Ages.
Seeing Milton's shadow, some daughters of Beulah trembling
Returned, but Ololon remained before the gates of the dead.

And Ololon looked down into the heavens of Ulro in fear.
They said: 'How are the wars of Man, which in great Eternity   170
Appear around, in the external spheres of visionary life,
Here rendered deadly, within the life and interior vision!
How are the beasts and fishes and plants and minerals
Here fixed into a frozen bulk, subject to decay and death!
Those visions of human life, and shadows of wisdom and
    knowledge,

Are here frozen to unexpansive, deadly, destroying terrors.
And war and hunting, the two fountains of the River of Life,°
Are become fountains of bitter death and of corroding hell;
Till brotherhood is changed into a curse and a flattery
By differences between ideas, that ideas themselves (which are   180
The divine members) may be slain in offerings for sin.
Oh dreadful loom of death! Oh piteous female forms, compelled
To weave the woof of death! On Camberwell Tirzah's courts,
Malah's on Blackheath, Rahab and Noah dwell on Windsor's
    heights,
Where once the cherubs of Jerusalem spread to Lambeth's vale.
Milcah's pillars shine from Harrow to Hampstead, where Hoglah
On Highgate's heights magnificent weaves, over trembling
    Thames,
To Shooter's Hill and thence to Blackheath, the dark woof! Loud,
Loud roll the weights and spindles over the whole earth, let down
On all sides round to the four quarters of the world, eastward on   190
Europe to Euphrates and Hindu, to Nile and back, in clouds
Of death across the Atlantic to America North and South.'

So spake Ololon in reminiscence, astonished; but they
Could not behold Golgonooza without passing the polypus
(A wondrous journey not passable by immortal feet, and none
But the divine Savour can pass it without annihilation).
For Golgonooza cannot be seen till, having passed the polypus,
It is viewed on all sides round by a fourfold vision,
Or till you become mortal and vegetable in sexuality;
Then you behold its mighty spires, and domes of ivory and gold.   200

And Ololon examined all the couches of the dead,
Even of Los and Enitharmon, and all the sons of Albion,
And his four Zoas terrified and on the verge of death.
In midst of these was Milton's couch, and when they saw eight
Immortal starry ones guarding the couch in flaming fires,
They thunderous uttered all a universal groan, falling down
Prostrate before the starry eight, asking with tears forgiveness,
Confessing their crime with humiliation and sorrow.

Oh how the starry eight rejoiced to see Ololon descended!
And now that a wide road was open to Eternity,°   210
By Ololon's descent through Beulah to Los and Enitharmon.
For mighty were the multitudes of Ololon, vast the extent

Of their great sway, reaching from Ulro to Eternity,
Surrounding the Mundane Shell outside in its caverns
And through Beulah; and all silent forbore to contend
With Ololon, for they saw the Lord in the clouds of Ololon.

There is a moment in each day that Satan cannot find,
Nor can his watch-fiends find it. But the industrious find
This moment and it multiply, and when it once is found
It renovates every moment of the day if rightly placed.   220
In this moment Ololon descended to Los and Enitharmon,
Unseen beyond the Mundane Shell, southward in Milton's track.

Just in this moment, when the morning odours rise abroad,
And first from the wild thyme, stands a fountain in a rock°
Of crystal flowing into two streams. One flows through Golgo-
    nooza
And through Beulah to Eden, beneath Los's western wall;
The other flows through the aerial void and all the churches,
Meeting again in Golgonooza beyond Satan's seat.

The wild thyme is Los's messenger to Eden, a mighty demon.
Terrible, deadly and poisonous his presence in Ulro dark;   230
Therefore he appears only a small root creeping in grass,
Covering over the rock of odours his bright purple mantle
Beside the fount, above the lark's nest, in Golgonooza.
Luvah slept here in death, and here is Luvah's empty tomb.
Ololon sat beside this fountain on the rock of odours.

Just at the place to where the lark mounts is a crystal gate;
It is the entrance of the first heaven, named Luther. For
The lark is Los's messenger through the twenty-seven churches,
That the seven eyes of God, who walk even to Satan's seat
Through all the twenty-seven heavens, may not slumber nor
    sleep.°   240
But the lark's nest is at the gate of Los, at the eastern
Gate of wide Golgonooza, and the lark is Los's messenger.
When on the highest lift of his light pinions he arrives
At that bright gate, another lark meets him, and back to back
They touch their pinions, tip tip, and each descend
To their respective earths, and there all night consult with angels
Of providence, and with the eyes of God all night in slumbers
Inspired; and at the dawn of day send out another lark

Into another heaven to carry news upon his wings.
Thus are the messengers dispatched till they reach the earth again   250
In the east gate of Golgonooza, and the twenty-eighth bright
Lark met the female Ololon descending into my garden.
(Thus it appears to mortal eyes and those of the Ulro heavens,
But not thus to Immortals; the lark is a mighty angel.)

For Ololon stepped into the polypus within the Mundane Shell.
They could not step into vegetable worlds without becoming
The enemies of humanity, except in a female form,
And as one female; Ololon and all its mighty hosts
Appeared, a virgin of twelve years. Nor time nor space was
To the perception of the virgin Ololon, but as the   260
Flash of lightning, but more quick; the virgin in my garden
Before my cottage stood (for the Satanic space is delusion).

For when Los joined with me he took me in his fiery whirlwind;
My vegetated portion was hurried from Lambeth's shades.
He set me down in Felpham's vale and prepared a beautiful
Cottage for me, that in three years I might write all these visions
To display Nature's cruel holiness, the deceits of natural religion.
Walking in my cottage garden sudden I beheld
The virgin Ololon, and addressed her as a daughter of Beulah:

'Virgin of providence, fear not to enter into my cottage.   270
What is thy message to thy friend? What am I now to do?
Is it again to plunge into deeper affliction? Behold me
Ready to obey, but pity thou my shadow of delight.
Enter my cottage, comfort her, for she is sick with fatigue.'

The virgin answered: 'Knowest thou of Milton, who descended
Driven from Eternity? Him I seek! Terrified at my act
In great Eternity, which thou knowest, I come him to seek.'

So Ololon uttered in words distinct the anxious thought.
Mild was the voice, but more distinct than any earthly—
That Milton's shadow heard, and, condensing all his fibres   280
Into a strength impregnable of majesty and beauty infinite,
I saw he was the covering cherub, and within him Satan
And Rahab, in an outside which is fallacious (within,
Beyond the outline of identity, in the selfhood deadly).
And he appeared the wicker man of Scandinavia, in whom°
Jerusalem's children consume in flames among the stars.

Descending down into my garden, a human wonder of God
Reaching from heaven to earth, a cloud and human form,
I beheld Milton with astonishment, and in him beheld°
The monstrous churches of Beulah, the gods of Ulro dark.   290
Twelve monstrous dishumanized terrors, synagogues of Satan,
A double twelve and thrice nine: such their divisions.

And these their names and their places within the Mundane Shell.

In Tyre and Sidon I saw Baal and Ashtaroth; in Moab, Chemosh;°
In Ammon, Molech. (Loud his furnaces rage among the wheels
Of Og, and pealing loud the cries of the victims of fire,
And pale his priestesses enfold in veils of pestilence, bordered
With war, woven in looms of Tyre and Sidon by beautiful
     Ashtaroth.)
In Palestine Dagon, sea-monster worshipped o'er the sea;°
Thammuz in Lebanon, and Rimmon in Damascus curtained;°   300
Osiris, Isis, Orus in Egypt. (Dark their tabernacles on Nile,°
Floating with solemn songs, and on the lakes of Egypt nightly
With pomp, even till morning break, and Osiris appear in the sky.)
But Belial of Sodom and Gomorrha, obscure demon of bribes°
And secret assassinations, not worshipped nor adored, but
With the finger on the lips and the back turned to the light;
And Saturn, Jove and Rhea of the isles of the sea remote.°
These twelve gods are the twelve spectre sons of the Druid Albion.

And these the names of the twenty-seven heavens and their
     churches:
Adam, Seth, Enos, Cainan, Mahalaleel, Jared, Enoch,°   310
Methuselah, Lamech (these are giants mighty, hermaphroditic),°
Noah, Shem, Arphaxad, Cainan the second, Salah, Heber,
Peleg, Reu, Serug, Nahor, Terah (these are the female-males,
A male within a female hid as in an ark and curtains),
Abraham, Moses, Solomon, Paul, Constantine, Charlemagne,
Luther (these seven are the male-females, the dragon forms,
Religion hid in war, a dragon red and hidden harlot).°

All these are seen in Milton's shadow, who is the covering cherub,
The spectre of Albion, in which the spectre of Luvah inhabits,
In the Newtonian voids between the substances of creation.°   320
For the chaotic voids outside of the stars are measured by

The stars, which are the boundaries of kingdoms, provinces
And empires of chaos, invisible to the vegetable man.
The kingdom of Og is in Orion: Sihon is in Ophiuchus;
Og has twenty-seven districts; Sihon's districts twenty-one.
From star to star, mountains and valleys, terrible dimension
Stretched out, compose the Mundane Shell, a mighty incrustation
Of forty-eight deformed human wonders of the Almighty,
With caverns whose remotest bottoms meet again beyond
The Mundane Shell in Golgonooza. But the fires of Los rage   330
In the remotest bottoms of the caves, that none can pass
Into Eternity that way, but all descend to Los,
To Bowlahoola and Allamanda, and to Entuthon Benython.

The heavens are the cherub, the twelve gods are Satan.

And the forty-eight starry regions are cities of the Levites,°
The heads of the great polypus, fourfold twelve enormity
In mighty and mysterious commingling, enemy with enemy,
Woven by Urizen into sexes from his mantle of years.
And Milton, collecting all his fibres into impregnable strength,
Descended down a paved work of all kinds of precious stones°   340
Out from the eastern sky, descending down into my cottage
Garden, clothed in black; severe and silent he descended.

The Spectre of Satan stood upon the roaring sea and beheld
Milton within his sleeping humanity. Trembling and shudd'ring
He stood upon the waves, a twenty-sevenfold mighty demon,
Gorgeous and beautiful. Loud roll his thunders against Milton;
Loud Satan thundered, loud and dark upon mild Felpham shore.
Not daring to touch one fibre he howled round upon the sea.

I also stood in Satan's bosom and beheld its desolations,
A ruined man, a ruined building of God not made with hands:°   350
Its plains of burning sand, its mountains of marble terrible,
Its pits and declivities flowing with molten ore, and fountains
Of pitch and nitre, its ruined palaces and cities and mighty works,
Its furnaces of affliction, in which his angels and Emanations
Labour with blackened visages among its stupendous ruins,
Arches and pyramids and porches, colonnades and domes,
In which dwells Mystery, Babylon. Here is her secret place;
From hence she comes forth on the churches in delight;

Here is her cup filled with its poisons, in these horrid vales,
And here her scarlet veil woven in pestilence and war.   360
Here is Jerusalem bound in chains, in the dens of Babylon.

In the eastern porch of Satan's universe Milton stood and said:

'Satan, my Spectre, I know my power thee to annihilate
And be a greater in thy place, and be thy tabernacle,
A covering for thee to do thy will: till one greater comes°
And smites me as I smote thee, and becomes my covering.
Such are the laws of thy false heavens, but laws of Eternity
Are not such. Know thou, I come to self-annihilation.
Such are the laws of Eternity: that each shall mutually
Annihilate himself for other's good, as I for thee.   370
Thy purpose and the purpose of thy priests and of thy churches
Is to impress on men the fear of death; to teach
Trembling and fear, terror, constriction, abject selfishness.
Mine is to teach men to despise death, and to go on
In fearless majesty annihilating self, laughing to scorn
Thy laws and terrors, shaking down thy synagogues as webs.
I come to discover before Heaven and Hell the self-righteousness
In all its hypocritic turpitude, opening to every eye
These wonders of Satan's holiness, showing to the earth
The idol-virtues of the natural heart, and Satan's seat   380
Explore, in all its selfish natural virtue, and put off
In self-annihilation all that is not of God alone—
To put off self and all I have, ever and ever. Amen.'

Satan heard, coming in a cloud with trumpets and flaming fire,
Saying: 'I am God, the judge of all, the living and the dead.
Fall therefore down and worship me; submit thy supreme
Dictate to my eternal will, and to my dictate bow.
I hold the balances of right and just, and mine the sword.
Seven angels bear my name and in those seven I appear;°
But I alone am God, and I alone in Heaven and Earth,   390
Of all that live, dare utter this. Others tremble and bow
Till all things become one great Satan, in holiness
Opposed to mercy, and the divine delusion, Jesus, be no more.'

Suddenly around Milton on my path the starry seven
Burned terrible. My path became a solid fire, as bright

As the clear sun, and Milton silent came down on my path.
And there went forth from the starry limbs of the seven forms
Human, with trumpets innumerable, sounding articulate
As the seven spake; and they stood in a mighty column of fire,
Surrounding Felpham's vale, reaching to the Mundane Shell,   400
  saying:

'Awake, Albion, awake! Reclaim thy reasoning Spectre. Subdue
Him to the divine mercy; cast him down into the lake°
Of Los, that ever burneth with fire, ever and ever. Amen!
Let the four Zoas awake from slumbers of six thousand years.'

Then loud the furnaces of Los were heard, and seen as seven
  heavens
Stretching from south to north over the mountains of Albion.

Satan heard; trembling round his body, he encircled it.
He trembled with exceeding great trembling and astonishment,
Howling in his Spectre round his body, hungering to devour,
But fearing for the pain; for if he touches a vital,   410
His torment is unendurable. Therefore he cannot devour,
But howls round it as a lion round his prey continually.
Loud Satan thundered, loud and dark upon mild Felpham's shore,
Coming in a cloud with trumpets and with fiery flame,
An awful form eastward, from midst of a bright paved-work
Of precious stones by cherubim surrounded (so permitted—
Lest he should fall apart in his eternal death—to imitate
The eternal great Humanity Divine, surrounded by
His cherubim and seraphim in ever-happy Eternity).
Beneath sat Chaos, Sin on his right hand, Death on his left;°   420
And Ancient Night spread over all the heaven his mantle of laws.
He trembled with exceeding great trembling and astonishment.

Then Albion rose up in the night of Beulah on his couch
Of dread repose, seen by the visionary eye; his face is toward
The east, toward Jerusalem's gates; groaning he sat above
His rocks. London and Bath and Legions and Edinburgh°
Are the four pillars of his throne. His left foot near London
Covers the shades of Tyburn; his instep from Windsor
To Primrose Hill, stretching to Highgate and Holloway.
London is between his knees, its basements fourfold.°   430

His right foot stretches to the sea on Dover cliffs, his heel
On Canterbury's ruins. His right hand covers lofty Wales,
His left Scotland. His bosom girt with gold involves
York, Edinburgh, Durham and Carlisle, and on the front
Bath, Oxford, Cambridge, Norwich. His right elbow
Leans on the rocks of Erin's land, Ireland, ancient nation.
His head bends over London. He sees his embodied Spectre
Trembling before him, with exceeding great trembling and fear.
He views Jerusalem and Babylon; his tears flow down.
He moved his right foot to Cornwall, his left to the rocks of
    Bognor.   440
He strove to rise to walk into the deep, but strength failing
Forbade, and down with dreadful groans he sunk upon his couch
In moony Beulah. Los, his strong guard, walks round beneath the
    moon.

Urizen faints in terror striving among the brooks of Arnon
With Milton's spirit. As the ploughman or artificer or shepherd,
While in the labours of his calling, sends his thought abroad
To labour in the ocean or in the starry heaven, so Milton
Laboured in chasms of the Mundane Shell—though here before
My cottage 'midst the starry seven, where the virgin Ololon
Stood trembling in the porch. Loud Satan thundered on the   450
    stormy sea,
Circling Albion's cliffs in which the fourfold world resides,
Though seen in fallacy outside, a fallacy of Satan's churches.
Before Ololon Milton stood, and perceived the eternal form
Of that mild vision. Wondrous were their acts, by me unknown
Except remotely; and I heard Ololon say to Milton:

'I see thee strive upon the brooks of Arnon. There a dread
And awful man I see, o'ercovered with the mantle of years.
I behold Los and Urizen, I behold Orc and Tharmas,
The four Zoas of Albion, and thy spirit with them striving,
In self-annihilation giving thy life to thy enemies.   460
Are those who contemn religion and seek to annihilate it
Become in their feminine portions the causes and promoters
Of these religions? How is that thing, this Newtonian phantasm,
This Voltaire and Rousseau, this Hume and Gibbon and
    Bolingbroke,°
This natural religion, this impossible absurdity?

Is Ololon the cause of this? Oh, where shall I hide my face?
These tears fall for the little ones, the children of Jerusalem,
Lest they be annihilated in any annihilation.'

No sooner she had spoke but Rahab Babylon appeared
Eastward upon the paved-work across Europe and Asia,   470
Glorious as the midday sun, in Satan's bosom glowing:
A female hidden in a male, religion hidden in war,
Named 'Moral Virtue', cruel twofold monster shining bright,
A dragon red and hidden harlot which John in Patmos saw.°

And all beneath the nations innumerable of Ulro
Appeared: the seven kingdoms of Canaan and five Baalim°
Of Philistia, into twelve divided, called after the names
Of Israel, as they are in Eden mountain, river and plain,
City and sandy desert intermingled beyond mortal ken.
But turning toward Ololon in terrible majesty, Milton   480
Replied: 'Obey thou the words of the inspired man.
All that can be annihilated must be annihilated,
That the children of Jerusalem may be saved from slavery.
There is a negation, and there is a contrary.
The negation must be destroyed to redeem the contraries.
The negation is the Spectre, the reasoning power in man.
This is a false body, an incrustation over my immortal
Spirit, a selfhood which must be put off and annihilated alway.

'To cleanse the face of my spirit by self-examination,
To bathe in the waters of life, to wash off the not-human,   490
I come in self-annihilation and the grandeur of inspiration.
To cast off rational demonstration by faith in the Saviour;
To cast off the rotten rags of memory by inspiration;
To cast off Bacon, Locke and Newton from Albion's covering;
To take off his filthy garments, and clothe him with imagination;
To cast aside from poetry all that is not inspiration—
That it no longer shall dare to mock with the aspersion of madness,
Cast on the inspired by the tame high-finisher of paltry blots,
Indefinite or paltry rhymes, or paltry harmonies,
Who creeps into state government like a caterpillar to destroy.   500
To cast off the idiot questioner; who is always questioning
But never capable of answering; who sits with a sly grin
Silent plotting when to question, like a thief in a cave;

Who publishes doubt and calls it knowledge; whose science is
   despair,
Whose pretence to knowledge is envy; whose whole science is
To destroy the wisdom of ages to gratify ravenous envy,
That rages round him like a wolf day and night without rest.
He smiles with condescension; he talks of benevolence and virtue;
And those who act with benevolence and virtue, they murder time
   on time.
These are the destroyers of Jerusalem; these are the murderers   510
Of Jesus, who deny the faith and mock at eternal life,
Who pretend to poetry, that they may destroy imagination
By imitation of Nature's images drawn from remembrance.
These are the sexual garments, the abomination of desolation,
Hiding the human lineaments as with an ark and curtains:°
Which Jesus rent and now shall wholly purge away with fire,
Till generation is swallowed up in regeneration.'

Then trembled the virgin Ololon and replied in clouds of despair:

'Is this our feminine portion, the sixfold Miltonic female?
Terribly this portion trembles before thee, O awful man.   520
Although our human power can sustain the severe contentions
Of friendship, our sexual cannot, but flies into the Ulro.
Hence arose all our terrors in Eternity! And now remembrance
Returns upon us! Are we contraries, O Milton, thou and I?
O Immortal! how were we led to war the wars of death?
Is this the void outside of existence, which if entered into
Becomes a womb, and is this the death-couch of Albion?
Thou goest to eternal death, and all must go with thee!'

So saying the virgin divided sixfold, and with a shriek ·
Dolorous that ran through all creation, a double sixfold wonder,   530
Away from Ololon she divided and fled into the depths
Of Milton's shadow, as a dove upon the stormy sea.

Then as a moony ark Ololon descended to Felpham's vale
In clouds of blood, in streams of gore, with dreadful thunderings,
Into the fires of intellect that rejoiced in Felpham's vale
Around the starry eight. With one accord the starry eight became
One man, Jesus the Saviour, wonderful! Round his limbs
The clouds of Ololon folded as a garment dipped in blood,°

Written within and without in woven letters; and the writing°
Is the divine revelation in the literal expression,°
A garment of war. I heard it named the woof of six thousand years.

And I beheld the twenty-four cities of Albion
Arise upon their thrones to judge the nations of the earth;
And the immortal four in whom the twenty-four appear fourfold
Arose around Albion's body. Jesus wept and walked forth°
From Felpham's vale, clothed in clouds of blood, to enter into
Albion's bosom, the bosom of death, and the four surrounded him
In the column of fire in Felpham's vale. Then to their mouths the
    four
Applied their four trumpets, and them sounded to the four winds.

Terror struck in the vale. I stood at that immortal sound;   550
My bones trembled. I fell outstretched upon the path
A moment, and my soul returned into its mortal state,
To resurrection and judgement in the vegetable body.
And my sweet shadow of delight stood trembling by my side.

Immediately the lark mounted with a loud trill from Felpham's
    vale,
And the wild thyme from Wimbledon's green and empurpled
    hills;
And Los and Enitharmon rose over the hills of Surrey.
Their clouds roll over London with a south wind; soft Oothoon
Pants in the vales of Lambeth, weeping o'er her human harvest.
Los listens to the cry of the poor man, his cloud   560
Over London in volume terrific, low bended in anger.

Rintrah and Palamabron view the human harvest beneath.
Their winepresses and barns stand open; the ovens are prepared,
The waggons ready; terrific, lions and tigers sport and play;
All animals upon the earth are prepared in all their strength
To go forth to the great harvest and vintage of the nations.

                          Finis

Jerusalem
The Emanation of the Giant Albion (*1804–18*)

[*see above, p. 72*]

Μόνℴς δ ᾽Ιησοῦς°

CHAPTER I

TO THE PUBLIC

Of the sleep of Ulro! and of the passage through°
Eternal death! and of the awaking to eternal life!

This theme calls me in sleep night after night, and ev'ry morn
Awakes me at sunrise; then I see the Saviour over me,
Spreading his beams of love and dictating the words of this mild
    song.

'Awake! Awake O sleeper of the land of shadows; wake! expand!
I am in you and you in me, mutual in love divine:°
Fibres of love from man to man through Albion's pleasant land.
In all the dark Atlantic vale, down from the hills of Surrey,
A black water accumulates. Return, Albion! Return!   10
Thy brethren call thee; and thy fathers and thy sons,
Thy nurses and thy mothers, thy sisters and thy daughters
Weep at thy soul's disease, and the Divine Vision is darkened.
Thy Emanation that was wont to play before thy face,
Beaming forth with her daughters into the divine bosom—
Where hast thou hidden thy Emanation, lovely Jerusalem,
From the vision and fruition of the Holy One?
I am not a God afar off; I am a brother and friend.°
Within your bosoms I reside, and you reside in me.
Lo! we are one, forgiving all evil, not seeking recompense!   20
Ye are my members, O ye sleepers of Beulah, land of shades!'

But the perturbed man away turns down the valleys dark.

'Phantom of the overheated brain! Shadow of immortality,
Seeking to keep my soul a victim to thy love! which binds
Man, the enemy of Man, into deceitful friendships.
Jerusalem is not! Her daughters are indefinite.
By demonstration Man alone can live, and not by faith.

My mountains are my own, and I will keep them to myself;°
The Malvern and the Cheviot, the Wolds, Plinlimmon and
    Snowdon
Are mine. Here will I build my laws of moral virtue.   30
Humanity shall be no more, but war and princedom and victory!'

So spoke Albion in jealous fears, hiding his Emanation
Upon the Thames and Medway, rivers of Beulah, dissembling
His jealousy before the throne divine, darkening, cold!
The banks of the Thames are clouded! The ancient porches of
    Albion are
Darkened! They are drawn through unbounded space, scattered
    upon
The void in incoherent despair! Cambridge and Oxford and
    London
Are driven among the starry wheels, rent away and dissipated,
In chasms and abysses of sorrow, enlarged without dimension,
    terrible.
Albion's mountains run with blood; the cries of war and of tumult   40
Resound into the unbounded night. Every human perfection
Of mountain and river and city are small and withered and
    darkened.
Cam is a little stream! Ely is almost swallowed up!
Lincoln and Norwich stand trembling on the brink of Udan-Adan!
Wales and Scotland shrink themselves to the west and to the north!
Mourning for fear of the warriors in the vale of Entuthon
    Benython,
Jerusalem is scattered abroad like a cloud of smoke through non-
    entity.
Moab and Ammon and Amalek and Canaan and Egypt and Aram
Receive her little ones for sacrifices and the delights of cruelty.

Trembling I sit day and night. My friends are astonished at me,   50
Yet they forgive my wanderings. I rest not from my great task!
To open the eternal worlds, to open the immortal eyes
Of Man inwards into the worlds of thought, into Eternity
Ever expanding in the bosom of God, the human imagination.
O Saviour, pour upon me thy spirit of meekness and love;
Annihilate the selfhood in me; be thou all my life!
Guide thou my hand which trembles exceedingly upon the Rock of
    Ages,

While I write of the building of Golgonooza and of the terrors of
    Entuthon;
Of Hand and Hyle and Coban; of Gwantok, Peachey, Brereton,
    Slayd and Huttn;°
Of the terrible sons and daughters of Albion and their generations.   60

Skofield, Kox, Kotope and Bowen revolve most mightily upon°
The furnace of Los, before the eastern gate bending their fury.
They war to destroy the furnaces, to desolate Golgonooza,
And to devour the sleeping humanity of Albion in rage and
    hunger.
They revolve into the furnaces southward and are driven forth
    northward,
Divided into male and female forms time after time.
From these twelve all the families of England spread abroad.°

The male is a furnace of beryl; the female is a golden loom.°
I behold them and their rushing fires overwhelm my soul,
In London's darkness; and my tears fall day and night   70
Upon the Emanations of Albion's sons, the daughters of Albion!
Names anciently remembered but now contemned as fictions,
Although in every bosom they control our vegetative powers.

These are united into Tirzah and her sisters on Mount Gilead:°
Cambel and Gwendolen and Conwenna and Cordella and
    Ignoge.°
And these united into Rahab in the covering cherub on Euph-
    rates:°
Gwineverra and Gwinefred and Gonorill and Sabrina beautiful,°
Estrild, Mehetabel and Ragan, lovely daughters of Albion.
They are the beautiful Emanations of the twelve sons of Albion.

The starry wheels revolved heavily over the furnaces,   80
Drawing Jerusalem in anguish of maternal love
Eastward: a pillar of a cloud, with Vala upon the mountains
Howling in pain, redounding from the arms of Belulah's daughters
Out from the furnaces of Los above the head of Los;
A pillar of smoke writhing afar from non-entity, redounding
Till the cloud reaches afar, outstretched among the starry wheels
Which revolve heavily in the mighty void above the furnaces.

Oh what avail the loves and tears of Beulah's lovely daughters?
They hold the immortal form in gentle bands and tender tears,
But all within is opened into the deeps of Entuthon Benython,   90
A dark and unknown night, indefinite, unmeasurable, without end,
Abstract philosophy warring in enmity against imagination
(Which is the divine body of the Lord Jesus, blessed for ever),
And there Jerusalem wanders with Vala upon the mountains.
Attracted by the revolutions of those wheels, the cloud of smoke
Immense, and Jerusalem and Vala weeping in the cloud,
Wander away into the chaotic void, lamenting with her shadow
Among the daughters of Albion, among the starry wheels,
Lamenting for her children, for the sons and daughters of Albion.

Los heard her lamentations in the deeps afar! His tears fall   100
Incessant before the furnaces, and his Emanation divided in pain,
Eastward toward the starry wheels. But westward a black horror:
His Spectre driven by the starry wheels of Albion's sons, black and
Opaque, divided from his back, he labours and he mourns!

For as his Emanation divided, his Spectre also divided
In terror of those starry wheels. And the Spectre stood over Los
Howling in pain, a black'ning shadow, black'ning dark and opaque,
Cursing the terrible Los, bitterly cursing him for his friendship
To Albion, suggesting murderous thoughts against Albion.

Los raged and stamped the earth in his might and terrible wrath!   110
He stood and stamped the earth! Then he threw down his hammer
    in rage and
In fury! Then he sat down and wept, terrified! Then arose
And chanted his song, labouring with the tongs and hammer;
But still the Spectre divided, and still his pain increased.

In pain the Spectre divided! in pain of hunger and thirst,
To devour Los's human perfection. But when he saw that Los
Was living, panting like a frighted wolf and howling
He stood over the Immortal in the solitude and darkness,
Upon the dark'ning Thames, across the whole island westward,
A horrible shadow of death among the furnaces, beneath   120
The pillar of folding smoke. And he sought by other means
To lure Los: by tears, by arguments of science and by terrors,
Terrors in every nerve, by spasms and extended pains,
While Los answered unterrified to the opaque blackening fiend.

And thus the Spectre spoke: 'Wilt thou still go on to destruction,
Till thy life is all taken away by this deceitful friendship?
He drinks thee up like water! Like wine he pours thee
Into his tuns. Thy daughters are trodden in his vintage.
He makes thy sons the trampling of his bulls; they are ploughed
And harrowed for his profit. Lo! thy stolen Emanation   130
Is his garden of pleasure! All the Spectres of his sons mock thee.
Look how they scorn thy once admired palaces, now in ruins
Because of Albion, because of deceit and friendship! For lo!
Hand has peopled Babel and Nineveh; Hyle, Asshur and Aram;°
Coban's son is Nimrod; his son Cush is adjoined to Aram
By the daughter of Babel in a woven mantle of pestilence and war.
They put forth their spectrous cloudy sails, which drive their
    immense
Constellations over the deadly deeps of indefinite Udan-Adan.
Kox is the father of Shem and Ham and Japheth; he is the Noah
Of the flood of Udan-Adan. Huttn is the father of the seven   140
From Enoch to Adam. Skofield is Adam who was new
Created in Edom. I saw it indignant, and thou art not moved!
This has divided thee in sunder, and wilt thou still forgive?
Oh! thou seest not what I see: what is done in the furnaces.
Listen, I will tell thee what is done in moments to thee unknown.
Luvah was cast into the furnaces of affliction and sealed,°
And Vala fed in cruel delight the furnaces with fire.
Stern Urizen beheld, urged by necessity to keep
The evil day afar, and if perchance with iron power
He might avert his own despair. In woe and fear he saw   150
Vala encircle round the furnaces where Luvah was closed.
With joy she heard his howlings, and forgot he was her Luvah,
With whom she lived in bliss in times of innocence and youth!
Vala comes from the furnace in a cloud, but wretched Luvah
Is howling in the furnaces, in flames among Albion's Spectres,
To prepare the Spectre of Albion to reign over thee, O Los,
Forming the Spectres of Albion according to his rage;
To prepare the Spectre sons of Adam, who is Skofield, the ninth
Of Albion's sons, and the father of all his brethren in the shadowy
Generation. Cambel and Gwendolen wove webs of war and of   160
Religion to involve all Albion's sons, and when they had
Involved eight, their webs rolled outwards into darkness;
And Skofield the ninth remained on the outside of the eight,
And Kox, Kotope, and Bowen, one in him, a fourfold wonder,

Involved the eight. Such are the generations of the giant Albion:
To separate a law of sin, to punish thee in thy members.'

Los answered: 'Although I know not this, I know far worse than
   this.
I know that Albion hath divided me and that thou, O my Spectre,
Hast just cause to be irritated. But look steadfastly upon me;
Comfort thyself in my strength. The time will arrive   170
When all Albion's injuries shall cease, and when we shall
Embrace him tenfold bright, rising from his tomb in immortality.
They have divided themselves by wrath; they must be united by
Pity. Let us therefore take example and warning, O my Spectre.
Oh that I could abstain from wrath! Oh that the Lamb
Of God would look upon me and pity me in my fury,
In anguish of regeneration, in terrors of self-annihilation.
Pity must join together those whom wrath has torn in sunder,
And the religion of generation, which was meant for the
   destruction
Of Jerusalem, become her covering till the time of the end.   180
Oh holy generation! Image of regeneration!
Oh point of mutual forgiveness between enemies!
Birthplace of the Lamb of God incomprehensible!
The dead despise and scorn thee, and cast thee out as accursed,
Seeing the Lamb of God in thy gardens and thy palaces,
Where they desire to place the abomination of desolation.
Hand sits before his furnace; scorn of others and furious pride
Freeze round him to bars of steel and to iron rocks beneath
His feet. Indignant self-righteousness like whirlwinds of the north
Rose up against me thundering from the brook of Albion's river,   190
From Ranelagh and Strumbolo, from Cromwell's Gardens and
   Chelsea,°
The place of wounded soldiers. But when he saw my mace
Whirled round from heaven to earth trembling he sat; his cold
Poisons rose up, and his sweet deceits covered them all over
With a tender cloud. As thou art now, such was he, O Spectre.
I know thy deceit and thy revenges, and unless thou desist
I will certainly create an eternal hell for thee. Listen;
Be attentive; be obedient. Lo, the furnaces are ready to receive
   thee.
I will break thee into shivers and melt thee in the furnaces of death;
I will cast thee into forms of abhorrence and torment if thou   200

Desist not from thine own will, and obey not my stern command.
I am closed up from my childen; my Emanation is dividing
And thou my Spectre art divided against me. But mark:
I will compel thee to assist me in my terrible labours. To beat
These hypocritic selfhoods on the anvils of bitter death
I am inspired. I act not for myself; for Albion's sake
I now am what I am, a horror and an astonishment;
Shudd'ring the heavens to look upon me. Behold what cruelties
Are practised in Babel and Shinar, and have approached to Zion's
    hill.'°

While Los spoke the terrible Spectre fell shudd'ring before him,   210
Watching his time with glowing eyes to leap upon his prey.
Los opened the furnaces in fear. The Spectre saw to Babel and
    Shinar,°
Across all Europe and Asia; he saw the tortures of the victims;
He saw now from the outside what he before saw and felt from
    within;
He saw that Los was the sole, uncontrolled lord of the furnaces.
Groaning he kneeled before Los's iron-shod feet on London
    Stone,
Hung'ring and thirsting for Los's life, yet pretending obedience,
While Los pursued his speech in threat'nings loud and fierce:

'Thou art my pride and self-righteousness. I have found thee out;
Thou art revealed before me in all thy magnitude and power.   220
Thy uncircumcised pretences to chastity must be cut in sunder!
Thy holy wrath and deep deceit cannot avail against me,
Nor shalt thou ever assume the triple form of Albion's Spectre,
For I am one of the living. Dare not to mock my inspired fury.
If thou wast cast forth from my life, if I was dead upon the
    mountains,
Thou mightest be pitied and loved, but now I am living. Unless
Thou abstain ravening I will create an eternal hell for thee.°
Take thou this hammer, and in patience heave the thundering
    bellows;
Take thou these tongs, strike thou alternate with me; labour
    obedient.
Hand and Hyle and Coban, Skofield, Kox and Kotope labour   230
    mightily
In the wars of Babel and Shinar; all their Emanations were
Condensed. Hand has absorbed all his brethren in his might;

All the infant loves and graces were lost, for the mighty Hand
Condensed his Emanations into hard opaque substances,
And his infant thoughts and desires into old, dark cliffs of death.
His hammer of gold he seized, and his anvil of adamant.
He seized the bars of condensed thoughts to forge them
Into the sword of war, into the bow and arrow,
Into the thundering cannon and into the murdering gun.
I saw the limbs formed for exercise contemned, and the beauty of 240
Eternity looked upon as deformity, and loveliness as a dry tree.°
I saw disease forming a body of death around the Lamb
Of God, to destroy Jerusalem and to devour the body of Albion,
By war and stratagem to win the labour of the husbandman:

'Awkwardness armed in steel, folly in a helmet of gold,
Weakness with horns and talons, ignorance with a rav'ning beak,
Every emanative joy forbidden as a crime.
And the Emanations buried alive in the earth with pomp of
   religion,
Inspiration denied, genius forbidden by laws of punishment,
I saw terrified. I took the sighs and tears and bitter groans; 250
I lifted them into my furnaces, to form the spiritual sword
That lays open the hidden heart; I drew forth the pang
Of sorrow red-hot; I worked it on my resolute anvil.
I heated it in the flames of Hand and Hyle and Coban
Nine times. Gwendolen and Cambel and Gwineverra
Are melted into the gold, the silver, the liquid ruby,
The crysolite, the topaz, the jacinth, and every precious stone.
Loud roar my furnaces and loud my hammer is heard.
I labour day and night. I behold the soft affections
Condense beneath my hammer into forms of cruelty, 260
But still I labour in hope, though still my tears flow down:
That he who will not defend truth may be compelled to defend
A lie; that he may be snared and caught, and snared and taken;
That enthusiasm and life may not cease. Arise, Spectre, arise!'

Thus they contended among the furnaces with groans and tears.
Groaning, the Spectre heaved the bellows, obeying Los's frowns,
Till the spaces of Erin were perfected in the furnaces
Of affliction. And Los drew them forth, compelling the harsh
  Spectre
Into the furnaces, and into the valleys of the anvils of death,

And into the mountains of the anvils and of the heavy hammers,   270
Till he should bring the sons and daughters of Jerusalem to be
The sons and daughters of Los, that he might protect them from
Albion's dread Spectres. Storming, loud, thunderous and mighty
The bellows and the hammers move, compelled by Los's hand.

And this the manner of the sons of Albion in their strength.
They take the two contraries which are called Qualities, with
    which
Every substance is clothed; they name them Good and Evil.
From them they make an abstract, which is a negation
Not only of the substance from which it is derived,
A murderer of its own body, but also a murderer   280
Of every divine member. It is the reasoning power,
An abstract objecting power that negatives everything.
This is the Spectre of Man, the holy reasoning power;
And in its holiness is closed the abomination of desolation.

Therefore Los stands in London building Golgonooza,
Compelling his Spectre to labours mighty; trembling in fear
The Spectre weeps, but Los unmoved by tears or threats remains.

'I must create a system, or be enslaved by another man's.
I will not reason and compare; my business is to create.'

So Los, in fury and strength, in indignation and burning wrath. 290
Shudd'ring the Spectre howls; his howlings terrify the night.
He stamps around the anvil, beating blows of stern despair.
He curses heaven and earth, day and night, and sun and moon;
He curses forest, spring and river, desert and sandy waste,
Cities and nations, families and peoples, tongues and laws,
Driven to desperation by Los's terrors and threat'ning fears.

Los cries: 'Obey my voice and never deviate from my will,
And I will be merciful to thee. Be thou invisible to all
To whom I make thee invisible, but chief to my own children,
O Spectre of Urthona. Reason not against their dear approach,   300
Nor them obstruct with thy temptations of doubt and despair.
O shame, O strong and mighty shame, I break thy brazen fetters.
If you refuse thy present torments will seem southern breezes
To what thou shalt endure if thou obey not my great will.'

The Spectre answered: 'Art thou not ashamed of those thy sins
That thou callest thy children? Lo, the law of God commands
That they be offered upon his altar. Oh cruelty and torment,
For thine are also mine! I have kept silent hitherto
Concerning my chief delight; but thou has broken silence.
Now I will speak my mind. Where is my lovely Enitharmon,   310
O thou my enemy, where is my great sin? She is also thine.
I said: "Now is my grief at worst, incapable of being
Surpassed", but every moment it accumulates more and more;
It continues accumulating to eternity! The joys of God advance,
For he is righteous; he is not a being of pity and compassion;
He cannot feel distress; he feeds on sacrifice and offering,
Delighting in cries and tears, and clothed in holiness and solitude.
But my griefs advance also, for ever and ever without end.
Oh that I could cease to be! Despair! I am despair,
Created to be the great example of horror and agony. Also my   320
Prayer is vain. I called for compassion; compassion mocked;
Mercy and pity threw the gravestone over me, and with lead
And iron bound it over me for ever. Life lives on my
Consuming, and the Almighty hath made me his contrary:
To be all evil, all reversed and for ever dead, knowing
And seeing life, yet living not. How can I then behold
And not tremble? How can I be beheld and not abhorred?'

So spoke the Spectre shudd'ring, and dark tears ran down his
    shadowy face,
Which Los wiped off, but comfort none could give, or beam of
    hope!
Yet ceased he not from labouring at the roarings of his forge,   330
With iron and brass building Golgonooza in great contendings
Till his sons and daughters came forth from the furnaces
At the sublime labours. For Los compelled the invisible Spectre
To labours mighty, with vast strength, with his mighty chains,
In pulsations of time, and extensions of space, like urns of Beulah,
With great labour upon his anvils. And in his ladles the ore
He lifted, pouring it into the clay ground prepared with art,
Striving with systems to deliver individuals from those systems,
That, whenever any Spectre began to devour the dead,
He might feel the pain as if a man gnawed his own tender nerves. 340
Then Erin came forth from the furnaces, and all the daughters of
    Beulah

Came from the furnaces, by Los's mighty power for Jerusalem's
Sake, walking up and down among the spaces of Erin.
And the sons and daughters of Los came forth in perfection lovely!
And the spaces of Erin reached from the starry height to the starry
    depth.

Los wept with exceeding joy, and all wept with joy together.
They feared they never more should see their father, who
Was built in from eternity in the cliffs of Albion.

But, when the joy of meeting was exhausted in loving embrace,
Again they lament: 'Oh what shall we do for lovely Jerusalem,   350
To protect the Emanations of Albion's mighty ones from cruelty?
Sabrina and Ignoge begin to sharpen their beamy spears
Of light and love; their little children stand with arrows of gold.
Ragan is wholly cruel. Skofield is bound in iron armour;
He is like a mandrake in the earth before Reuben's gate;
He shoots beneath Jerusalem's walls to undermine her founda-
    tions.
Vala is but thy shadow, O thou loveliest among women!
A shadow animated by thy tears, O mournful Jerusalem!

'Why wilt thou give to her a body whose life is but a shade?
Her joy and love a shade, a shade of sweet repose.   360
But animated and vegetated she is a devouring worm.
What shall we do for thee, O lovely mild Jerusalem?'

And Los said: 'I behold the finger of God in terrors!
Albion is dead! His Emanation is divided from him!
But I am living! Yet I feel my Emanation also dividing.
Such thing was never known! Oh pity me, thou all-piteous
    one!
What shall I do, or how exist, divided from Enitharmon?
Yet why despair? I saw the finger of God go forth
Upon my furnaces from within the wheels of Albion's sons,
Fixing their systems permanent, by mathematic power   370
Giving a body to falsehood that it may be cast off for ever,
With demonstrative science piercing Apollyon with his own
    bow!°
God is within and without! He is even in the depths of Hell!'

Such were the lamentations of the labourers in the furnaces!

And they appeared within and without, encircling on both sides
The starry wheels of Albion's sons, with spaces for Jerusalem,
And for Vala the shadow of Jerusalem, the ever-mourning shade;
On both sides within and without beaming gloriously!

Terrified at the sublime wonder, Los stood before his furnaces.
And they stood around, terrified with admiration at Erin's spaces,   380
For the spaces reached from the starry height to the starry depth;
And they builded Golgonooza, terrible eternal labour!

What are those golden builders doing? Where was the burying-
    place
Of soft Ethinthus? Near Tyburn's fatal tree? Is that°
Mild Zion's hill's most ancient promontory, near mournful,
Ever-weeping Paddington? Is that Calvary and Golgotha
Becoming a building of pity and compassion? Lo!
The stones are pity and the bricks well-wrought affections,
Enamelled with love and kindness, and the tiles engraven gold,
Labour of merciful hands. The beams and rafters are forgiveness,   390
The mortar and cement of the work, tears of honesty. The nails
And the screws and iron braces are well-wrought blandishments
And well-contrived words, firm fixing, never forgotten,
Always comforting the remembrance; the floors, humility;
The ceilings, devotion; the hearths, thanksgiving
(Prepare the furniture, O Lambeth, in thy pitying looms);
The curtains, woven tears and sighs, wrought into lovely forms
For comfort. There the secret furniture of Jerusalem's chamber
Is wrought. Lambeth! the bride, the Lamb's wife loveth thee.
Thou art one with her and knowest not of self in thy supreme joy.   400
Go on, builders, in hope; though Jerusalem wanders far away,
Without the gate of Los, among the dark Satanic wheels.

Fourfold the sons of Los in their divisions, and fourfold
The great city of Golgonooza. Fourfold toward the north,
And toward the south fourfold, and fourfold toward the east and
    west,
Each within other toward the four points: that toward
Eden, and that toward the world of generation,
And that toward Beulah, and that toward Ulro.

Ulro is the space of the terrible starry wheels of Abion's sons;
But that toward Eden is walled up till time of renovation;   410
Yet it is perfect in its building, ornaments and perfection.

And the four points are thus beheld in great Eternity:°
West, the circumference; south, the zenith; north,
The nadir; east, the centre, unapproachable for ever.
These are the four faces towards the four worlds of humanity
In every man; Ezekiel saw them by Chebar's flood.
And the eyes are the south, and the nostrils are the east,
And the tongue is the west, and the ear is the north.

And the north gate of Golgonooza toward generation
Has four sculptured bulls terrible, before the gate of iron,   420
And iron the bulls. And that which looks toward Ulro,
Clay-baked and enamelled, eternal glowing as four furnaces,
Turning upon the wheels of Albion's sons with enormous power.
And that toward Beulah four, gold, silver, brass, and iron.
And that toward Eden four, formed of gold, silver, brass, and iron.

The south, a golden gate, has four lions, terrible, living;
That toward generation four, of iron carved wondrous;
That toward Ulro four, clay-baked laborious workmanship;
That toward Eden four, immortal gold, silver, brass and iron.

The western gate fourfold is closed, having four cherubim   430
Its guards, living, the work of elemental hands (laborious task!)
Like men hermaphroditic, each winged with eight wings.
That towards generation, iron; that toward Beulah, stone;
That toward Ulro, clay; that toward Eden, metals.
But all closed up till the last day, when the graves shall yield their
    dead.

The eastern gate fourfold, terrible and deadly its ornaments,
Taking their forms from the wheels of Albion's sons, as cogs
Are formed in a wheel to fit the cogs of the adverse wheel.

That toward Eden eternal ice, frozen in seven folds
Of forms of death. And that toward Beulah stone;   440
The seven diseases of the earth are carved, terrible.
And that toward Ulro forms of war, seven enormities.
And that toward generation, seven generative forms.

And every part of the city is fourfold, and every inhabitant
  fourfold.
And every pot and vessel and garment and utensil of the houses,
And every house, fourfold; but the third gate in every one
Is closed as with a threefold curtain of ivory and fine linen and
  ermine.
And Luban stands in middle of the city; a moat of fire
Surrounds Luban, Los's palace and the golden looms of Cathed-
  ron.

And sixty-four thousand genii guard the eastern gate;   450
And sixty-four thousand gnomes guard the northern gate;
And sixty-four thousand nymphs guard the western gate;
And sixty-four thousand fairies guard the southern gate.

Around Golgonooza lies the land of death eternal, a land
Of pain and misery and despair and ever-brooding melancholy,
In all the twenty-seven heavens, numbered from Adam to Luther,
From the blue Mundane Shell reaching to the vegetative earth.

The vegetative universe opens like a flower from the earth's centre,
In which is Eternity. It expands in stars to the Mundane Shell,
And there it meets Eternity again, both within and without,   460
And the abstract voids between the stars are the Satanic wheels.

There is the cave, the rock, the tree, the lake of Udan-Adan,
The forest, and the marsh, and the pits of bitumen deadly,
The rocks of solid fire, the ice valleys, the plains
Of burning sand, the rivers, cataract and lakes of fire,
The islands of the fiery lakes, the trees of malice, revenge,
And black anxiety, and the cities of the salamandrine men°
(But whatever is visible to the generated man
Is a creation of mercy and love from the Satanic void).
The land of darkness, flamed, but no light and no repose;   470
The land of snows, of trembling, and of iron hail incessant;
The land of earthquakes and the land of woven labyrinths;
The land of snares and traps and wheels and pit-falls and dire
  mills;
The voids, the solids; and the land of clouds and regions of
  waters—
With their inhabitants in the twenty-seven heavens beneath
  Beulah,

Self-righteousnesses conglomerating against the Divine Vision.
A concave earth, wondrous, chasmal, abyssal, incoherent,
Forming the Mundane Shell above, beneath, on all sides
    surrounding
Golgonooza. Los walks round the walls night and day.

He views the city of Golgonooza and its smaller cities,   480
The looms and mills and prisons and work-houses of Og and Anak,
The Amalekite, the Canaanite, the Moabite, the Egyptian,
And all that has existed in the space of six thousand years,
Permanent, and not lost, not lost nor vanished. And every little act,
Word, work and wish that has existed, all remaining still,
In those churches ever-consuming and ever-building, by the
    Spectres
Of all the inhabitants of earth wailing to be created,
Shadowy to those who dwell not in them, mere possibilities;
But to those who enter into them they seem the only substances.
For everything exists, and not one sigh nor smile nor tear,   490
One hair nor particle of dust, not one can pass away.

He views the cherub at the Tree of Life, also the serpent,
Orc the first-born, coiled in the south, the dragon Urizen,
Tharmas the vegetated tongue, even the devouring tongue;
A threefold region, a false brain, a false heart,
And false bowels, altogether composing the false tongue
Beneath Beulah, as a wat'ry flame revolving every way,
And as dark roots and stems, a forest of affliction, growing
In seas of sorrow. Los also views the four females:
Ahania and Enion and Vala and Enitharmon lovely,   500
And from them all the lovely beaming daughters of Albion.
Ahania and Enion and Vala are three evanescent shades;
Enitharmon is a vegetated mortal wife of Los,
His Emanation, yet his wife till the sleep of death is past.

Such are the buildings of Los and such are the woofs of
    Enitharmon.

And Los beheld his sons and he beheld his daughters,
Every one a translucent wonder, a universe within,
Increasing inwards into length and breadth and height,
Starry and glorious; and they every one in their bright loins

Have a beautiful golden gate which opens into the vegetative   510
    world;
And every one a gate of rubies and all sorts of precious stones
In their translucent hearts which opens into the vegetative world;
And every one a gate of iron dreadful and wonderful
In their translucent heads which opens into the vegetative world.
And every one has the three regions: childhood, manhood and age.
But the gate of the tongue, the western gate, in them is closed,
Having a wall builded against it, and thereby the gates
Eastward and southward and northward are encircled with
    flaming fires.
And the north is breadth; the south is height and depth;
The east is inwards; and the west is outwards every way.   520

And Los beheld the mild Emanation Jerusalem, eastward bending
Her revolutions toward the starry wheels in maternal anguish,
Like a pale cloud arising from the arms of Beulah's daughters,
In Entuthon Benython's deep vales beneath Golgonooza.

And Hand and Hyle rooted into Jerusalem by a fibre
Of strong revenge, and Skofield vegetated by Reuben's gate
In every nation of the earth, till the twelve sons of Albion
Enrooted into every nation: a mightly polypus growing
From Albion over the whole earth. Such is my awful vision.

I see the fourfold man, the humanity, in deadly sleep,   530
And its fallen Emanation, the Spectre and its cruel shadow.
I see the past, present and future, existing all at once
Before me. O Divine Spirit, sustain me on thy wings!
That I may awake Albion from his long and cold repose.
For Bacon and Newton, sheathed in dismal steel, their terrors
    hang
Like iron scourges over Albion; reasonings like vast serpents
Enfold around my limbs, bruising my minute articulations.

I turn my eyes to the schools and universities of Europe,
And there behold the loom of Locke whose woof rages dire,
Washed by the water-wheels of Newton. Black the cloth   540
In heavy wreaths folds over every nation. Cruel works
Of many wheels I view, wheel without wheel, with cogs tyrannic
Moving by compulsion each other; not as those in Eden, which
Wheel within wheel in freedom revolve, in harmony and peace.

I see, in deadly fear, in London Los raging round his anvil
Of death, forming an axe of gold. The four sons of Los
Stand round him, cutting the fibres from Albion's hills
That Albion's sons may roll apart over the nations,
While Reuben enroots his brethren in the narrow Canaanite
From the limit Noah to the limit Abram, in whose loins   550
Reuben in his twelvefold majesty and beauty shall take refuge,
As Abraham flees from Chaldea shaking his gory locks.
But first Albion must sleep, divided from the nations.

I see Albion sitting upon his rock in the first winter;
And thence I see the chaos of Satan and the world of Adam
When the Divine Hand went forth on Albion in the midwinter,
And at the place of death when Albion sat in eternal death
Among the furnaces of Los in the valley of the son of Hinnom.°

Hampstead, Highgate, Finchley, Hendon, Muswell Hill rage loud
Before Bromion's iron tongs and glowing poker reddening fierce. 560
Hertfordshire glows with fierce vegetation; in the forests
The oak frowns terrible; the beech and ash and elm enroot
Among the spiritual fires. Loud the cornfields thunder along
The soldier's fife, the harlot's shriek, the virgin's dismal groan,
The parent's fear, the brother's jealousy, the sister's curse,
Beneath the storms of Theotormon; and the thund'ring bellows
Heaves in the hand of Palamabron, who in London's darkness
Before the anvil watches the bellowing flames. Thundering
The hammer loud rages in Rintrah's strong grasp, swinging loud
Round from heaven to earth, down falling with heavy blow   570
Dead on the anvil, where the red-hot wedge groans in pain.
He quenches it in the black trough of his forge. London's river
Feeds the dread forge, trembling and shuddering along the valleys.

Humber and Trent roll dreadful before the seventh furnace,
And Tweed and Tyne anxious give up their souls for Albion's
    sake.
Lincolnshire, Derbyshire, Nottinghamshire, Leicestershire,
From Oxfordshire to Norfolk on the lake of Udan-Adan,
Labour within the furnaces, walking among the fires
With ladles huge and iron pokers over the island white.

Scotland pours out his sons to labour at the furnaces;   580
Wales give his daughters to the looms; England nursing mothers

Gives to the children of Albion and to the children of Jerusalem.
From the blue Mundane Shell even to the earth of vegetation,
Throughout the whole creation which groans to be delivered,
Albion groans in the deep slumbers of death upon his rock.

Here Los fixed down the fifty-two counties of England and Wales,
The thirty-six of Scotland, and the thirty-four of Ireland
With mighty power, when they fled out at Jerusalem's gates
Away from the conflict of Luvah and Urizen, fixing the gates
In the twelve counties of Wales; and thence gates looking every   590
    way
To the four points conduct to England and Scotland and Ireland,°
And thence to all the kingdoms and nations and families of the
    earth.
The gate of Reuben in Carmarthenshire, the gate of Simeon in
Cardiganshire, and the gate of Levi in Montgomeryshire;
The gate of Judah, Merionethshire; the gate of Dan, Flintshire;
The gate of Naphtali, Radnorshire; the gate of Gad, Pembroke-
    shire;
The gate of Asher, Carnarvonshire; the gate of Issachar, Breck-
    nockshire;
The gate of Zebulun in Anglesey and Sodor. So is Wales divided.
The gate of Joseph Denbighshire, the gate of Benjamin Glamor-
    ganshire,
For the protection of the twelve Emanations of Albion's sons.   600

And the forty counties of England are thus divided: in the gates
Of Reuben—Norfolk, Suffolk, Essex; Simeon—Lincoln, York,
    Lancashire;
Levi—Middlesex, Kent, Surrey; Judah—Somerset, Gloucester,
    Wiltshire;
Dan—Cornwall, Devon, Dorset; Naphtali—Warwick, Leicester,
    Worcester;
Gad—Oxford, Bucks, Hertford; Asher—Sussex, Hampshire,
    Berkshire;
Issachar—Northampton, Rutland, Nottingham; Zebulun—Bed-
    ford, Huntingdon, Cambridge;
Joseph—Stafford, Shropshire, Hereford; Benjamin—Derby,
    Cheshire, Monmouth;
And Cumberland, Northumberland, Westmorland and Durham
    are
Divided in the gates of Reuben, Judah, Dan and Joseph.

And the thirty-six counties of Scotland divided in the gates   610
Of Reuben (Kincardine, Haddington, Forfar), Simeon (Ayr,
    Argyll, Banff),
Levi (Edinburgh, Roxburgh, Ross), Judah (Aberdeen, Berwick,
    Dumfries),
Dan (Bute, Caithness, Clackmannan), Naphtali (Nairn, Inverness,
    Linlithgow),
Gad (Peebles, Perth, Renfrew), Asher (Sutherland, Stirling,
    Wigtown),
Issachar (Selkirk, Dunbarton, Glasgow), Zebulun (Orkney, Shet-
    land, Skye)
Joseph (Elgin, Lanark, Kinross), Benjamin (Cromarty, Moray,
    Kirkcudbright)—
Governing all by the sweet delights of secret amorous glances
In Enitharmon's halls builded by Los and his mighty children.

All things acted on earth are seen in the bright sculptures of
Los's halls and every age renews its powers from these works,   620
With every pathetic story possible to happen, from hate or
Wayward love; and every sorrow and distress is carved here.
Every affinity of parents, marriages and friendships are here
In all their various combinations wrought with wondrous art,
All that can happen to Man in his pilgrimage of seventy years.
Such is the divine written law of Horeb and Sinai,
And such the holy gospel of Mount Olivet and Calvary.
His Spectre divides and Los in fury compels it to divide,
To labour in the fire, in the water, in the earth, in the air,
To follow the daughters of Albion as the hound follows the scent   630
Of the wild inhabitant of the forest, to drive them from his own,
To make a way for the children of Los to come from the furnaces.
But Los himself against Albion's sons his fury bends, for he
Dare not approach the daughters openly, lest he be consumed
In the fires of their beauty and perfection, and be vegetated
    beneath
Their looms, in a generation of death and resurrection to
    forgetfulness.
They woo Los continually to subdue his strength; he continually
Shows them his Spectre, sending him abroad over the four points
    of heaven
In the fierce desires of beauty and in the tortures of repulse! He is
The Spectre of the living pursuing the Emanations of the dead.   640

Shudd'ring they flee; they hide in the Druid temples in cold
    chastity,
Subdued by the Spectre of the living and terrified by undisguised
    desire.

For Los said: 'Though my Spectre is divided, as I am a living man
I must compel him to obey me wholly, that Enitharmon may not
Be lost, and lest he should devour Enitharmon. Ah me!
Piteous image of my soft desires and loves, O Enitharmon!
I will compel my Spectre to obey; I will restore to thee thy
    children.
No one bruises or starves himself to make himself fit for labour!

'Tormented with sweet desire for these beauties of Albion,
They would never love my power if they did not seek to destroy   650
Enitharmon. Vala would never have sought and loved Albion
If she had not sought to destroy Jerusalem. Such is that false
And generating love: a pretence of love to destroy love,
Cruel hypocrisy, unlike the lovely delusions of Beulah,
And cruel forms, unlike the merciful forms of Beulah's night.

'They know not why they love nor wherefore they sicken and die,
Calling that Holy Love which is envy, revenge and cruelty,
Which separated the stars from the mountains, the mountains
    from Man,
And left Man, a little grovelling root, outside of himself.
Negations are not contraries. Contraries mutually exist,   660
But negations exist not. Exceptions and objections and unbeliefs
Exist not, nor shall they ever be organised, for ever and ever.
If thou separate from me, thou art a negation, a mere
Reasoning and derogation from me, an objecting and cruel spite,
And malice and envy; but my Emanation, alas! will become
My contrary. O thou negation, I will continually compel
Thee to be invisible to any but whom I please, and when
And where and how I please, and never, never shalt thou be
    organized
But as a distorted and reversed reflection in the darkness
And in the non-entity! Nor shall that which is above   670
Ever descend into thee, but thou shalt be a non-entity for ever.
And if any enter into thee, thou shalt be an unquenchable fire,°
And he shall be a never-dying worm, mutually tormented by
Those that thou tormentest, a hell and despair for ever and ever.'

So Los in secret with himself communed, and Enitharmon heard
In her darkness and was comforted; yet still she divided away
In gnawing pain from Los's bosom in the deadly night,
First as a red globe of blood trembling beneath his bosom.
Suspended over her he hung, he enfolded her in his garments
Of wool; he hid her from the Spectre, in shame and confusion of   680
Face. In terrors and pains of Hell and eternal death the
Trembling globe shot forth self-living, and Los howled over it,
Feeding it with his groans and tears day and night without ceasing.
And the spectrous darkness from his back divided in temptations,
And in grinding agonies, in threats, stiflings, and direful strug-
    glings.

'Go thou to Skofield. Ask him if he is Bath or if he is Canterbury.
Tell him to be no more dubious. Demand explicit words.
Tell him I will dash him into shivers, where and at what time
I please. Tell Hand and Skofield they are my ministers of evil
To those I hate; for I can hate also as well as they!'   690

From every one of the four regions of human majesty
There is an outside spread without and an outside spread within,
Beyond the outline of identity both ways, which meet in one:
An orbed void of doubt, despair, hunger, and thirst, and sorrow.
Here the twelve sons of Albion, joined in dark assembly,
Jealous of Jerusalem's children, ashamed of her little ones
(For Vala produced the bodies, Jerusalem gave the souls)
Became as three immense wheels, turning upon one another
Into non-entity, and their thunders hoarse appal the dead,
To murder their own souls, to build a kingdom among the dead.   700

'Cast, cast ye Jeruslaem forth! The shadow of delusions!°
The harlot daughter! Mother of pity and dishonourable forgive-
    ness,
Our father Albion's sin and shame! But father now no more!
Nor sons! Nor hateful peace and love, nor soft complacencies
With transgressors meeting in brotherhood around the table,
Or in the porch or garden! No more the sinful delights
Of age and youth, and boy and girl, and animal and herb,
And river and mountain, and city and village, and house and
    family,
Beneath the oak and palm, beneath the vine and fig-tree

In self-denial! But war and deadly contention between   710
Father and son, and light and love! All bold asperities
Of haters met in deadly strife, rending the house and garden,
The unforgiving porches, the tables of enmity, and beds
And chambers of trembling and suspicion, hatreds of age and
      youth,
And boy and girl, and animal and herb, and river and mountain,
And city and village, and house and family. That the perfect
May live in glory, redeemed by sacrifice of the Lamb
And of his children before sinful Jerusalem, to build
Babylon, the city of Vala, the goddess virgin-mother.
She is our mother! Nature! Jerusalem is our harlot sister   720
Returned with children of pollution, to defile our house
With sin and shame. Cast, cast her into the potter's field!°
Her little ones she must slay upon our altars, and her aged
Parents must be carried into captivity, to redeem her soul,
To be for a shame and a curse, and to be our slaves for ever.'

So cry Hand and Hyle, the eldest of the fathers of Albion's
Little ones; to destroy the divine Saviour, the friend of sinners,
Building castles in desolated places and strong fortifications.
Soon Hand mightily devoured and absorbed Albion's twelve sons.
Out from his bosom a mighty polypus, vegetating in darkness,   730
And Hyle and Coban were his two chosen ones, for emissaries
In war. Forth from his bosom they went and returned
Like wheels from a great wheel reflected in the deep.
Hoarse turned the starry wheels, rending a way in Albion's loins
Beyond the night of Beulah; in a dark and unknown night,
Outstretched his giant beauty on the ground in pain and tears.

His children exiled from his breast pass to and fro before him.
His birds are silent on his hills; flocks die beneath his branches.
His tents are fallen; his trumpets and the sweet sound of his harp
Are silent on his clouded hills, that belch forth storms and fire.   740
His milk of cows, and honey of bees, and fruit of golden harvest,
Is gathered in the scorching heat and in the driving rain.
Where once he sat he weary walks in misery and pain,
His giant beauty and perfection fallen into dust;
Till from within his withered breast, grown narrow with his woes,
The corn is turned to thistles and the apples into poison,
The birds of song to murderous crows, his joys to bitter groans,

The voices of children in his tents to cries of helpless infants.
And, self-exiled from the face of light and shine of morning,
In the dark world, a narrow house, he wanders up and down,   750
Seeking for rest and finding none! And hidden far within
His eon weeping in the cold and desolated earth.

All his affections now appear withoutside. All his sons,
Hand, Hyle and Coban, Gwantok, Peachey, Brereton, Slayd and
    Huttn,
Skofield, Kox, Kotope and Bowen, his twelve sons, Satanic mill,
Who are the Spectres of the twenty-four, each double-formed,
Revolve upon his mountains: groaning in pain beneath
The dark incessant sky, seeking for rest and finding none,
Raging against their human natures, ravening to gormandize
The human majesty and beauty of the twenty-four,   760
Condensing them into solid rocks with cruelty and abhorrence.
Suspicion and revenge, and the seven diseases of the soul,
Settled around Albion and around Luvah in his secret cloud.
Willing the friends endured, for Albion's sake and for
Jerusalem his Emanation shut within his bosom;
Which hardened against them more and more as he builded
    onwards
On the gulf of death in self-righteousness, that rolled
Before his awful feet, in pride of virtue for victory.
And Los was roofed in from Eternity in Albion's cliffs,
Which stand upon the ends of Beulah, and withoutside all   770
Appeared a rocky form against the Divine Humanity.

Albion's circumference was closed. His centre began dark'ning
Into the night of Beulah, and the moon of Beulah rose
Clouded with storms. Los his strong guard walked round beneath
    the moon,
And Albion fled inward among the currents of his rivers.

He found Jerusalem upon the river of his city, soft reposed
In the arms of Vala, assimilating in one with Vala
The lily of Havilah. And they sang soft through Lambeth's vales,°
In a sweet moony night and silence that they had created,
With a blue sky spread over with wings and a mild moon,   780
Dividing and uniting into many female forms, Jerusalem
Trembling! Then in one commingling in eternal tears,
Sighing to melt his giant beauty on the moony river.

But when they saw Albion fallen upon mild Lambeth's vale,
Astonished, terrified they hovered over his giant limbs.
Then thus Jerusalem spoke, while Vala wove the veil of tears,
Weeping in pleadings of love in the web of despair:

'Wherefore has thou shut me into the winter of human life,
And closed up the sweet regions of youth and virgin innocence,
Where we live, forgetting error, not pondering on evil,   790
Among my lambs and brooks of water, among my warbling birds,
Where we delight in innocence before the face of the Lamb,
Going in and out before him in his love and sweet affection?'

Vala replied weeping and trembling, hiding in her veil:

'When winter rends the hungry family and the snow falls
Upon the ways of men, hiding the paths of man and beast,
Then mourns the wanderer; then he repents his wanderings and
     eyes
The distant forest; then the slave groans in the dungeon of stone,
The captive in the mill of the stranger, sold for scanty hire.
They view their former life; they number moments over and over,   800
Stringing them on their remembrance as on a thread of sorrow.
Thou art my sister and my daughter; thy shame is mine also.
Ask me not of my griefs! Thou knowest all my griefs.'

Jerusalem answered with soft tears over the valley:

'O Vala, what is sin that thou shudderest and weepest
At sight of thy once-loved Jerusalem? What is sin but a little
Error and fault that is soon forgiven? But mercy is not a sin,
Nor pity, nor love, nor kind forgiveness. Oh, if I have sinned
Forgive and pity me! Oh, unfold thy veil in mercy and love!

'Slay not my little ones, beloved virgin daughter of Babylon,   810
Slay not my infant loves and graces, beautiful daughter of Moab.
I cannot put off the human form; I strive but strive in vain.
When Albion rent thy beautiful net of gold and silver twine
Thou hadst woven it with art, thou hadst caught me in the bands
Of love; thou refusedst to let me go. Albion beheld thy beauty,
Beautiful through our love's comeliness, beautiful through pity;
The veil shone with thy brightness in the eyes of Albion
Because it enclosed pity and love, because we loved one another.

Albion loved thee! He rent thy veil! He embraced thee; he loved
      thee!
Astonished at his beauty and perfection, thou forgavest his furious   820
      love.
I redounded from Albion's bosom in my virgin loveliness.
The Lamb of God received me in his arms; he smiled upon us.
He made me his bride and wife; he gave thee to Albion.
Then was a time of love. Oh, why is it passed away?'

Then Albion broke silence and with groans replied:

'O Vala! O Jerusalem! Do you delight in my groans?
You, O lovely forms, you have prepared my death-cup.
The disease of shame covers me from head to feet: I have no hope.
Every boil upon my body is a separate and deadly sin.
Doubt first assailed me, then shame took possession of me.   830
Shame divides families; shame hath divided Albion in sunder.
First fled my sons, and then my daughters, then my wild
      animations;
My cattle next, last even the dog of my gate. The forests fled,
The corn-fields, and the breathing gardens outside separated,
The sea, the stars, the sun, the moon: driven forth by my disease.
All is eternal death unless you can weave a chaste
Body over an unchaste mind! Vala! Oh that thou wert pure!
That the deep wound of sin might be closed up with the needle
And with the loom, to cover Gwendolen and Ragan with costly
      robes
Of natural virtue; for their spiritual forms without a veil   840
Wither in Luvah's sepulchre. I thrust him from my presence,
And all my children followed his loud howlings into the deep.
Jerusalem! Dissembler Jerusalem! I look into thy bosom,
I discover thy secret places. Cordella, I behold
Thee, whom I thought pure as the heavens in innocence and fear,
Thy tabernacle taken down, thy secret cherubim disclosed.
Art thou broken? Ah me, Sabrina, running by my side
In childhood, what were thou? Unutterable anguish! Conwenna,
Thy cradled infancy is most piteous. Oh hide, oh hide!
Their secret gardens were made paths to the traveller;   850
I knew not of their secret loves with those I hated most,
Nor that their every thought was sin and secret appetite.
Hyle sees in fear, he howls in fury over them. Hand sees

In jealous fear; in stern accusation with cruel stripes
He drives them through the streets of Babylon before my face.
Because they taught Luvah to rise into my clouded heavens,
Battersea and Chelsea mourn for Cambel and Gwendolen!
Hackney and Holloway sicken for Estrild and Ignoge!—
Because the Peak, Malvern and Cheviot reason in cruelty,
Penmaenmawr and Dhinas-bran demonstrate in unbelief,   860
Manchester and Liverpool are in tortures of doubt and despair,
Maldon and Colchester demonstrate. I hear my children's voices;
I see their piteous faces gleam out upon the cruel winds
From Lincoln and Norwich, from Edinburgh and Monmouth;
I see them distant from my bosom scourged along the roads,
Then lost in clouds. I hear their tender voices! Clouds divide;
I see them die beneath the whips of the captains. They are taken
In solemn pomp into Chaldea across the breadths of Europe.
Six months they lie embalmed in silent death, worshipped,
Carried in arks of oak before the armies in the spring.   870
Bursting their arks they rise again to life; they play before
The armies. I hear their loud cymbals and their deadly cries.
Are the dead cruel? Are those who are enfolded in moral law
Revengeful? Oh that death and annihiliation were the same!'

Then Vala answered, spreading her scarlet veil over Albion:

'Albion, thy fear has made me tremble; thy terrors have
    surrounded me.
Thy sons have nailed me on the gates, piercing my hands and feet,
Till Skofield's Nimrod, the mighty huntsman Jehovah, came°
With Cush his son, and took me down. He in a golden ark
Bears me before his armies, though my shadow hovers here.   880
The flesh of multitudes fed and nourished me in my childhood;
My morn and evening food were prepared in battles of men.
Great is the cry of the hounds of Nimrod along the valley
Of vision; they scent the odour of war in the valley of vision.
All love is lost! Terror succeeds, and hatred, instead of love,
And stern demands of right and duty instead of liberty.
Once thou wast to me the loveliest son of heaven; but now
Where shall I hide from thy dread countenance and searching
    eyes?
I have looked into the secret soul of him I loved,
And in the dark recesses found sin and can never return.'   890

Albion again uttered his voice beneath the silent moon:

'I brought love into light of day, to pride in chaste beauty;
I brought love into light, and fancied innocence is no more.'

Then spoke Jerusalem: 'O Albion, my father Albion!
Why wilt thou number every little fibre of my soul,
Spreading them out before the sun like stalks of flax to dry?
The infant joy is beautiful, but its anatomy
Horrible, ghast and deadly! Nought shalt thou find in it°
But dark despair and everlasting brooding melancholy!'

Then Albion turned his face toward Jerusalem and spoke: 900

'Hide thou, Jerusalem, in impalpable voidness, not to be
Touched by the hand nor seen with the eye. O Jerusalem,
Would thou wert not and that thy place might never be found.°
But come, O Vala, with knife and cup! Drain my blood
To the last drop! Then hide me in thy scarlet tabernacle.
For I see Luvah whom I slew; I behold him in my Spectre
As I behold Jerusalem in thee, O Vala, dark and cold.'

Jerusalem then stretched her hand toward the moon and spoke:

'Why should punishment weave the veil with iron wheels of war,
When forgiveness might it weave with wings of cherubim?'   910

Loud groaned Albion from mountain to mountain and replied:

'Jerusalem! Jerusalem! deluding shadow of Albion!
Daughter of my fantasy! unlawful pleasure! Albion's curse!
I came here with intention to annihilate thee! But
My soul is melted away, enwoven within the veil.
Hast thou again knitted the veil of Vala, which I for thee
Pitying rent in ancient times? I see it whole and more
Perfect, and shining with beauty!'
                                            'But thou, O wretched father!'
Jerusalem replied, like a voice heard from a sepulchre,
'Father, once piteous! Is pity a sin? Embalmed in Vala's bosom   920
In an eternal death for Albion's sake, our best beloved,
Thou art my father and my brother; why hast thou hidden me
Remote from the Divine Vision, my Lord and Saviour?'

Trembling stood Albion at her words in jealous dark despair.

He felt that love and pity are the same: a soft repose,
Inward complacency of soul, a self-annihilation!

'I have erred! I am ashamed! and will never return more.
I have taught my children sacrifices of cruelty. What shall I
    answer?
I will hide it from Eternals! I will give myself for my children!
Which way soever I turn, I behold humanity and pity!'   930

He recoiled; he rushed outwards; he bore the veil whole away.
His fires redound from his dragon altars in errors returning.
He drew the veil of moral virtue, woven for cruel laws,
And cast it into the Atlantic deep, to catch the souls of the dead.
He stood between the palm tree and the oak of weeping
Which stand upon the edge of Beulah, and there Albion sunk
Down in sick pallid langour. These were his last words, relapsing°
Hoarse from his rocks, from caverns of Derbyshire and Wales
And Scotland, uttered from the circumference into Eternity:

'Blasphemous sons of feminine delusion! God in the dreary void   940
Dwells from eternity, wide separated from the human soul.
But thou, deluding image—by whom imbued, the veil I rent—
Lo, here is Vala's veil whole, for a law, a terror and a curse!
And therefore God takes vengeance on me; from my clay-cold
    bosom
My children wander, trembling victims of his moral justice.
His snows fall on me and cover me, while in the veil I fold
My dying limbs. Therefore, O manhood, if thou art aught
But a mere fantasy, hear dying Albion's curse!
May God who dwells in this dark Ulro and voidness, vengeance
    take,
And draw thee down into this abyss of sorrow and torture,   950
Like me thy victim. Oh that death and annihilation were the same!

'What have I said? What have I done? O all-powerful human
    words!
You recoil back upon me in the blood of the Lamb slain in his
    children.
Two bleeding contraries, equally true, are his witnesses against
    me.

We reared mighty stones; we danced naked around them,
Thinking to bring love into light of day, to Jerusalem's shame
Displaying our giant limbs to all the winds of heaven! Sudden
Shame seized us; we could not look on one another for abhorrence.
  The blue
Of our immortal veins and all their hosts fled from our limbs,
And wandered distant in a dismal night clouded and dark.   960
The sun fled from the Briton's forehead, the moon from his mighty
    loins;
Scandinavia fled with all his mountains filled with groans.

'Oh, what is life and what is Man? Oh, what is death? Wherefore
Are you, my children, natives in the grave to where I go?
Or are you born to feed the hungry ravenings of destruction,
To be the sport of accident, to waste in wrath and love a weary
Life, in brooding cares and anxious labours that prove but chaff?
Oh Jerusalem, Jerusalem, I have forsaken thy courts:°
Thy pillars of ivory and gold, thy curtains of silk and fine
Linen, thy pavements of precious stones, thy walls of pearl   970
And gold, thy gates of thanksgiving, thy windows of praise,
Thy clouds of blessing, thy cherubims of tender mercy,
Stetching their wings sublime over the little ones of Albion.
O human imagination, O divine body I have crucified!
I have turned my back upon thee into the wastes of moral law;
There Babylon is builded in the waste, founded in human
    desolation.
O Babylon, thy watchman stands over thee in the night;
Thy severe judge all the day long proves thee, O Babylon,
With provings of destruction, with giving thee thy heart's desire.
But Albion is cast forth to the potter, his children to the builders 980
To build Babylon, because they have forsaken Jerusalem.
The walls of Babylon are souls of men, her gates the groans
Of nations, her towers are the miseries of once happy families.
Her streets are paved with destruction, her houses built with
    death,
Her palaces with Hell and the grave, her synagogues with torments
Of ever-hardening despair, squared and polished with cruel skill.
Yet thou wast lovely as the summer cloud upon my hills,
When Jerusalem was thy heart's desire in times of youth and love.
Thy sons came to Jerusalem with gifts; she sent them away

With blessings on their hands and on their feet, blessings of gold   990
And pearl and diamond. Thy daughters sang in her courts;
They came up to Jerusalem; they walked before Albion.
In the exchanges of London every nation walked,
And London walked in every nation, mutual in love and harmony.
Albion covered the whole earth; England encompassed the
    nations;
Mutual each within other's bosom in visions of regeneration.
Jerusalem covered the Atlantic mountains and the Erythrean,
From bright Japan and China to Hesperia, France and England.°
Mount Zion lifted his head in every nation under heaven,
And the Mount of Olives was beheld over the whole earth.   1000
The footsteps of the Lamb of God were there. But now no more,
No more shall I behold him; he is closed in Luvah's sepulchre.
Yet why these smitings of Luvah, the gentlest, mildest Zoa?
If God was merciful this could not be. O Lamb of God,
Thou art a delusion, and Jerusalem is my sin! O my children,
I have educated you in the crucifying cruelties of demonstration,
Till you have assumed the providence of God and slain your
    father.
Dost thou appear before me who liest dead in Luvah's sepulchre?
Dost thou forgive me, thou who wast dead and art alive?°
Look not so merciful upon me, O thou slain Lamb of God,°   1010
I die! I die in thy arms, though hope is banished from me.'

Thund'ring the veil rushes from his hand, vegetating knot by
Knot, day by day, night by night. Loud roll the indignant Atlantic
Waves and the Erythrean, turning up the bottoms of the deeps,
And there was heard a great lamenting in Beulah. All the regions
Of Beulah were moved as the tender bowels are moved, and they
    said:

'Why did you take vengeance, O ye sons of the mighty Albion,
Planting these oaken groves, erecting these dragon temples?
Injury the Lord heals, but vengeance cannot be healed.
As the sons of Albion have done to Luvah, so they have in him   1020
Done to the divine Lord and Saviour, who suffers with those that
    suffer.
For not one sparrow can suffer, and the whole universe not suffer
    also,°
In all its regions, and its father and Saviour not pity and weep.
But vengeance is the destroyer of grace and repentance in the
    bosom

Of the injurer, in which the Divine Lamb is cruelly slain.
Descend, O Lamb of God, and take away the imputation of sin°
By the creation of states and the deliverance of individuals,
    evermore. Amen.'

Thus wept they in Beulah over the four regions of Albion.
But many doubted and despaired, and imputed sin and righteous-
    ness
To individuals and not to states; and these slept in Ulro.   1030

CHAPTER II

TO THE JEWS

[see above, p. 31]

Every ornament of perfection, and every labour of love,
In all the Garden of Eden and in all the golden mountains,
Was become an envied horror and a remembrance of jealousy;
And every act a crime, and Albion the punisher and judge.

And Albion spoke from his secret seat and said:

'All these ornaments are crimes; they are made by the labours
Of loves, of unnatural consanguinities and friendships
Horrid to think of when enquired deeply into; and all
These hills and valleys are accursed witnesses of sin.
I therefore condense them into solid rocks, steadfast,   10
A foundation and certainty and demonstrative truth,
That man be separate from man; and here I plant my seat.'

Cold snows drifted around him; ice covered his loins around.
He sat by Tyburn's brook, and underneath his heel shot up
A deadly tree; he named it Moral Virtue and the Law
Of God, who dwells in chaos hidden from the human sight.

The tree spread over him its cold shadows (Albion groaned);
They bent down, they felt the earth and again enrooting
Shot into many a tree, an endless labyrinth of woe!

From willing sacrifice of self, to sacrifice of (miscalled) enemies 20
For atonement, Albion began to erect twelve altars,
Of rough unhewn rocks, before the potter's furnace.
He named them Justice and Truth. And Albion's sons

Must have become the first victims, being the first transgressors,
But they fled to the mountains to seek ransom, building a strong
Fortification against the Divine Humanity and Mercy,
In shame and jealousy to annihilate Jerusalem.

Turning his back to the Divine Vision, his spectrous
Chaos before his face appeared, an unformed memory.

Then spoke the spectrous chaos to Albion, dark'ning cold,   30
From the back and loins where dwell the spectrous dead:

'I am your rational power, O Albion, and that human form
You call divine is but a worm seventy inches long,
That creeps forth in a night and is dried in the morning sun
In fortuitous concourse of memories accumulated and lost.°
It ploughs the earth in its own conceit, it overwhelms the hills
Beneath its winding labyrinths, till a stone of the brook°
Stops it in midst of its pride among its hills and rivers.
Battersea and Chelsea mourn, London and Canterbury tremble;
Their place shall not be found as the wind passes over.°   40
The ancient cities of the earth remove as a traveller;
And shall Albion's cities remain when I pass over them
With my deluge of forgotten remembrances over the tablet?'

So spoke the Spectre to Albion. He is the great selfhood,
Satan, worshipped as God by the mighty ones of the earth,
Having a white dot called a centre from which branches out
A circle in continual gyrations. This became a heart,
From which sprang numerous branches varying their motions,
Producing many heads, three or seven or ten, and hands and feet
Innumerable at will of the unfortunate contemplator,   50
Who becomes his food. Such is the way of the devouring power.

And this is the cause of the appearance in the frowning chaos:
Albion's Emanation, which he had hidden in jealousy,
Appeared now in the frowning chaos, prolific upon the chaos,
Reflecting back to Albion in sexual reasoning hermaphroditic.

Albion spoke: 'Who art thou that appearest in gloomy pomp,
Involving the Divine Vision in colours of autumn ripeness?
I never saw thee till this time, nor beheld life abstracted,
Nor darkness immingled with light on my furrowed field.

Whence camest thou? Who art thou, O loveliest? The Divine   60
    Vision
Is as nothing before thee; faded is all life and joy.'

Vala replied in clouds of tears, Albion's garment embracing:

'I was a city and a temple built by Albion's children.
I was a garden planted with beauty. I allured on hill and valley
The River of Life to flow against my walls and among my trees.
Vala was Albion's bride and wife in great Eternity,
The loveliest of the daughters of Eternity, when in day-break
I emanated from Luvah over the towers of Jerusalem
And in her courts, among her little children offering up
The sacrifice of fanatic love. Why loved I Jerusalem?   70
Why was I one with her embracing in the vision of Jesus?
Wherefore did I loving create love, which never yet
Immingled God and Man, when thou and I hid the Divine
    Vision
In cloud of secret gloom, which, behold, involves me round
    about.
Know me now, Albion; look upon me. I alone am beauty;
The imaginative human form is but a breathing of Vala.
I breathe him forth into the heaven from my secret cave,
Born of the woman to obey the woman, O Albion the mighty.
For the divine appearance is brotherhood, but I am love,
Elevate into the region of brotherhood with my red fires.'°   80

'Art thou Vala?' replied Albion, 'image of my repose?
Oh how I tremble! How my members pour down milky fear!
A dewy garment covers me all over; all manhood is gone.
At thy word and at thy look death enrobes me about
From head to feet, a garment of death and eternal fear.
Is not that sun thy husband and that moon thy glimmering veil?
Are not the stars of heaven their children? Art thou not Babylon?
Art thou Nature, mother of all? Is Jerusalem thy daughter?
Why hast thou elevate inward, O dweller of outward chambers,
From grot and cave beneath the moon, dim region of death   90
Where I laid my plough in the hot noon, where my hot team fed,
Where implements of war are forged, the plough to go over the
    nations,

In pain girding me round like a rib of iron in heaven? O Vala,
In Eternity they neither marry nor are given in marriage.°
Albion the high cliff of the Atlantic is become a barren land.'

Los stood at his anvil; he heard the contentions of Vala.
He heaved his thund'ring bellows upon the valleys of Middlesex;
He opened his furnaces before Vala. Then Albion frowned in
    anger
On his rock, ere yet the starry heavens were fled away
From his awful members. And thus Los cried aloud   100
To the sons of Albion, and to Hand, the eldest son of Albion:

'I hear the screech of childbirth loud pealing, and the groans
Of death, in Albion's clouds dreadful uttered over all the earth.
What may Man be? Who can tell! but what may Woman be,
To have power over Man from cradle to corruptible grave?
There is a throne in every man; it is the throne of God.
This, Woman has claimed as her own, and Man is no more.
Albion is the tabernacle of Vala and her temple,
And not the tabernacle and temple of the most high.
O Albion, why wilt thou create a female will?   110
To hide the most evident God in a hidden covert, even
In the shadows of a woman and a secluded holy place,
That we may pry after him as after a stolen treasure
Hidden among the dead, and mured up from the paths of life.
Hand! Art thou not Reuben enrooting thyself into Bashan,
Till thou remainest a vaporous shadow in a void? O Merlin!
Unknown among the dead where never before existence came,
Is this the female will? O ye lovely daughters of Albion!—to
Converse concerning weight and distance in the wilds of Newton
    and Locke?'

So Los spoke, standing on Mam Tor, looking over Europe and   120
    Asia;°
The graves thunder beneath his feet from Ireland to Japan.

Reuben slept in Bashan like one dead in the valley,
Cut off from Albion's mountains and from all the earth's summits,
Between Succoth and Zaretan beside the Stone of Bohan,°
While the daughters of Albion divided Luvah into three bodies.
Los bended his nostrils down to the earth, then sent him over

Jordan to the land of the Hittite. Everyone that saw him
Fled. They fled at his horrible form; they hid in caves
And dens; they looked on one another and became what they
    beheld.

Reuben returned to Bashan; in despair he slept on the stone.   130
Then Gwendolen divided into Rahab and Tirzah in twelve
    portions.
Los rolled his eyes into two narrow circles, then sent him
Over Jordan. All terrified fled; they became what they beheld.

If perceptive organs vary, objects of perception seem to vary.
If the perceptive organs close, their objects seem to close also.
'Consider this, O mortal man, O worm of sixty winters,' said Los;
'Consider sexual organization and hide thee in the dust.'°

Then the divine hand found the two limits, Satan and Adam,
In Albion's bosom; for in every human bosom those limits stand.
And the divine voice came from the furnaces, as multitudes   140
    without
Number, the voices of the innumerable multitudes of Eternity.
And the appearance of a man was seen in the furnaces,
Saving those who have sinned from the punishment of the law
(In pity of the punisher whose state is eternal death),
And keeping them from sin by the mild counsels of his love:

'Albion goes to eternal death; in me all Eternity
Must pass through condemnation, and awake beyond the grave!
No individual can keep these laws, for they are death
To every energy of Man, and forbid the springs of life.
Albion hath entered the state Satan! Be permanent, O state!   150
And be thou for ever accursed, that Albion may arise again!
And be thou created into a state! I go forth to create
States, to deliver individuals evermore! Amen.'

So spoke the voice from the furnaces, descending into non-
    entity.

Reuben returned to his place; in vain he sought beautiful Tirzah,
For his eyelids were narrowed and his nostrils scented the
    ground.
And sixty winters Los raged in the divisions of Reuben,

Building the moon of Ulro, plank by plank and rib by rib.
Reuben slept in the cave of Adam, and Los folded his tongue
Between lips of mire and clay, then sent him forth over Jordan.   160
In the love of Tirzah he said: 'Doubt is my food day and night'.
All that beheld him fled howling and gnawed their tongues
For pain; they became what they beheld. In reasonings Reuben
    returned
To Heshbon; disconsolate he walked through Moab, and he
    stood°
Before the furnaces of Los in a horrible dreamful slumber,
On Mount Gilead looking toward Gilgal. And Los bended°
His ear in a spiral circle outward, then sent him over Jordan.

The seven nations fled before him; they became what they
    beheld.
Hand, Hyle and Coban fled; they became what they beheld.
Gwantok and Peachey hid in Damascus beneath Mount Leba-   170
    non,
Brereton and Slayd in Egypt. Huttn and Skofield and Kox
Fled over Chaldea in terror, in pains in every nerve.
Kotope and Bowen became what they beheld, fleeing over the
    earth,
And the twelve female Emanations fled with them agonising.

Jerusalem trembled, seeing her children driven by Los's hammer
In the visions of the dreams of Beulah on the edge of non-entity.
Hand stood between Reuben and Merlin, as the reasoning
    Spectre
Stands between the vegetative man and his immortal imagina-
    tion.

And the Four Zoas, clouded, rage east and west and north and
    south;
They change their situations in the universal man.   180
Albion groans; he sees the elements divide before his face,
And England, who is Britannia, divided into Jerusalem and Vala.
And Urizen assumes the east, Luvah assumes the south,
In his dark Spectre ravening from his open sepulchre.

And the Four Zoas, who are the four eternal senses of Man,
Became four elements, separating from the limbs of Albion;
These are their names in the vegetative generation.

And accident and chance were found hidden in length, breadth
   and height,
And they divided into four ravening deathlike forms,
Fairies and genii and nymphs and gnomes of the elements.   190
These are states permanently fixed by the divine power.
The Atlantic continent sunk round Albion's cliffy shore,
And the sea poured in amain upon the giants of Albion,
As Los bended the senses of Reuben. Reuben is Merlin
Exploring the three states of Ulro: creation, redemption, and
   judgement.

And many of the eternal ones laughed after their manner:

'Have you known the judgement that is arisen among the
Zoas of Albion, where a man dare hardly to embrace
His own wife, for the terrors of chastity that they call
By the name of morality? Their daughters govern all   200
In hidden deceit! They are vegetable, only fit for burning;
Art and science cannot exist but by naked beauty displayed.'

Then those in great Eternity who contemplate on death
Said thus: 'What seems to be, is!—to those to whom
It seems to be—and is productive of the most dreadful
Consequences to those to whom it seems to be, even of
Torments, despair, eternal death. But the divine mercy
Steps beyond and redeems Man in the body of Jesus. Amen.
And length, breadth, height again obey the Divine Vision.
   Hallelujah.'

And one stood forth from the divine family, and said: 210

'I feel my Spectre rising upon me! Albion! Arouse thyself!
Why dost thou thunder with frozen spectrous wrath against us?
The Spectre is, in giant man, insane and most deformed.
Thou wilt certainly provoke my Spectre against thine in fury!
He has a sepulchre hewn out of a rock ready for thee,
And a death of eight thousand years, forged by thyself, upon
The point of his spear!—if thou persistest to forbid with laws
Our Emanations, and to attack our secret supreme delights.'

So Los spoke, but when he saw blue death in Albion's feet
Again he joined the divine body, following merciful;   220

While Albion fled more indignant, revengeful, covering
His face and bosom with petrific hardness, and his hands
And feet, lest any should enter his bosom and embrace
His hidden heart. His Emanation wept and trembled within him,
Uttering not his jealousy, but hiding it as with
Iron and steel, dark and opaque, with clouds and tempests
    brooding.
His strong limbs shuddered upon his mountains high and dark.

Turning from universal love, petrific as he went,
His cold against the warmth of Eden raged, with loud
Thunders of deadly war (the fever of the human soul),   230
Fires and clouds of rolling smoke! But mild the Saviour followed
    him,
Displaying the eternal vision, the divine similitude,
In loves and tears of brothers, sisters, sons, fathers, and friends
(Which if Man ceases to behold, he ceases to exist),

Saying: 'Albion! Our wars are wars of life, and wounds of love,
With intellectual spears, and long-winged arrows of thought.
Mutual in one another's love and wrath all-renewing
We live as one man; for contracting our infinite senses
We behold multitude, or, expanding, we behold as one,
As one man all the universal family. And that one man   240
We call Jesus the Christ, and he in us, and we in him,
Live in perfect harmony in Eden, the land of life,
Giving, receiving, and forgiving each other's trespasses.
He is the good shepherd; he is the lord and master;
He is the shepherd of Albion; he is all in all,
In Eden, in the garden of God, and in heavenly Jerusalem.
If we have offended, forgive us, take not vengeance against us.'

Thus speaking, the divine family follow Albion;
I see them in the vision of God upon my pleasant valleys.

I behold London, a human awful wonder of God!   250
He says: 'Return, Albion, return! I give myself for thee;
My streets are my ideas of imagination.
Awake, Albion, awake! And let us awake up together.
My houses are thoughts walking within my blood-vessels,
The children of my thoughts walking within my blood-vessels,

Shut from my nervous form which sleeps upon the verge of
   Beulah
In dreams of darkness, while my vegetating blood, in veiny pipes,
Rolls dreadful through the furnaces of Los and the mills of Satan.
For Albion's sake, and for Jerusalem thy Emanation
I give myself, and these brethren give themselves for Albion.'   260

So spoke London, immortal guardian! I heard in Lambeth's
   shades.°
In Felpham I heard and saw the visions of Albion.
I write in South Molton Street, what I both see and hear
In regions of humanity, in London's opening streets.

I see thee, awful parent land in light, behold I see!
Verulam! Canterbury, venerable parent of men,
Generous immortal guardian, golden clad! For cities
Are men, fathers of multitudes, and rivers and mountains
Are also men; every thing is human, mighty! sublime!
In every bosom a universe expands as wings   270
Let down at will around, and called the universal tent.
York, crowned with loving kindness! Edinburgh! clothed
With fortitude, as with a garment of immortal texture
Woven in looms of Eden, in spiritual deaths of mighty men
Who give themselves, in Golgotha, victims to justice, where
There is in Albion a gate of precious stones and gold
Seen only by Emanations, by vegetations viewless.
Bending across the road of Oxford Street, it from Hyde Park
To Tyburn's deathful shades admits the wandering souls
Of multitudes who die from earth. This gate cannot be found   280
By Satan's watch-fiends, though they search numbering every
   grain
Of sand on earth every night; they never find this gate.
It is the gate of Los. Withoutside is the mill, intricate, dreadful
And filled with cruel tortures; but no mortal man can find the
   mill
Of Satan, in his mortal pilgrimage of seventy years,
For human beauty knows it not, nor can mercy find it! But
In the fourth region of humanity, Urthona named,
Mortality begins to roll the billows of eternal death
Before the gate of Los. Urthona here is named Los,
And here begins the system of moral virtue named Rahab.   290

Albion fled through the gate of Los, and he stood in the gate.

Los was the friend of Albion who most loved him. In Cambridgeshire,
His eternal station, he is the twenty-eighth and is fourfold.
Seeing Albion had turned his back against the Divine Vision,
Los said to Albion, 'Whither fleest thou?' Albion replied:

'I die! I go to eternal death. The shades of death
Hover within me and beneath, and spreading themselves outside
Like rocky clouds, build me a gloomy monument of woe.
Will none accompany me in my death, or be a ransom for me
In that dark valley? I have girded round my cloak, and on my feet   300
Bound these black shoes of death, and on my hands death's iron gloves.
God hath forsaken me, and my friends are become a burden,
A weariness to me, and the human footstep is a terror to me.'

Los answered, troubled, and his soul was rent in twain:
'Must the wise die for an atonement? Does mercy endure atonements?
No! It is moral severity, and destroys mercy in its victim';
So speaking, not yet infected with the error and illusion.

Los shuddered at beholding Albion, for his disease
Arose upon him pale and ghastly; and he called around
The friends of Albion. Trembling at the sight of eternal death   310
The four appeared with their Emanations in fiery
Chariots. Black their fires roll, beholding Albion's house of Eternity.
Damp couch the flames beneath, and silent, sick, stand shuddering
Before the porch of sixteen pillars. Weeping every one
Descended and fell down upon their knees round Albion's knees,
Swearing the oath of God with awful voice of thunders round
Upon the hills and valleys; the cloudy oath rolled far and wide.

'Albion is sick!' said every valley, every mournful hill
And every river. 'Our brother Albion is sick to death.
He hath leagued himself with robbers! He hath studied the arts   320
Of unbelief! Envy hovers over him! His friends are his abhorrence!

Those who give their lives for him are despised!
Those who devour his soul are taken into his bosom!
To destroy his Emanation is their intention.
Arise! Awake, O friends of the giant Albion!
They have persuaded him of horrible falsehoods!
They have sown errors over all his fruitful fields!'

The twenty-four heard! They came trembling on wat'ry chariots,
Borne by the living creatures of the third procession
Of human majesty; the living creatures wept aloud as they   330
Went along Albion's roads, till they arrived at Albion's house.

Oh, how the torments of eternal death waited on Man!
And the loud-rending bars of the creation ready to burst,
That the wide world might fly from its hinges, and the immortal
    mansion
Of Man for ever be possessed by monsters of the deeps,
And Man himself become a fiend, wrapped in an endless curse,
Consuming and consumed forever in flames of moral justice.

For had the body of Albion fall'n down, and from its dreadful
    ruins
Let loose the enormous Spectre on the darkness of the deep,
At enmity with the merciful and filled with devouring fire,   340
A netherworld must have received the foul enormous spirit,
Under pretence of moral virtue, filled with revenge and law:
There to eternity chained down and issuing in red flames
And curses, with his mighty arms brandished against the
    heavens,
Breathing cruelty, blood and vengeance, gnashing his teeth with
    pain,
Torn with black storms and ceaseless torrents of his own
    consuming fire,
Within his breast his mighty sons chained down and filled with
    cursings,
And his dark eon, that once fair crystal form divinely clear,
Within his ribs producing serpents whose souls are flames of fire.
But, glory to the merciful one, for he is of tender mercies!   350
And the divine family wept over him as one man.

And these the twenty-four in whom the divine family
Appeared; and they were one in him, a human vision!—
Human-divine, Jesus the Saviour, blessed for ever and ever.

Selsey, true friend who afterwards submitted to be devoured°
By the waves of despair, whose Emanation rose above
The flood, and was named Chichester, lovely, mild and gentle!
   Lo,
Her lambs bleat to the sea-fowls' cry, lamenting still for Albion!'

Submitting to be called the son of Los, the terrible vision,
Winchester stood devoting himself for Albion; his tents   360
Outspread with abundant riches, and his Emanations
Submitting to be called Enitharmon's daughters, and be born
In vegetable mould, created by the hammer and loom
In Bowlahoola and Allamanda, where the dead wail night and
   day.

I call them by their English names: English, the rough basement.
Los built the stubborn structure of the language, acting against
Albion's melancholy, who must else have been a dumb despair.

Gloucester and Exeter and Salisbury and Bristol, and benevolent
   Bath;
Bath who is Legions. He is the seventh, the physician and°
The poisoner, the best and worst in Heaven and Hell,   370
Whose Spectre first assimilated with Luvah in Albion's moun-
   tains.
A triple octave he took, to reduce Jerusalem to twelve,
To cast Jerusalem forth upon the wilds to Poplar and Bow,
To Maldon and Canterbury in the delights of cruelty.
The shuttles of death sing in the sky to Islington and Pancras
Round Marybone to Tyburn's river, weaving black melancholy
   as a net,
And despair as meshes closely wove over the west of London,
Where mild Jerusalem sought to repose in death and be no more.
She fled to Lambeth's mild vale and hid herself beneath
The Surrey hills where Rephaim terminates. Her sons are seized   380
For victims of sacrifice, but Jerusalem cannot be found, hid
By the daughters of Beulah, gently snatched away and hid in
   Beulah.

There is a grain of sand in Lambeth that Satan cannot find,
Nor can his watch-fiends find it; 'tis translucent and has many
  angles.
But he who finds it will find Oothoon's palace, for within,
Opening into Beulah, every angle is a lovely heaven.
But should the watch-fiends find it, they would call it Sin,
And lay its heavens and their inhabitants in blood of punishment.
Here Jerusalem and Vala were hid in soft slumberous repose,
Hid from the terrible east, shut up in the south and west.   390

The twenty-eight trembled in death's dark caves; in cold despair°
They kneeled around the couch of death, in deep humiliation
And tortures of self-condemnation, while their Spectres raged
  within.
The four Zoas in terrible combustion clouded rage,
Drinking the shuddering fears and loves of Albion's families,
Destroying by selfish affections the things that they most admire,
Drinking and eating, and pitying and weeping, as at a tragic scene
The soul drinks murder and revenge, and applauds its own
  holiness.

They saw Albion endeavouring to destroy their Emanations.
They saw their wheels rising up poisonous against Albion: 400
Urizen, cold and scientific; Luvah, pitying and weeping;
Tharmas, indolent and sullen; Urthona, doubting and despair-
  ing;
Victims to one another and dreadfully plotting against each other
To prevent Albion walking about in the four complexions.

They saw America closed out by the oaks of the western shore,
And Tharmas dashed on the rocks of the altars of victims in
  Mexico.
'If we are wrathful, Albion will destroy Jerusalem with rooty
  groves;
If we are merciful, ourselves must suffer destruction on his oaks.
Why should we enter into our Spectres to behold our own
  corruptions?
O God of Albion, descend! Deliver Jerusalem from the oaken   410
  groves!'
Then Los grew furious, raging: 'Why stand we here trembling
  around,

Calling on God for help and not ourselves in whom God dwells,
Stretching a hand to save the falling man? Are we not four,
Beholding Albion upon the precipice ready to fall into non-
    entity,
Seeing these heavens and hells conglobing in the void? Heavens
    over hells
Brooding in holy hypocritic lust, drinking the cries of pain
From howling victims of law, building heavens twenty-seven-
    fold;
Swelled and bloated general forms, repugnant to the Divine
Humanity, who is the only general and universal form,
To which all lineaments tend and seek with love and sympathy.   420
All broad and general principles belong to benevolence,
Who protects minute particulars, every one in their own identity;
But here the affectionate touch of the tongue is closed in by
    deadly teeth,
And the soft smile of friendship and the open dawn of
    benevolence
Become a net and a trap, and every energy rendered cruel,
Till the existence of friendship and benevolence is denied.
The wine of the spirit and the vineyards of the holy one
Here turn into poisonous stupor and deadly intoxication,
That they may be condemned by law and the Lamb of God be
    slain!
And the two sources of life in Eternity, hunting and war,   430
Are becoming the sources of dark and bitter death and of
    corroding hell.
The open heart is shut up in integuments of frozen silence,
That the spear that lights it forth may shatter the ribs and bosom.
A pretence of art to destroy art! A pretence of liberty
To destroy liberty, a pretence of religion to destroy religion.
Oshea and Caleb fight; they contend in the valleys of Peor°
In the terrible family contentions of those who love each other.
The armies of Balaam weep; no women come to the field.°
Dead corses lay before them, and not as in wars of old.
For the soldier who fights for truth, calls his enemy his brother;   440
They fight and contend for life, and not for eternal death!
But here the soldier strikes, and a dead corse falls at his feet;
Nor daughter nor sister nor mother come forth to embosom the
    slain,
But death! eternal death! remains in the valleys of Peor.

The English are scattered over the face of the nations. Are these
Jerusalem's children? Hark! Hear the giants of Albion cry at
   night:
"We smell the blood of the English! We delight in their blood on
   our altars.°
The living and the dead shall be ground in our rumbling mills
For bread of the sons of Albion, of the giants Hand and
   Skofield."
Skofield and Kox are let loose upon my Saxons! They accumu-   450
   late
A world in which man is by his nature the enemy of man,
In pride of selfhood unwieldy stretching out into non-entity,
Generalizing art and science till art and science is lost.

Bristol and Bath, listen to my words, and ye seventeen, give ear!
It is easy to acknowledge a man to be great and good while we
Derogate from him in the trifles and small articles of that
   goodness;
Those alone are his friends who admire his minutest powers.
Instead of Albion's lovely mountains and the curtains of
   Jerusalem
I see a cave, a rock, a tree deadly and poisonous, unimaginative.
Instead of the mutual forgivenesses, the minute particulars, I see   460
Pits of bitumen ever burning, artificial riches of the Canaanite,
Like lakes of liquid lead. Instead of heavenly chapels built
By our dear Lord I see worlds crusted with snows and ice.
I see a wicker idol woven round Jerusalem's children. I see
The Canaanite, the Amalekite, the Moabite, the Egyptian
By demonstrations, the cruel sons of quality and negation,
Driven on the void in incoherent despair into non-entity.
I see America closed apart, and Jerusalem driven in terror
Away from Albion's mountains, far away from London's spires.
I will not endure this thing! I alone withstand to death   470
This outrage! Ah me, how sick and pale you all stand round me!
Ah me, pitiable ones! Do you also go to death's vale?
All you my friends and brothers, all you my beloved companions,
Have you also caught the infection of sin and stern repentance?
I see disease arise upon you! Yet speak to me and give
Me some comfort. Why do you all stand silent? I alone
Remain in permanent strength. Or is all this goodness and pity
   only
That you may take the greater vengeance in your sepulchre?'

So Los spoke. Pale they stood around the house of death,
In the midst of temptations and despair, among the rooted oaks,   480
Among reared rocks of Albion's sons. At length they rose
With one accord in love sublime, and as on cherub's wings
They Albion surround with kindest violence, to bear him back
Against his will through Los's gate to Eden. Fourfold, loud,
Their wings! waving over the bottomless immense to bear
Their awful charge back to his native home. But Albion dark,
Repugnant, rolled his wheels backward into non-entity!
Loud roll the starry wheels of Albion into the world of death,
And all the gate of Los clouded with clouds redounding from
Albion's dread wheels, stretching out spaces immense between.   490
That every little particle of light and air became opaque,
Black and immense, a rock of difficulty and a cliff
Of black despair; that the immortal wings laboured against
Cliff after cliff, and over valleys of despair and death.
The narrow sea between Albion and the Atlantic continent,
Its waves of pearl, became a boundless ocean, bottomless,
Of grey obscurity, filled with clouds and rocks and whirling
    waters,
And Albion's sons ascending and descending in the horrid void.

But as the will must not be bended but in the day of divine
Power, silent, calm and motionless, in the mid-air sublime,   500
The family divine hover around the darkened Albion.

Such is the nature of the Ulro: that whatever enters
Becomes sexual, and is created, and vegetated, and born.
From Hyde Park spread their vegetating roots beneath Albion
In dreadful pain, the spectrous uncircumcized vegetation
Forming a sexual machine, an aged virgin form
In Erin's land toward the north, joint after joint, and burning
In love and jealousy immingled, and calling it Religion.
And feeling the damps of death they with one accord delegated
    Los,
Conjuring him by the highest that he should watch over them,   510
Till Jesus shall appear. And they gave their power to Los,
Naming him the Spirit of Prophecy, calling him Elijah.

Stricken with Albion's disease they become what they behold;
They assimilate with Albion in pity and compassion.

Their Emanations return not; their Spectres rage in the deep.
The slumbers of death came over them around the couch of
    death,
Before the gate of Los and in the depths of non-entity,
Among the furnaces of Los, among the oaks of Albion.

Man is adjoined to man by his emanative portion,
Who is Jerusalem in every individual man; and her   520
Shadow is Vala, builded by the reasoning power in Man.
Oh search and see. Turn your eyes inward. Open, O thou world
Of love and harmony in Man; expand thy ever-lovely gates.

They wept into the deeps a little space. At length was heard
The voice of Bath, faint as the voice of the dead in the house of
    death,
Bath, healing city, whose wisdom in midst of poetic
Fervour mild spoke through the western porch, in soft gentle
    tears:

'O Albion, mildest son of Eden! Closed is thy western gate.
Brothers of Eternity! This man, whose great example
We all admired and loved, whose all-benevolent countenance,   530
    seen
In Eden, in lovely Jerusalem, drew even from envy
The tear, and the confession of honesty, open and undisguised,
From mistrust and suspicion—the man is himself become
A piteous example of oblivion. To teach the sons
Of Eden that however great and glorious, however loving
And merciful the individuality, however high
Our palaces and cities, and however fruitful are our fields
In selfhood, we are nothing, but fade away in morning's breath.
Our mildness is nothing; the greatest mildness we can use
Is incapable and nothing. None but the Lamb of God can heal   540
This dread disease, none but Jesus. O Lord, descend and save!
Albion's western gate is closed; his death is coming apace.
Jesus alone can save him; for alas, we none can know
How soon his lot may be our own. When Africa in sleep
Rose in the night of Beulah, and bound down the sun and moon,
His friends cut his strong chains and overwhelmed his dark
Machines in fury and destruction, and the man reviving
    repented.

He wept before his wrathful brethren, thankful and considerate
For their well-timed wrath. But Albion's sleep is not
Like Africa's, and his machines are woven with his life.   550
Nothing but mercy can save him! Nothing but mercy, interpos-
   ing
Lest he should slay Jerusalem in his fearful jealousy.
O God, descend! Gather our brethren; deliver Jerusalem.
But that we may omit no office of the friendly spirit,
Oxford, take thou these leaves of the Tree of Life. With
   eloquence°
That thy immortal tongue inspires, present them to Albion.
Perhaps he may receive them, offered from thy loved hands.'

So spoke, unheard by Albion, the merciful Son of Heaven
To those whose western gates were open, as they stood weeping
Around Albion. But Albion heard him not, obdurate! hard!   560
He frowned on all his friends, counting them enemies in his
   sorrow.

And the seventeen—conjoining with Bath, the seventh,
In whom the other ten shone manifest, a Divine Vision—
Assimilated and embraced eternal death for Albion's sake.

And these the names of the eighteen combining with those ten.°
Bath, mild physician of Eternity, mysterious power,
Whose springs are unsearchable and knowledge infinite.
Hereford, ancient guardian of Wales, whose hands
Builded the mountain palaces of Eden, stupendous works.
Lincoln, Durham and Carlisle, counsellors of Los,   570
And Ely, scribe of Los, whose pen no other hand
Dare touch. Oxford, immortal bard! With eloquence
Divine he wept over Albion, speaking the words of God
In mild persuasion, bringing leaves of the Tree of Life:

'Thou art in error, Albion, the land of Ulro.
One error not removed will destroy a human soul.
Repose in Beulah's night till the error is removed.
Reason not on both sides. Repose upon our bosoms,
Till the plough of Jehovah, and the harrow of Shaddai
Have passed over the dead to awake the dead to judgement.'   580
But Albion turned away, refusing comfort.

Oxford trembled while he spoke, then fainted in the arms
Of Norwich, Peterborough, Rochester, Chester awful, Worces-
ter,
Lichfield, Saint David's, Llandaff, Asaph, Bangor, Sodor,
Bowing their heads devoted. And the furnaces of Los
Began to rage. Thundering loud the storms began to roar
Upon the furnaces, and loud the furnaces rebellow beneath.

And these the four in whom the twenty-four appeared fourfold:
Verulam, London, York, Edinburgh, mourning one towards
another.
Alas, the time will come when a man's worst enemies   590
Shall be those of his own house and family, in a religion
Of generation, to destroy by sin and atonement happy Jerusalem,
The bride and wife of the Lamb. O God, thou art not an
avenger!'

Thus Albion sat, studious of others in his pale disease,
Brooding on evil. But when Los opened the furnaces before him
He saw that the accursed things were his own affections,
And his own beloved's. Then he turned sick; his soul died within
him.
Also Los sick and terrified beheld the furnaces of death
And must have died, but the divine Saviour descended
Among the infant loves and affections, and the Divine Vision   600
wept
Like evening dew on every herb upon the breathing ground.

Albion spoke in his dismal dreams: 'O thou deceitful friend,
Worshipping mercy and beholding thy friend in such affliction.
Los! Thou now discoverest thy turpitude to the heavens.
I demand righteousness and justice, O thou ingratitude!
Give me my Emanations back, food for my dying soul.
My daughters are harlots! My sons are accursed before me.
Enitharmon is my daughter, accursed with a father's curse.
Oh, I have utterly been wasted! I have given my daughters to
devils.'

So spoke Albion in gloomy majesty, and deepest night   610
Of Ulro rolled round his skirts from Dover to Cornwall.

Los answered: 'Righteousness and justice I give thee in return
For thy righteousness! But I add mercy also, and bind
Thee from destroying these little ones. Am I to be only
Merciful to thee, and cruel to all that thou hatest?
Thou wast the image of God surrounded by the four Zoas.
Three thou hast slain! I am the fourth; thou canst not destroy me.
Thou art in error; trouble me not with thy righteousness.
I have innocence to defend and ignorance to instruct.
I have no time for seeming, and little arts of compliment   620
In morality and virtue, in self-glorying and pride.
There is a limit of opaqueness and a limit of contraction
In every individual man, and the limit of opaqueness
Is named Satan, and the limit of contraction is named Adam.
But when Man sleeps in Beulah the Saviour in mercy takes
Contraction's limit, and of the limit he forms Woman: that
Himself may in process of time be born, Man to redeem.
But there is no limit of expansion! There is no limit of
    translucence
In the bosom of Man for ever from eternity to eternity!
Therefore I break thy bonds of righteousness; I crush thy   630
    messengers,
That they may not crush me and mine. Do thou be righteous,
And I will return it; otherwise I defy thy worst revenge.
Consider me as thine enemy; on me turn all thy fury,
But destroy not these little ones, nor mock the Lord's anointed.°
Destroy not by moral virtue the little ones whom he hath chosen!
The little ones whom he hath chosen in preference to thee.
He hath cast thee off for ever; the little ones he hath anointed!
Thy selfhood is for ever accursed from the divine presence.'

So Los spoke, then turned his face and wept for Albion.

Albion replied: 'Go, Hand and Hyle! Seize the abhorred friend,   640
As you have seized the twenty-four rebellious ingratitudes,
To atone for you, for spiritual death. Man lives by deaths of men.
Bring him to justice before Heaven, here upon London Stone,
Between Blackheath and Hounslow, between Norwood and
    Finchley.
All that they have is mine; from my free gen'rous gift,
They now hold all they have. Ingratitude to me,
To me their benefactor, calls aloud for vengeance deep!'

Los stood before his furnaces awaiting the fury of the dead;
And the divine hand was upon him, strengthening him mightily.

The Spectres of the dead cry out from the deeps beneath 650
Upon the hills of Albion. Oxford groans in his iron furnace,
Winchester in his den and cavern. They lament against
Albion; they curse their human kindness and affection;
They rage like wild beasts in the forests of affliction.
In the dreams of Ulro they repent of their human kindness.

'Come up; build Babylon, Rahab is ours, and all her multitudes
With her in pomp and glory of victory. Depart,
Ye twenty-four, into the deeps; let us depart to glory!'

Their human majestic forms sit up upon their couches
Of death; they curb their Spectres as with iron curbs. 660
They enquire after Jerusalem in the regions of the dead
With the voices of dead men, low, scarcely articulate;
And with tears cold on their cheeks they weary repose.

'Oh, when shall the morning of the grave appear, and when
Shall our salvation come? We sleep upon our watch;
We cannot awake! And our Spectres rage in the forests.
O God of Albion, where art thou? Pity the watchers!'

Thus mourn they. Loud the furnaces of Los thunder upon
The clouds of Europe and Asia, among the serpent temples.

And Los drew his seven furnaces around Albion's altars, 670
And as Albion built his frozen altars, Los built the Mundane
  Shell
In the four regions of humanity, east and west and north and
  south,
Till Norwood and Finchley and Blackheath and Hounslow
  covered the whole earth.
This is the net and veil of Vala among the souls of the dead.

Then the Divine Vision like a silent sun appeared above°
Albion's dark rocks, setting behind the gardens of Kensington
On Tyburn's river, in clouds of blood, where was mild Zion hill's
Most ancient promontory; and in the sun a human form
  appeared.
And thus the voice divine went forth upon the rocks of Albion:

'I elected Abion for my glory; I gave to him the nations   680
Of the whole earth. He was the angel of my presence, and all
The sons of God were Albion's sons and Jerusalem was my joy.
The reactor hath hid himself through envy. I behold him;
But you cannot behold him till he be revealed in his system.
Albion's reactor must have a place prepared; Albion must sleep
The sleep of death, till the man of sin and repentance be revealed.
Hidden in Albion's forests he lurks; he admits of no reply
From Albion, but hath founded his reaction into a law°
Of action, for obedience to destroy the contraries of man.
He hath compelled Albion to become a punisher, and hath   690
  possessed
Himself of Albion's forests and wilds! And Jerusalem is taken!
The city of the woods in the forest of Ephratah is taken!°
London is a stone of her ruins; Oxford is the dust of her walls;
Sussex and Kent are her scattered garments, Ireland her holy
  place;
And the murdered bodies of her little ones are Scotland and
  Wales.
The cities of the nations are the smoke of her consummation;
The nations are her dust! Ground by the chariot wheels
Of her lordly conquerors, her palaces levelled with the dust.
I come that I may find a way for my banished ones to return.
Fear not, O little flock; I come. Albion shall rise again.'°   700

So saying, the mild sun enclosed the human family.

Forthwith from Albion's dark'ning rocks came two immortal
  forms,
Saying: 'We alone are escaped, O merciful Lord and Saviour.°
We flee from the interiors of Albion's hills and mountains,
From his valleys eastward, from Amalek, Canaan and Moab,
Beneath his vast ranges of hills surrounding Jerusalem!

'Albion walked on the steps of fire before his halls,
And Vala walked with him in dreams of soft deluding slumber.
He looked up and saw the prince of light with splendour faded.
Then Albion ascended mourning into the porches of his palace.   710
Above him rose a shadow from his wearied intellect
Of living gold, pure, perfect, holy; in white linen pure he
  hovered,

A sweet entrancing self-delusion, a wat'ry vision of Albion,
Soft exulting in existence, all the man absorbing!

'Albion fell upon his face prostrate before the wat'ry shadow,
Saying: "O Lord, whence is this change? Thou knowest I am
   nothing?"
And Vala trembled and covered her face! And her locks were
   spread on the pavement.

'We heard, astonished at the vision, and our hearts trembled
   within us;
We heard the voice of slumberous Albion, and thus he spake,
Idolatrous to his own shadow, words of eternity uttering: 720

' "Oh, I am nothing when I enter into judgement with thee!°
If thou withdraw thy breath I die and vanish into Hades;°
If thou dost lay thine hand upon me, behold I am silent;
If thou withhold thine hand, I perish like a fallen leaf.
Oh I am nothing, and to nothing must return again.
If thou withdraw thy breath, behold I am oblivion."

'He ceased. The shadowy voice was silent; but the cloud hovered
   over their heads
In golden wreaths, the sorrow of Man, and the balmy drops fell
   down.
And lo! that son of Man, that shadowy spirit of mild Albion,
Luvah, descended from the cloud; in terror Albion rose.   730
Indignant rose the awful man, and turned his back on Vala.

'We heard the voice of Albion starting from his sleep:

' "Whence is this voice crying, *Enion*! that soundeth in my ears?
O cruel pity! O dark deceit! Can love seek for dominion?"

'And Luvah strove to gain dominion over Albion.
They strove together above the body where Vala was enclosed,
And the dark body of Albion left prostrate upon the crystal
   pavement,
Covered with boils from head to foot, the terrible smitings of
   Luvah.°

'Then frowned the fall'n man and put forth Luvah from his
  presence,
Saying: "Go and die the death of Man for Vala the sweet   740
  wanderer.
I will turn the volutions of your ears outward, and bend your
  nostrils
Downward, and your fluxile eyes englobed roll round in fear;
Your with'ring lips and tongue shrink up into a narrow circle,
Till into narrow forms you creep. Go take your fiery way,
And learn what 'tis to absorb the man, you spirits of pity and
  love."

'They heard the voice and fled swift as the winter's setting sun.
And now the human blood foamed high; the spirits Luvah and
  Vala
Went down the human heart where paradise and its joys
  abounded,
In jealous fears and fury and rage, and flames roll round their
  fervid feet.
And the vast form of Nature like a serpent played before them.   750
And as they fled in folding fires and thunders of the deep,
Vala shrunk in like the dark sea that leaves its slimy banks,
And from her bosom Luvah fell far as the east and west.
And the vast form of Nature like a serpent rolled between,
Whether of Jerusalem's or Vala's ruins congenerated we know
  not:
All is confusion, all is tumult, and we alone are escaped.'
So spoke the fugitives; they joined the divine family, trembling.

And the two that escaped were the Emanation of Los and his
Spectre, for wherever the Emanation goes, the Spectre
Attends her as her guard. And Los's Emanation is named   760
Enitharmon, and his Spectre is named Urthona. They knew
Not where to flee; they had been on a visit to Albion's children,
And they strove to weave a shadow of the Emanation
To hide themselves, weeping and lamenting for the vegetation
Of Albion's children, fleeing through Albion's vales in streams of
  gore.

Being not irritated by insult, bearing insulting benevolences,
They perceived that corporeal friends are spiritual enemies.

They saw the sexual religion in its embryon uncircumcision,
And the divine hand was upon them, bearing them through
    darkness
Back safe to their humanity as doves to their windows.   770
Therefore the sons of Eden praise Urthona's Spectre in songs,
Because he kept the Divine Vision in time of trouble.

They wept and trembled, and Los put forth his hand and took
    them in,
Into his bosom, from which Albion shrunk in dismal pain,
Rending the fibres of brotherhood, and in feminine allegories
Enclosing Los. But the Divine Vision appeared with Los,
Following Albion into his central void among his oaks.

And Los prayed and said: 'O Divine Saviour, arise
Upon the mountains of Albion as in ancient time. Behold!
The cities of Albion seek thy face; London groans in pain   780
From hill to hill; and the Thames laments along the valleys.
The little villages of Middlesex and Surrey hunger and thirst;
The twenty-eight cities of Albion stretch their hands to thee,
Because of the oppressors of Albion in every city and village.
They mock at the labourer's limbs! They mock at his starved
    children;
They buy his daughters that they may have power to sell his sons.
They compel the poor to live upon a crust of bread by soft mild
    arts.
They reduce the man to want, then give with pomp and
    ceremony.
The praise of Jehovah is chanted from lips of hunger and thirst.
Humanity knows not of sex; wherefore are sexes in Beulah?   790
In Beulah the female lets down her beautiful tabernacle;
Which the male enters magnificent between her cherubim,°
And becomes one with her, mingling, condensing in self-love
The rocky law of condemnation and double generation and
    death.°
Albion hath entered the loins, the place of the Last Judgement,
And Luvah hath drawn the curtains around Albion in Vala's
    bosom.
The dead awake to generation! Arise, O Lord, and rend the veil!'

So Los in lamentations followed Albion. Albion covered
His western heaven with rocky clouds of death and despair.

Fearing that Albion should turn his back against the Divine   800
    Vision,
Los took his globe of fire to search the interiors of Albion's
Bosom in all the terrors of friendship, entering the caves
Of despair and death to search the tempters out, walking among
Albion's rocks and precipices, caves of solitude and dark despair,
And saw every minute particular of Albion degraded and
    murdered—
But saw not by whom. They were hidden within in the minute
    particulars
Of which they had possessed themselves, and there they take up
The articulations of man's soul, and laughing throw it down°
Into the frame, then knock it out upon the plank; and souls are
    baked
In bricks to build the pyramids of Heber and Terah. But Los   810
Searched in vain; closed from the minutia he walked difficult.
He came down from Highgate through Hackney and Holloway
    towards London
Till he came to Old Stratford, and thence to Stepney and the Isle
Of Leutha's Dogs, thence through the narrows of the river's side;
And saw every minute particular, the jewels of Albion, running
    down
The kennels of the streets and lanes as if they were abhorred.°
Every universal form was become barren mountains of moral
Virtue, and every minute particular hardened into grains of sand,
And all the tenderness of the soul cast forth as filth and mire
Among the winding places of deep contemplation intricate.   820
To where the Tower of London frowned dreadful over
    Jerusalem,
A building of Luvah, builded in Jerusalem's eastern gate to be
His secluded court; thence to Bethlehem, where was builded°
Dens of despair in the house of bread, enquiring in vain
Of stones and rocks he took his way, for human form was none.
And thus he spoke, looking on Albion's city with many tears:

'What shall I do? What could I do if I could find these criminals?
I could not dare to take vengeance; for all things are so
    constructed

And builded by the divine hand that the sinner shall always
    escape,
And he who takes vengeance alone is the criminal of providence.   830
If I should dare to lay my finger on a grain of sand
In way of vengeance, I punish the already punished. Oh, whom
Should I pity if I pity not the sinner who is gone astray?
O Albion, if thou takest vengeance, if thou revengest thy wrongs,
Thou art for ever lost! What can I do to hinder the sons
Of Albion from taking vengeance, or how shall I them persuade?'

So spoke Los, travelling through darkness and horrid solitude.
And he beheld Jerusalem in Westminster and Marybone
Among the ruins of the temple, and Vala who is her shadow,°
Jerusalem's shadow, bent northward over the island white.   840
At length he sat on London Stone and heard Jerusalem's voice:

'Albion, I cannot be thy wife. Thine own minute particulars
Belong to God alone, and all thy little ones are holy.
They are of faith and not of demonstration, wherefore is Vala
Clothed in black mourning upon my river's currents. Vala,
    awake!
I hear thy shuttles sing in the sky, and round my limbs
I feel the iron threads of love and jealousy and despair.'

Vala replied: 'Albion is mine! Luvah gave me to Albion
And now receives reproach and hate. Was it not said of old,
"Set your son before a man and he shall take you and your sons   850
For slaves: but set your daugher before a man and she
Shall make him and his sons and daughters your slaves for ever"?
And is this faith? Behold, the strife of Albion and Luvah
Is great in the east; their spears of blood rage in the eastern
    heaven.
Urizen is the champion of Albion; they will slay my Luvah.
And thou, O harlot daughter, daughter of despair, art all
This cause of these shakings of my towers on Euphrates.
Here is the house of Albion, and here is thy secluded place,
And here we have found thy sins. And hence we turn thee forth
For all to avoid thee, to be astonished at thee for thy sins;   860
Because thou art the impurity and the harlot, and thy children
Children of whoredoms, born for sacrifice, for the meat and
    drink

Offering, to sustain the glorious combat and the battle and war,
That Man may be purified by the death of thy delusions.'

So saying, she her dark threads cast over the trembling river
And over the valleys: from the hills of Hertfordshire to the hills
Of Surrey, across Middlesex and across Albion's house
Of eternity. Pale stood Albion at his eastern gate,
Leaning against the pillars, and his disease rose from his skirts.
Upon the precipice he stood, ready to fall into non-entity.   870

Los was all astonishment and terror; he trembled, sitting on the
     Stone
Of London. But the interiors of Albion's fibres and nerves were
     hidden
From Los; astonished he beheld only the petrified surfaces,
And saw his furnaces in ruins (for Los is the demon of the
     furnaces).
He also saw the four points of Albion reversed inwards.
He seized his hammer and tongs, his iron poker and his bellows,
Upon the valleys of Middlesex, shouting loud for aid divine.

In stern defiance came from Albion's bosom Hand, Hyle, Coban,
Gwantok, Peachey, Brereton, Slayd, Huttn, Skofield, Kox,
     Kotope,
Bowen, Albion's sons. They bore him a golden couch into the   880
     porch
And on the couch reposed his limbs, trembling from the bloody
     field,
Rearing their Druid patriarchal rocky temples around his limbs
(All things begin and end in Albion's ancient Druid rocky shore.)

From Camberwell to Highgate, where the mighty Thames
     shudders along,
Where Los's furnaces stand, where Jerusalem and Vala howl,
Luvah tore forth from Albion's loins, in fibrous veins, in rivers
Of blood over Europe: a vegetating root in grinding pain
Animating the dragon temples, soon to become that holy fiend,
The wicker man of Scandinavia, in which, cruelly consumed,
The captives reared to heaven howl in flames among the stars.   890
Loud the cries of war on the Rhine and Danube with Albion's
     sons;

Away from Beulah's hills and vales break forth the souls of the
    dead,
With cymbal, trumpet, clarion and the scythed chariots of
    Britain.

And the veil of Vala is composed of the Spectres of the dead.

Hark, the mingling cries of Luvah with the sons of Albion!
Hark, and record the terrible wonder!—that the punisher
Mingles with his victim's Spectre, enslaved and tormented
To him whom he has murdered, bound in vengeance and enmity.
Shudder not, but write, and the hand of God will assist you!
Therefore I write Albion's last words: 'Hope is banished from   900
    me.'

These were his last words, and the merciful Saviour in his arms
Received him, in the arms of tender mercy, and reposed
The pale limbs of his eternal individuality
Upon the Rock of Ages. Then, surrounded with a cloud,
In silence the divine Lord builded with immortal labour
Of gold and jewels a sublime ornament, a couch of repose,
With sixteen pillars, canopied with emblems and written verse,
Spiritual verse, ordered and measured—from whence time shall
    reveal
The five books of the Decalogue, the books of Joshua and
    Judges,°
Samuel, a double book, and Kings, a double book, the Psalms   910
    and Prophets,
The fourfold Gospel, and the Revelations everlasting.
Eternity groaned and was troubled at the image of eternal death!

Beneath the bottoms of the graves, which is earth's central joint,
There is a place where contrarieties are equally true
(To protect from the giant blows in the sports of intellect,
Thunder in the midst of kindness, and love that kills its beloved,
Because death is for a period, and they renew tenfold).
From this sweet place maternal love awoke Jerusalem;
With pangs she forsook Beulah's pleasant lovely shadowy
    universe,
Where no dispute can come, created for those who sleep.   920

Weeping was in all Beulah, and all the daughters of Beulah
Wept for their sister the daughter of Albion, Jerusalem,
When out of Beulah the Emanation of the sleeper descended
With solemn mourning, out of Beulah's moony shades and hills
Within the human heart, whose gates closed with solemn sound.

And this the manner of the terrible separation.
The Emanations of the grievously afflicted friends of Albion
Concentre in one female form, an aged pensive woman.
Astonished, lovely, embracing the sublime shade, the daughters
    of Beulah
Beheld her with wonder! With awful hands she took   930
A moment of time, drawing it out with many tears and afflictions
And many sorrows oblique across the Atlantic vale
(Which is the vale of Rephaim dreadful from east to west,
Where the human harvest waves abundant in the beams of
    Eden),
Into a rainbow of jewels and gold, a mild reflection from
Albion's dread tomb, eight thousand and five hundred years
In its extension; every two hundred years has a door to Eden.
She also took an atom of space, with dire pain opening it, a centre
Into Beulah. Trembling the daughters of Beulah dried
Her tears; she ardent embraced her sorrows, occupied in labours   940
Of sublime mercy in Rephaim's vale. Perusing Albion's tomb
She sat; she walked among the ornaments solemn mourning.
The daughters attended her shudderings, wiping the death-
    sweat.

Los also saw her in his seventh furnace; he also terrified
Saw the finger of God go forth upon his seventh furnace,
Away from the starry wheels to prepare Jerusalem a place—
When with a dreadful groan the Emanation mild of Albion
Burst from his bosom in the tomb like a pale snowy cloud,
Female and lovely, struggling to put off the human form,
Writhing in pain. The daughters of Beulah in kind arms received   950
Jerusalem, weeping over her among the spaces of Erin,
In the ends of Beulah, where the dead wail night and day.

And thus Erin spoke to the daughters of Beulah, in soft tears:

'Albion the vortex of the dead! Albion the generous!
Albion the mildest son of Heaven! The place of holy sacrifice,
Where friends die for each other!—will become the place

Of murder, and unforgiving, never-awaking sacrifice of enemies.
The children must be sacrificed (a horror never known
Till now in Beulah!), unless a refuge can be found
To hide them from the wrath of Albion's law, that freezes sore   960
Upon his sons and daughters, self-exiled from his bosom.
Draw ye Jerusalem away from Albion's mountains
To give a place for redemption; let Sihon and Og
Remove eastward to Bashan and Gilead, and leave
The secret coverts of Albion and the hidden places of America.
Jerusalem, Jerusalem! why wilt thou turn away?
Come ye, O daughters of Beulah, lament for Og and Sihon
Upon the lakes of Ireland from Rathlin to Baltimore;
Stand ye upon the Dargle from Wicklow to Drogheda;
Come and mourn over Albion, the white cliff of the Atlantic,   970
The mountain of giants. All the giants of Albion are become
Weak, withered, darkened! And Jerusalem is cast forth from
    Albion.
They deny that they ever knew Jerusalem, or ever dwelt in
    Shiloh.°
The gigantic roots and twigs of the vegetating sons of Albion,
Filled with the little ones, are consumed in the fires of their altars.
The vegetating cities are burned and consumed from the earth,
And the bodies in which all animals and vegetations, the earth
    and heaven,
Were contained in the all-glorious imagination are withered and
    darkened.
The golden gate of Havilah, and all the garden of God,
Was caught up with the sun in one day of fury and war.   980
The lungs, the heart, the liver shrunk away far distant from Man,
And left a little slimy substance floating upon the tides.
In one night the Atlantic continent was caught up with the moon,
And became an opaque globe far distant, clad with moony
    beams.
The visions of Eternity, by reason of narrowed perceptions,
Are become weak visions of time and space, fixed into furrows of
    death,
Till deep dissimulation is the only defence an honest man has
    left.
O polypus of death! O Spectre over Europe and Asia,
Withering the human form by laws of sacrifice for sin!
By laws of chastity and abhorrence I am withered up,   990

Striving to create a heaven in which all shall be pure and holy
In their own selfhoods, in natural selfish chastity; to banish pity
And dear mutual forgiveness, and to become one great Satan,
Enslaved to the most powerful selfhood; to murder the Divine
   Humanity,
In whose sight all are as the dust, and who chargeth his angels
   with folly.°
Ah, weak and wide astray! Ah, shut in narrow doleful form!
Creeping in reptile flesh upon the bosom of the ground!
The eye of Man, a little narrow orb, closed up and dark,
Scarcely beholding the great light, conversing with the void;
The ear, a little shell, in small volutions shutting out   1000
True harmonies, and comprehending great as very small;
The nostrils, bent down to the earth and closed with senseless
   flesh,
That odours cannot them expand, nor joy on them exult;
The tongue, a little moisture fills, a little food it cloys,
A little sound it utters, and its cries are faintly heard.
Therefore they are removed; therefore they have taken root
In Egypt and Philistia, in Moab and Edom and Aram;
In the Erythrean Sea their uncircumcision in heart and loins
Be lost for ever and ever. Then they shall arise from self
By self-annihilation, into Jerusalem's courts and into Shiloh,   1010
Shiloh the masculine Emanation among the flowers of Beulah.
Lo, Shiloh dwells over France, as Jerusalem dwells over Albion.
Build and prepare a wall and curtain for America's shore!
Rush on, rush on, rush on, ye vegetating sons of Albion!
The sun shall go before you in day, the moon shall go°
Before you in night. Come on, come on, come on! The Lord
Jehovah is before, behind, above, beneath, around.
He has builded the arches of Albion's tomb, binding the stars
In merciful order, bending the laws of cruelty to peace.
He hath placed Og and Anak, the giants of Albion, for their   1020
   guards,
Building the body of Moses in the valley of Peor, the body
Of divine analogy; and Og and Sihon in the tears of Balaam,
The son of Beor, have given their power to Joshua and Caleb.°
Remove from Albion, far remove these terrible surfaces.
They are beginning to form heavens and hells in immense
Circles, the hells for food to the heavens, food of torment,
Food of despair; they drink the condemned soul and rejoice

In cruel holiness in their heavens of chastity and uncircumcision.
Yet they are blameless, and iniquity must be imputed only
To the state they are entered into that they may be delivered.   1030
Satan is the state of death and not a human existence;
But Luvah is named Satan, because he has entered that state,
A world where man is by nature the enemy of man,
Because the evil is created into a state, that men
May be delivered time after time evermore. Amen.
Learn therefore, O sisters, to distinguish the eternal human
That walks about among the stones of fire, in bliss and woe°
Alternate, from those states or worlds in which the spirit travels;
This is the only means to forgiveness of enemies.
Therefore remove from Albion these terrible surfaces,   1040
And let wild seas and rocks close up Jerusalem away from
The Atlantic mountains: where giants dwelt in intellect,
Now given to stony Druids and allegoric generation,
To the twelve gods of Asia, the Spectres of those who sleep,
Swayed by a providence opposed to the divine Lord Jesus,
A murderous providence!—a creation that groans, living on
    death,
Where fish and bird and beast and man and tree and metal and
    stone
Live by devouring, going into eternal death continually.
Albion is now possessed by the war of blood! The sacrifice
Of envy Albion is become, and his Emanation cast out.   1050
Come, Lord Jesus, Lamb of God, descend! For if, O Lord!°
If thou hadst been here, our brother Albion had not died.°
Arise, sisters! Go ye and meet the Lord, while I remain.
Behold the foggy mornings of the dead on Albion's cliffs,
Ye know that if the Emanation remains in them
She will become an eternal death, an avenger of sin,
A self-righteousness: the proud virgin-harlot! mother of war!
And we also, and all Beulah, consume beneath Albion's curse.'

So Erin spoke to the daughters of Beulah. Shuddering
With their wings they sat in the furnace, in a night   1060
Of stars; for all the sons of Albion appeared distant stars,
Ascending and descending into Albion's sea of death.°
And Erin's lovely bow enclosed the wheels of Albion's sons.
Expanding on wing, the daughters of Beulah replied in sweet
    response:

'Come, O thou Lamb of God, and take away the remembrance of
    sin.°
To sin and to hide the sin in sweet deceit is lovely!
To sin in the open face of day is cruel and pitiless! But
To record the sin for a reproach, to let the sun go down,°
In a remembrance of the sin, is a woe and a horror,
A brooder of an evil day, and a sun rising in blood!   1070
Come then, O Lamb of God, and take away the remembrance of
    sin.'

CHAPTER III

TO THE DEISTS

[see above, p. 32]

But Los, who is the vehicular form of strong Urthona,
Wept vehemently over Albion where Thames currents spring
From the rivers of Beulah: pleasant river, soft, mild, parent
    stream!
And the roots of Albion's tree entered the soul of Los
As he sat before his furnaces clothed in sackcloth of hair,
In gnawing pain, dividing him from his Emanation,
Enclosing all the children of Los time after time,
Their giant forms condensing into nations and peoples and
    tongues,
Translucent the furnaces, of beryl and emerald immortal;
And sevenfold each within other, incomprehensible   10
To the vegetated mortal eye's perverted and single vision.
The bellows are the animal lungs; the hammers, the animal heart;
The furnaces, the stomach or digestion. Terrible their fury,
Like seven burning heavens ranged from south to north.

Here, on the banks of the Thames, Los builded Golgonooza,
Outside of the gates of the human heart, beneath Beulah,
In the midst of the rocks of the altars of Albion. In fears
He builded it, in rage and in fury. It is the spiritual fourfold
London, continually building and continually decaying desolate!
In eternal labours, loud the furnaces and loud the anvils   20
Of death thunder incessant around the flaming couches of
The twenty-four friends of Albion and around the awful four,
For the protection of the twelve Emanations of Albion's sons,
The mystic union of the Emanation in the Lord. Because

Man divided from his Emanation is a dark Spectre,
His Emanation is an ever-weeping melancholy shadow.
But she is made receptive of generation through mercy
In the potter's furnace, among the funeral urns of Beulah,
From Surrey hills, through Italy and Greece, to Hinnom's vale.

In great Eternity, every particular form gives forth or emanates   30
Its own peculiar light, and the form is the Divine Vision,
And the light is his garment. This is Jerusalem in every man,
A tent and tabernacle of mutual forgiveness, male and female
    clothings;
And Jerusalem is called Liberty among the children of Albion.

But Albion fell down, a rocky fragment from Eternity hurled
By his own Spectre, who is the reasoning power in every man,
Into his own chaos, which is the memory between man and man.

The silent broodings of deadly revenge, springing from the
All-powerful parental affection, fills Albion from head to foot.
Seeing his sons assimilate with Luvah, bound in the bonds   40
Of spiritual hate, from which springs sexual love as iron chains,
He tosses like a cloud outstretched among Jerusalem's ruins
Which overspread all the earth; he groans among his ruined
    porches.

But the Spectre like a hoar-frost and a mildew rose over Albion,
Saying: 'I am God, O sons of men! I am your rational power!
Am I not Bacon and Newton and Locke, who teach humility to
    Man,
Who teach doubt and experiment? And my two wings, Voltaire,
    Rousseau?
Where is that friend of sinners, that rebel against my laws,
Who teaches belief to the nations and an unknown eternal life?
Come hither into the desert and turn these stones to bread.°   50
Vain foolish man! Wilt thou believe without experiment?
And build a world of fantasy upon my great abyss,
A world of shapes, in craving lust and devouring appetite?'

So spoke the hard, cold, constrictive Spectre (he is named
    Arthur)
Constricting into Druid rocks round Canaan, Agag and Aram
    and Pharaoh.°

Then Albion drew England into his bosom in groans and tears.
But she stretched out her starry night in spaces against him, like
A long serpent, in the abyss of the Spectre, which augmented
The night with dragon wings covered with stars; and in the wings
Jerusalem and Vala appeared, and above, between the wings   60
    magnificent,
The Divine Vision dimly appeared in clouds of blood, weeping.

When those who disregard all mortal things saw a mighty one
Among the flowers of Beulah still retain his awful strength,
They wondered. Checking their wild flames, and many gathering
Together into an assembly, they said: 'Let us go down
And see these changes!' Others said: 'If you do so, prepare
For being driven from our fields; what have we to do with the
    dead?
To be their inferiors or superiors we equally abhor;
Superior, none we know; inferior none. All equal share°
Divine benevolence and joy, for the Eternal Man   70
Walketh among us, calling us his brothers and his friends,
Forbidding us that veil which Satan puts between Eve and
    Adam,
By which the princes of the dead enslave their votaries,
Teaching them to form the serpent of precious stones and gold;
To seize the sons of Jerusalem and plant them in one man's loins;
To make one family of contraries that Joseph may be sold
Into Egypt, for negation, a veil the Saviour born and dying
    rends.'°

But others said: 'Let us to him who only is, and who
Walketh among us, give decision. Bring forth all your fires!'

So saying an eternal deed was done; in fiery flames   80
The universal concave raged, such thunderous sounds as never
Were sounded from a mortal cloud, not on Mount Sinai old,
Nor in Havilah where the cherub rolled his redounding flame.

Loud, loud, the mountains lifted up their voices, loud the forests!
Rivers thundered against their banks; loud winds furious fought.
Cities and nations contended in fires and clouds and tempests.
The seas raised up their voices and lifted their hands on high;
The stars in their courses fought, the sun, moon, heaven, earth°

Contending!—for Albion and for Jerusalem his Emanation,
And for Shiloh the Emanation of France, and for lovely Vala.   90

Then far the greatest number were about to make a separation,
And they elected seven, called the Seven Eyes of God:
Lucifer, Molech, Elohim, Shaddai, Pahad, Jehovah, Jesus.
They named the eighth; he came not, he hid in Albion's forests.
But first they said (and their words stood in chariots in array,
Curbing their tigers with golden bits and bridles of silver and
  ivory):

'Let the human organs be kept in their perfect integrity,
At will contracting into worms or expanding into gods,
And then, behold, what are these Ulro visions of chastity?
Then, as the moss upon the tree or dust upon the plough,   100
Or as the sweat upon the labouring shoulder, or as the chaff
Of the wheat-floor, or as the dregs of the sweet wine-press,
Such are these Ulro visions. For though we sit down within
The ploughed furrow, listening to the weeping clods till we
Contract or expand space at will, or if we raise ourselves
Upon the chariots of the morning, contracting or expanding
  time,
Everyone knows we are one family, one man blessed for ever!'

Silence remained, and every one resumed his human majesty.
And many conversed on these things as they laboured at the
  furrow,
Saying: 'It is better to prevent misery than to release from   110
  misery.
It is better to prevent error than to forgive the criminal!
Labour well the minute particulars, attend to the little ones,
And those who are in misery cannot remain so long,
If we do but our duty. Labour well the teeming earth.'

They ploughed in tears; the trumpets sounded before the golden
  plough,
And the voices of the living creatures were heard in the clouds of
  heaven,
Crying: 'Compel the reasoner to demonstrate with unhewn
  demonstrations.
Let the indefinite be explored, and let every man be judged

By his own works. Let all indefinities be thrown into demon-
    strations,
To be pounded to dust and melted in the furnaces of affliction.   120
He who would do good to another must do it in minute
    particulars.
"General good" is the plea of the scoundrel, hypocrite and
    flatterer,
For art and science cannot exist but in minutely organised
    particulars,
And not in generalizing demonstrations of the rational power.
The infinite alone resides in definite and determinate identity.
Establishment of truth depends on destruction of falsehood
    continually,
On circumcision, not on virginity, O reasoners of Albion.'

So cried they at the plough. Albion's rock frowned above,
And the great voice of eternity rolled above, terrible in clouds,
Saying: 'Who will go forth for us, and who shall we send before   130
    our face?'°

Then Los heaved his thund'ring bellows on the valley of
    Middlesex,
And thus he chanted his song. The daughters of Albion reply:

'What may Man be? Who can tell? But what may Woman be,
To have power over Man from cradle to corruptible grave?
He who is an infant, and whose cradle is a manger,
Knoweth the infant sorrow, whence it came, and where it goeth,
And who weave it a cradle of the grass that withereth away.
This world is all a cradle for the erred wandering phantom,
Rocked by year, month, day and hour, and every two moments
Between dwells a daughter of Beulah to feed the human   140
    vegetable.
Entune, daughters of Albion, your hymning chorus mildly!
Cord of affection thrilling ecstatic on the iron reel,
To the golden loom of love, to the moth-laboured woof,
A garment and cradle weaving for the infantine terror,
For fear. At entering the gate into our world of cruel
Lamentation, it flees back and hides in non-entity's dark wild,
Where dwells the Spectre of Albion, destroyer of definite form.
The sun shall be a scythed chariot of Britain; the moon, a ship

In the British ocean created by Los's hammer!—measured out
Into days and nights and years and months, to travel with my feet   150
Over these desolate rocks of Albion. O daughters of despair!
Rock the cradle, and in mild melodies tell me where found
What you have enwoven with so much tears and care, so much
Tender artifice, to laugh, to weep, to learn, to know?
Remember, recollect, what dark befel in wintry days!'

'Oh, it was lost for ever! And we found it not; it came
And wept at our wintry door. Look, look, behold! Gwendolen
Is become a clod of clay! Merlin is a worm of the valley!'

Then Los uttered with hammer and anvil: 'Chant! Revoice!
I mind not your laugh, and your frown I not fear! And   160
You must my dictate obey from your gold-beamed looms. Trill
Gentle to Albion's watchman on Albion's mountains; re-echo
And rock the cradle while! Ah me, of that Eternal Man,°
And of the cradled infancy in his bowels of compassion!—
Who fell beneath his instruments of husbandry and became
Subservient to the clods of the furrow. The cattle and even
The emmet and earthworms are his superiors and his lords.'

Then the response came warbling from trilling looms in Albion:
'We women tremble at the light therefore, hiding fearful
The Divine Vision with curtain and veil and fleshy tabernacle.'   170
Los uttered, swift as the rattling thunder upon the mountains,
'Look back into the Church Paul! Look! Three women around
The cross! O Albion, why didst thou a female will create?'

And the voices of Bath and Canterbury and York and Edinburgh
    cry
Over the plough of nations in the strong hand of Albion,
    thundering along
Among the fires of the Druid and the deep black rethundering
    waters
Of the Atlantic, which poured in impetuous, loud, loud, louder
    and louder.
And the great voice of the Atlantic howled over the Druid altars,
Weeping over his children in Stonehenge, in Maldon and
    Colchester,
Round the rocky Peak of Derbyshire, London Stone and   180
    Rosamond's bower:°

'What is a wife and what is a harlot? What is a church and what
Is a theatre? Are they two and not one? Can they exist separate?
Are not religion and politics the same thing? Brotherhood is
    religion;
Oh demonstrations of reason, dividing families in cruelty and
    pride!'

But Albion fled from the Divine Vision, with the plough of
    nations enflaming.
The living creatures maddened, and Albion fell into the furrow,
    and
The plough went over him, and the living was ploughed in
    among the dead.
But his Spectre rose over the starry plough. Albion fled beneath
    the plough
Till he came to the Rock of Ages, and he took his seat upon the
    rock.

Wonder seized all in Eternity to behold the Divine Vision open   190
The centre into an expanse! And the centre rolled out into an
    expanse.

In beauty the daughters of Albion divide and unite at will,
Naked and drunk with blood, Gwendolen dancing to the timbrel
Of war, reeling up the street of London. She divides in twain,
Among the inhabitants of Albion. The people fall around;
The daughters of Albion divide and unite in jealousy and cruelty.
The inhabitants of Albion at the harvest and the vintage
Feel their brain cut round beneath the temples, shrieking,
Bonifying into a skull, the marrow exuding in dismal pain.°
They flee over the rocks bonifying; horses, oxen feel the knife.   200
And while the sons of Albion by severe war and judgement
    bonify,
The hermaphroditic condensations are divided by the knife,
The obdurate forms are cut asunder by jealousy and pity.

Rational philosophy and mathematic demonstration
Is divided in the intoxications of pleasure and affection.
Two contraries war against each other in fury and blood,
And Los fixes them on his anvil; incessant his blows.

He fixes them with strong blows, placing the stones and timbers
To create a world of generation from the world of death,
Dividing the masculine and feminine; for the commingling   210
Of Albion's and Luvah's spectres was hermaphroditic.

Urizen wrathful strode above, directing the awful building,
As a mighty temple, delivering form out of confusion.
Jordan sprang beneath its threshold, bubbling from beneath
Its pillars; Euphrates ran under its arches; white sails
And silver oars reflect on its pillars and sound on its echoing
Pavements, where walk the sons of Jerusalem who remain
    ungenerate.
But the revolving sun and moon pass through its porticoes;
Day and night, in sublime majesty and silence, they revolve
And shine glorious within. Hand and Coban arched over the sun   220
In the hot noon, as he travelled through his journey; Hyle and
    Skofield
Arched over the moon at midnight, and Los fixed them there
With his thunderous hammer. Terrified the Spectres rage and
    flee.
Canaan is his portico; Jordan is a fountain in his porch,
A fountain of milk and wine to relieve the traveller.
Egypt is the eight steps within; Ethiopia supports his pillars;
Lybia and the lands unknown are the ascent without;
Within is Asia and Greece, ornamented with exquisite art;
Persia and Media are his halls; his inmost hall is great Tartary;
China and India and Serbia are his temples for entertainment,   230
Poland and Russia and Sweden his soft retired chambers;
France and Spain and Italy and Denmark and Holland and
    Germany
Are the temples among his pillars. Britain is Los's forge;
America, North and South, are his baths of living waters.

Such is the ancient world of Urizen in the Satanic void,
Created from the valley of Middlesex by London's river
From Stonehenge and from London Stone, from Cornwall to
    Caithness.
The four Zoas rush around on all sides in dire ruin;
Furious in pride of selfhood, the terrible Spectres of Albion
Rear their dark rocks among the stars of God, stupendous   240
Works! A world of generation continually creating out of

The hermaphroditic Satanic world of rocky destiny,
And formed into four precious stones, for entrance from Beulah.

For the veil of Vala, which Albion cast into the Atlantic deep
To catch the souls of the dead, began to vegetate and petrify
Around the earth of Albion, among the roots of his tree.
This Los formed into the gates and mighty wall, between the oak
Of weeping and the palm of suffering beneath Albion's tomb.
Thus in process of time it became the beautiful Mundane Shell,
The habitation of the Spectres of the dead, and the place   250
Of redemption and of awaking again into Eternity.

For four universes round the mundane egg remain chaotic.
One to the north, Urthona; one to the south, Urizen;
One to the east, Luvah; one to the west, Tharmas.
They are the four Zoas that stood around the throne divine.°
Verulam, London, York and Edinburgh their English names.
But when Luvah assumed the world of Urizen southward,
And Albion was slain upon his mountain and in his tent,
All fell towards the centre, sinking downwards in dire ruin.
In the south remains a burning fire; in the east, a void;   260
In the west, a world of raging waters; in the north, solid darkness
Unfathomable, without end. But in the midst of these
Is built eternally the sublime universe of Los and Enitharmon.

And in the north gate, in the west of the north, toward Beulah,
Cathedron's looms are builded, and Los's furnaces in the south.
A wondrous golden building immense with ornaments sublime
Is bright Cathedron's golden hall, its courts, towers and
    pinnacles.

And one daughter of Los sat at the fiery reel, and another
Sat at the shining loom with her sisters attending round;
Terrible their distress, and their sorrow cannot be uttered.   270
And another daughter of Los sat at the spinning-wheel.
Endless their labour, with bitter food, void of sleep;
Though hungry they labour. They rouse themselves anxious,
Hour after hour labouring at the whirling wheel,
Many wheels—and as many lovely daughters sit weeping.
Yet the intoxicating delight that they take in their work
Obliterates every other evil; none pities their tears,
Yet they regard not pity and they expect no one to pity.

For they labour for life and love, regardles of anyone
But the poor Spectres that they work for, always, incessantly.   280

They are mocked by everyone that passes by. They regard not;
They labour. And when their wheels are broken by scorn and
    malice
They mend them sorrowing with many tears and afflictions.

Other daughters weave on the cushion and pillow network fine,
That Rahab and Tirzah may exist and live and breathe and love.
Ah, that it could be as the daughters of Beulah wish!

Other daughters of Los, labouring at looms less fine,
Create the silk-worm and the spider and the caterpillar
To assist in their most grievous work of pity and compassion.
And others create the woolly lamb and the downy fowl   290
To assist in the work. The lamb bleats; the sea-fowl cries.
Men understand not the distress and the labour and sorrow
That in the interior worlds is carried on in fear and trembling,
Weaving the shudd'ring fears and loves of Albion's families.
Thunderous rage the spindles of iron, and the iron distaff
Maddens in the fury of their hands, weaving in bitter tears
The veil of goats-hair, and purple and scarlet and fine-twined
    linen.°

The clouds of Albion's Druid temples rage in the eastern heaven,
While Los sat terrified beholding Albion's Spectre, who is
    Luvah,
Spreading in bloody veins in torments over Europe and Asia: 300
Not yet formed, but a wretched torment unformed and abyssal
In flaming fire. Within the furnaces the Divine Vision appeared
On Albion's hills; often, walking from the furnaces in clouds
And flames among the Druid temples and the starry wheels,
Gathered Jerusalem's children in his arms and bore them like
A shepherd, in the night of Albion which overspread all the
    earth.

'I gave thee liberty and life, O lovely Jerusalem,
And thou hast bound me down upon the stems of vegetation.
I gave thee sheep-walks upon the Spanish mountains, Jerusalem;
I gave thee Priam's city and the isles of Grecia lovely.
I gave thee Hand and Skofield and the counties of Albion.   310
They spread forth like a lovely root into the garden of God;

They were as Adam before me, united into one man.
They stood in innocence and their skiey tent reached over Asia
To Nimrod's tower, to Ham and Canaan walking with Mizraim°
Upon the Egyptian Nile, with solemn songs to Grecia
And sweet Hesperia, even to great Chaldea and Tesshina,°
Following thee as a shepherd by the four rivers of Eden.
Why wilt thou rend thyself apart, Jerusalem?
And build this Babylon, and sacrifice in secret groves   320
Among the gods of Asia, among the fountains of pitch and nitre!
Therefore thy mountains are become barren, Jerusalem!
Thy valleys, plains of burning sand; thy rivers, waters of death.
Thy villages die of the famine, and thy cities
Beg bread from house to house, lovely Jerusalem.
Why wilt thou deface thy beauty and the beauty of thy little ones
To please thy idols, in the pretended chastities of uncircum-
  cision?
Thy sons are lovelier than Egypt or Assyria; wherefore
Dost thou blacken their beauty by a secluded place of rest,
And a peculiar tabernacle, to cut the integuments of beauty   330
Into the veils of tears and sorrows, O lovely Jerusalem?
They have persuaded thee to this; therefore their end shall come.
And I will lead thee through the wilderness in shadow of my
  cloud,
And in my love I will lead thee, lovely shadow of sleeping
  Albion.'

This is the song of the Lamb, sung by slaves in evening time.

But Jerusalem faintly saw him, closed in the dungeons of
  Babylon.
Her form was held by Beulah's daughters, but all within unseen
She sat at the mills, her hair unbound, her feet naked,
Cut with the flints. Her tears run down; her reason grows like
The wheel of Hand, incessant turning day and night without   340
  rest.
Insane she raves upon the winds, hoarse, inarticulate.
All night Vala hears; she triumphs in pride of holiness
To see Jerusalem deface her lineaments with bitter blows
Of despair, while the Satanic holiness triumphed in Vala,
In a religion of chastity and uncircumcised selfishness,
Both of the head and the heart and loins, closed up in moral
  pride.

But the Divine Lamb stood beside Jerusalem; oft she saw
The lineaments divine and oft the voice heard, and oft she said:

'O Lord and Saviour, have the gods of the heathen pierced thee?
Or hast thou been pierced in the house of thy friends?   350
Art thou alive, and livest thou for evermore? Or art thou
Not—but a delusive shadow, a thought that liveth not?
Babel mocks, saying there is no God nor Son of God,
That thou, O human imagination, O divine body, art all
A delusion. But I know thee, O Lord, when thou arisest upon
My weary eyes, even in this dungeon and this iron mill.
The stars of Albion cruel rise; thou bindest to sweet influences,°
For thou also sufferest with me although I behold thee not.
And, although I sin and blaspheme thy holy name, thou pitiest
    me,
Because thou knowest I am deluded by the turning mills,   360
And by these visions of pity and love, because of Albion's death.'

Thus spake Jerusalem, and thus the divine voice replied:

'Mild shade of Man, pitiest thou these visions of terror and woe!
Give forth thy pity and love; fear not! Lo, I am with thee always.°
Only believe in me that I have power to raise from death
Thy brother who sleepeth in Albion! Fear not, trembling shade.°
Behold; in the visions of Elohim Jehovah behold Joseph and
    Mary,°
And be comforted, O Jerusalem, in the visions of Jehovah
    Elohim.'

She looked, and saw Joseph the carpenter in Nazareth and Mary
His espoused wife. And Mary said: 'If thou put me away from   370
    thee,
Dost thou not murder me?' Joseph spoke in anger and fury:
    'Should I
Marry a harlot and an adulteress?' Mary answered: 'Art thou
    more pure°
Than thy maker, who forgiveth sins and calls again her that is
    lost?
Though she hates, he calls her again in love. I love my dear
    Joseph,
But he driveth me away from his presence. Yet I hear the voice of
    God

In the voice of my husband; though he is angry for a moment, he will not

Utterly cast me away. If I were pure, never could I taste the sweets

Of the forgiveness of sins. If I were holy, I never could behold the tears

Of love of him who loves me in the midst of his anger in furnace of fire!'

'Ah, my Mary,' said Joseph, weeping over and embracing her closely in   380

His arms: 'Doth he forgive Jerusalem, and not exact purity from her who is

Polluted? I heard his voice in my sleep, and his angel in my dream,

Saying: "Doth Jehovah forgive a debt only on condition that it shall

Be paid? Doth he forgive pollution only on conditions of purity?

That debt is not forgiven! That pollution is not forgiven!

Such is the forgiveness of the gods, the moral virtues of the

Heathen, whose tender mercies are cruelty. But Jehovah's salvation

Is without money and without price, in the continual forgiveness of sins,°

In the perpetual mutual sacrifice in great Eternity! For behold!

There is none that liveth and sinneth not! And this is the covenant°   390

Of Jehovah: *If you forgive one another, so shall Jehovah forgive you,*

That he himself may dwell among you. Fear not then to take

To thee Mary thy wife, for she is with child by the Holy Ghost." '

Then Mary burst forth into a song! She flowed like a river of

Many streams in the arms of Joseph and gave forth her tears of joy

Like many waters—and emanating into gardens and palaces upon

Euphrates, and to forests and floods, and animals wild and tame from°

Gihon to Hiddekel, and to cornfields and villages and inhabitants

Upon Pison and Arnon and Jordan. And I heard the voice among
The reapers saying: 'Am I Jerusalem the lost adulteress? Or am I   400
Babylon come up to Jerusalem?' And another voice answered,
    saying:
'Does the voice of my Lord call me again? Am I pure through his
    mercy
And pity? Am I become lovely as a virgin in his sight, who am
Indeed a harlot drunken with the sacrifice of idols? Does he
Call her pure as he did in the days of her infancy, when she
Was cast out to the loathing of her person? The Chaldean took°
Me from my cradle. The Amalekite stole me away upon his
    camels
Before I had ever beheld with love the face of Jehovah, or known
That there was a God of mercy. O mercy, O Divine Humanity!
O forgiveness and pity and compassion! If I were pure I should   410
    never
Have known thee; if I were unpolluted I should never have
Glorified thy holiness, or rejoiced in thy great salvation.'

Mary leaned her side against Jerusalem; Jerusalem received
The infant into her hands in the visions of Jehovah. Times
    passed on.

Jerusalem fainted over the cross and sepulchre. She heard the
    voice:
'Wilt thou make Rome thy patriarch Druid, and the kings of
    Europe his
Horsemen? Man in the resurrection changes his sexual garments
    at will.
Every harlot was once a virgin, every criminal an infant love.
Repose on me till the morning of the grave. I am thy life.'

Jerusalem replied: 'I am an outcast; Albion is dead.   420
I am left to the trampling foot and the spurning heel.
A harlot I am called; I am sold from street to street.
I am defaced with blows and with the dirt of the prison.
And wilt thou become my husband, O my Lord and Saviour?
Shall Vala bring thee forth? Shall the chaste be ashamed also?
I see the maternal line, I behold the seed of the woman!°
Cainah and Ada and Zillah and Naamah wife of Noah,
Shuah's daughter and Tamar, and Rahab the Canaanitess,
Ruth the Moabite, and Bathsheba of the daughters of Heth,
Naamah the Ammonite, Zibeah the Philistine, and Mary.   430

These are the daughters of Vala, mother of the body of death;
But I thy Magdalen behold thy spiritual risen body.
Shall Albion arise? I know he shall arise at the last day!°
I know that in my flesh I shall see God; but Emanations°
Are weak, they know not whence they are, nor whither tend.'

Jesus replied: 'I am the resurrection and the life.°
I die and pass the limits of possibility, as it appears
To individual perception. Luvah must be created
And Vala; for I cannot leave them in the gnawing grave,
But will prepare a way for my banished ones to return.   440
Come now with me into the villages, walk through all the cities.
Though thou art taken to prison and judgement, starved in the
    streets,
I will command the cloud to give thee food, and the hard rock
To flow with milk and wine. Though thou seest me not a season,°
Even a long season and a hard journey and a howling wilderness,
Though Vala's cloud hide thee and Luvah's fires follow thee,
Only believe and trust in me. Lo, I am always with thee!'

So spoke the Lamb of God, while Luvah's cloud reddening
    above
Burst forth in streams of blood upon the heavens, and dark night
Involved Jerusalem. And the wheels of Albion's sons turned   450
    hoarse
Over the mountains, and the fires blazed on Druid altars,
And the sun set in Tyburn's brook, where victims howl and cry.

But Los beheld the Divine Vision among the flames of the
    furnaces.
Therefore he lived and breathed in hope, but his tears fell
    incessant
Because his children were closed from him apart, and Enithar-
    mon
Dividing in fierce pain. Also the vision of God was closed in
    clouds
Of Albion's Spectres, that Los in despair oft sat and often
    pondered
On death eternal, in fierce shudders upon the mountains of
    Albion
Walking, and in the vales in howling fierce. Then, to his anvils
Turning, anew began his labours, though in terrible pains.   460

Jehovah stood among the Druids in the valley of Annandale,
When the four Zoas of Albion, the four living creatures, the
    cherubim
Of Albion, tremble before the Spectre in the starry harness of the
    plough
Of nations. And their names are Urizen and Luvah and Tharmas
    and Urthona.

Luvah slew Tharmas, the angel of the tongue, and Albion
    brought him
To justice in his own city of Paris, denying the resurrection.
Then Vala, the wife of Albion, who is the daughter of Luvah,
Took vengeance twelve-fold among the chaotic rocks of the
    Druids,
Where the human victims howl to the moon, and Thor and
    Friga°
Dance the dance of death, contending with Jehovah among the   470
    cherubim.
The chariot wheels filled with eyes rage along the howling valley°
In the dividing of Reuben and Benjamin bleeding from Chester's
    river.

The giants and the witches and the ghosts of Albion dance with
Thor and Friga, and the fairies lead the moon along the valley of
    cherubim,
Bleeding in torrents from mountain to mountain, a lovely victim.
And Jehovah stood in the gates of the victim, and he appeared
A weeping infant in the gates of birth in the midst of heaven.

The cities and villages of Albion became rock and sand
    unhumanized,
The Druid sons of Albion and the heavens a void around
    unfathomable:
No human form, but sexual, and a little weeping infant pale,   480
    reflected
Multitudinous in the looking-glass of Enitharmon, on all sides
Around in the clouds of the female on Albion's cliffs of the dead.

Such the appearance in Cheviot in the divisions of Reuben,
When the cherubim hid their heads under their wings in deep
    slumbers,
When the Druids demanded chastity from Woman and all was
    lost.

'How can the female be chaste, O thou stupid Druid,' cried Los,
'Without the forgiveness of sins in the merciful clouds of
    Jehovah,
And without the baptism of repentance to wash away calumnies
    and
The accusations of sin, that each may be pure in their
    neighbour's sight?
O when shall Jehovah give us victims from his flock and herds,   490
Instead of human victims by the daughters of Albion and
    Canaan?'

Then laughed Gwendolen, and her laughter shook the nations
    and families of
The dead beneath Beulah, from Tyburn to Golgotha, and from
Ireland to Japan. Furious her lions and tigers and wolves sport
    before
Los on the Thames and Medway; London and Canterbury groan
    in pain.

Los knew not yet what was done; he thought it was all in vision,
In visions of the dreams of Beulah among the daughters of
    Albion.
Therefore the murder was put apart in the looking-glass of
    Enitharmon.

He saw in Vala's hand the Druid knife of revenge and the poison
    cup°
Of jealousy, and thought it a poetic vision of the atmospheres,°   500
Till Canaan rolled apart from Albion across the Rhine, along the
    Danube.

And all the land of Canaan suspended over the valley of Cheviot,
From Bashan to Tyre and from Troy to Gaza of the Amalekite.
And Reuben fled with his head downwards among the caverns
Of the Mundane Shell, which froze on all sides round Canaan on
The vast expanse, where the daughters of Albion weave the web
Of ages and generations, folding and unfolding it like a veil of
    cherubim.
And sometimes it touches the earth's summits, and sometimes
    spreads
Abroad into the indefinite Spectre, who is the rational power.

Then all the daughters of Albion became one before Los, even 510
    Vala.

And she put forth her hand upon the looms in dreadful howlings
Till she vegetated into a hungry stomach and a devouring
    tongue.
Her hand is a court of justice; her feet, two armies in battle;
Storms and pestilence in her locks; and in her loins earthquake,
And fire, and the ruin of cities and nations and families and
    tongues.

She cries: 'The human is but a worm, and thou, O male, thou art
Thyself female, a male. A breeder of seed, a son and husband,
    and lo,
The human-divine is woman's shadow, a vapour in the summer's
    heat.
Go assume papal dignity, thou Spectre, thou male harlot!
    Arthur,
Divide into the kings of Europe in times remote, O woman-born, 520
And woman-nourished, and woman-educated, and woman-
    scorned!'

'Wherefore art thou living,' said Los, 'and Man cannot live in thy
    presence?
Art thou Vala, the wife of Albion, O thou lovely daughter of
    Luvah?
All quarrels arise from reasoning, the secret murder and
The violent man-slaughter, these are the Spectre's double cave:
The sexual death, living on accusation of sin and judgement,
To freeze love and innocence into the gold and silver of the
    merchant.
Without forgiveness of sin, love is itself eternal death.'

Then the Spectre drew Vala into his bosom, magnificent, terrific,
Glittering with precious stones and gold, with garments of blood 530
    and fire.
He wept in deadly wrath of the Spectre, in self-contradicting
    agony,
Crimson with wrath and green with jealousy, dazzling with love
And jealousy immingled; and the purple of the violet darkened
    deep
Over the plough of nations thund'ring in the hand of Albion's
    Spectre.

A dark hermaphrodite they stood, frowning upon London's
    river;
And the distaff and spindle in the hands of Vala, with the flax of
Human miseries, turned fierce with the lives of men along the
    valley,
As Reuben fled before the daughters of Albion, taxing the
    nations.

Derby Peak yawned a horrid chasm at the cries of Gwendolen
    and at
The stamping feet of Ragan upon the flaming treadles of her   540
    loom,
That drop with crimson gore, with the loves of Albion and
    Canaan,
Opening along the valley of Rephaim, weaving over the caves of
    Machpelah,°
To decide two worlds with a great decision: a world of mercy and
A world of justice (the world of mercy for salvation,
To cast Luvah into the wrath, and Albion into the pity,
In the two contraries of humanity, and in the four regions.)

For in the depths of Albion's bosom in the eastern heaven
They sound the clarions strong! They chain the howling
    captives;
They cast the lots into the helmet; they give the oath of blood in
    Lambeth.
They vote the death of Luvah, and they nailed him to Albion's   550
    tree in Bath.
They stained him with poisonous blue; they enwove him in cruel
    roots°
To die a death of six thousand years bound round with
    vegetation.
The sun was black and the moon rolled a useless globe through
    Britain.

Then left the sons of Urizen the plough and harrow, the loom,
The hammer and the chisel, and the rule and compasses; from
    London fleeing
They forged the sword on Cheviot, the chariot of war and the
    battle-axe,
The trumpet fitted to mortal battle, and the flute of summer in
    Annandale.

And all the arts of life they changed into the arts of death in
    Albion:
The hour-glass contemned, because its simple workmanship
Was like the workmanship of the ploughman, and the water-   560
    wheel
That raises water into cisterns broken and burned with fire,
Because its workmanship was like the workmanship of the
    shepherd.
And in their stead intricate wheels invented, wheel without
    wheel,
To perplex youth in their outgoings, and to bind to labours in
    Albion
Of day and night the myriads of Eternity; that they may grind
And polish brass and iron hour after hour, laborious task,
Kept ignorant of its use; that they might spend the days of
    wisdom
In sorrowful drudgery, to obtain a scanty pittance of bread,
In ignorance to view a small portion and think that all,
And call it Demonstration, blind to all the simple rules of life.   570

'Now, now the battle rages round thy tender limbs, O Vala.
Now smile among thy bitter tears: now put on all thy beauty.
Is not the wound of the sword sweet, and the broken bone
    delightful?
Wilt thou now smile among the scythes when the wounded groan
    in the field?
We were carried away in thousands from London, and in tens
Of thousands from Westminster and Marybone: in ships closed
    up,
Chained hand and foot, compelled to fight under the iron whips
Of our captains, fearing our officers more than the enemy.
Lift up thy blue eyes, Vala, and put on thy sapphire shoes.
O melancholy Magdalen, behold the morning over Maldon   580
    break.
Gird on thy flaming zone; descend into the sepulchre of
    Canterbury.
Scatter the blood from thy golden brow, the tears from thy silver
    looks;
Shake off the waters from thy wings, and the dust from thy white
    garments.

Remember all thy feigned terrors on the secret couch of
    Lambeth's vale,
When the sun rose in glowing morn, with arms of mighty hosts
Marching to battle—who was wont to rise with Urizen's harps,
Girt as a sower with his seed to scatter life abroad over Albion.
Arise, O Vala! Bring the bow of Urizen; bring the swift arrows of
    light.
How raged the golden horses of Urizen, compelled to the chariot
    of love!
Compelled to leave the plough to the ox, to snuff up the winds of   590
    desolation,
To trample the cornfields in boastful neighings. This is no gentle
    harp;
This is no warbling brook, nor shadow of a myrtle tree,
But blood and wounds and dismal cries, and shadows of the oak,
And hearts laid open to the light by the broad grisly sword,
And bowels hid in hammered steel ripped quivering on the
    ground.
Call forth thy smiles of soft deceit; call forth thy cloudy tears.
We hear thy sighs in trumpets shrill when morn shall blood
    renew.'

So sang the Spectre sons of Albion round Luvah's stone of trial,
Mocking and deriding at the writhings of their victim on
    Salisbury,
Drinking his Emanation in intoxicating bliss, rejoicing in giant   600
    dance.
For a Spectre has no Emanation but what he imbibes from
    deceiving
A victim; then he becomes her priest and she is his tabernacle
And his oak grove, till the victim rend the woven veil,
In the end of his sleep when Jesus calls him from his grave.

Howling the victims on the Druid altars yield their souls
To the stern warriors; lovely sport the daughters round their
    victims,
Drinking their lives in sweet intoxication. Hence arose from Bath
Soft deluding odours, in spiral volutions intricately winding
Over Albion's mountains, a feminine indefinite cruel delusion.
Astonished, terrified, and in pain and torment, sudden they   610
    behold

Their own parent, the Emanation of their murdered enemy,
Become their Emanation, and their temple and tabernacle.
They knew not this Vala was their beloved mother Vala, Albion's
    wife.

Terrified at the sight of the victim, at his distorted sinews,
The tremblings of Vala vibrate through the limbs of Albion's
    sons,
While they rejoice over Luvah in mockery and bitter scorn.
Sudden they become like what they behold, in howlings and
    deadly pain.
Spasms smite their features, sinews and limbs; pale they look on
    one another.
They turn, contorted; their iron necks bend unwilling towards
Luvah; their lips tremble; their muscular fibres are cramped and   620
    smitten.
They become like what they behold! Yet immense in strength
    and power,
In awful pomp and gold, in all the precious unhewn stones of
    Eden,
They build a stupendous building on the Plain of Salisbury: with
    chains
Of rocks round London Stone, of reasonings, of unhewn
    demonstrations
In labyrinthine arches (mighty Urizen the architect), through
    which
The heavens might revolve and Eternity be bound in their chain.
Labour unparalleled! A wondrous rocky world of cruel destiny,
Rocks piled on rocks reaching the stars, stretching from pole to
    pole.
The building is natural religion and its altars natural morality,
A building of eternal death, whose proportions are eternal   630
    despair.

Here Vala stood turning the iron spindle of destruction
From heaven to earth, howling, invisible! But not invisible
Her two covering cherubs, afterwards named Voltaire and
    Rousseau:
Two frowning rocks on each side of the cove and stone of torture,
Frozen sons of the feminine tabernacle of Bacon, Newton and
    Locke.
For Luvah is France, the victim of the Spectres of Albion.

Los beheld in terror; he poured his loud storms on the furnaces.
The daughters of Albion, clothed in garments of needlework,
Strip them off from their shoulders and bosoms. They lay aside
Their garments; they sit naked upon the stone of trial.   640
The knife of flint passes over the howling victim; his blood°
Gushes and stains the fair side of the fair daughters of Albion.
They put aside his curls; they divide his seven locks upon
His forehead; they bind his forehead with thorns of iron.°
They put into his hand a reed; they mock, saying: 'Behold
The King of Canaan, whose are seven hundred chariots of iron!'
They take off his vesture whole with their knives of flint,
But they cut asunder his inner garments, searching with
Their cruel fingers for his heart. And there they enter in pomp,
In many tears; and there they erect a temple and an altar.   650
They pour cold water on his brain in front, to cause
Lids to grow over his eyes in veils of tears, and caverns
To freeze over his nostrils, while they feed his tongue from cups
And dishes of painted clay. Glowing with beauty and cruelty
They obscure the sun and the moon; no eye can look upon them.

Ah! Alas! At the sight of the victim, and at sight of those who are
    smitten,
All who see become what they behold; their eyes are covered
With veils of tears, and their nostrils and tongues shrunk up,
Their ears bent outwards. As their victim, so are they in the
    pangs
Of unconquerable fear, amidst delights of revenge earth-shaking!   660
And as their eye and ear shrunk, the heavens shrunk away;
The Divine Vision became first a burning flame, then a column
Of fire, then an awful fiery wheel surrounding earth and heaven,
And then a globe of blood wandering distant in an unknown
    night.
Afar into the unknown night the mountains fled away:
Six months of mortality, a summer; and six months of mortality,
    a winter.
The human form began to be altered by the daughters of Albion,
And the perceptions to be dissipated into the indefinite,
    becoming
A mighty polypus named Albion's Tree. They tie the veins
And nerves into two knots, and the seed into a double knot.   670
They look forth; the sun is shrunk, the heavens are shrunk

Away in the far remote, and the trees and mountains withered
Into indefinite cloudy shadows in darkness and separation.
By invisible hatreds adjoined, they seem remote and separate
From each other, and yet are a mighty polypus in the deep!
As the mistletoe grows on the oak, so Albion's tree on Eternity.
   Lo,
He who will not commingle in love, must be adjoined by hate!

They look forth from Stonehenge; from the cove round London
   Stone
They look on one another. The mountain calls out to the
   mountain;
Plinlimmon shrunk away; Snowdon trembled. The mountains   680
Of Wales and Scotland beheld the descending war, the routed
   flying.
Red run the streams of Albion; Thames is drunk with blood,
As Gwendolen cast the shuttle of war, as Cambel returned the
   beam.
The Humber and the Severn are drunk with the blood of the
   slain.
London feels his brain cut round; Edinburgh's heart is circums-
   cribed!
York and Lincoln hide among the flocks because of the griding
   knife;
Worcester and Hereford, Oxford and Cambridge, reel and
   stagger,
Overwearied with howling. Wales and Scotland alone sustain the
   fight!
The inhabitants are sick to death; they labour to divide into days
And nights the uncertain periods, and into weeks and months. In   690
   vain
They send the dove and raven, and in vain the serpent over the
   mountains,°
And in vain the eagle and lion over the fourfold wilderness.
They return not, but generate in rocky places desolate.
They return not, but build a habitation separate from Man.
The sun forgets his course like a drunken man; he hesitates
Upon the Chisledon hills, thinking to sleep on the Severn.
In vain: he is hurried afar into an unknown night.
He bleeds in torrents of blood as he rolls through heaven above;
He chokes up the paths of the sky. The moon is leprous as snow,

Trembling and descending down, seeking to rest upon high   700
    Mona,°
Scattering her leprous snows in flakes of disease over Albion.
The stars flee remote; the heaven is iron, the earth is sulphur,
And all the mountains and hills shrink up like a withering gourd;
As the senses of men shrink together under the knife of flint
In the hands of Albion's daughters, among the Druid temples,
By those who drink their blood and the blood of their covenant.°

And the twelve daughters of Albion united in Rahab and Tirzah,
A double female; and they drew out from the rocky stones
Fibres of life to weave. For every female is a golden loom;
The rocks are opaque hardnesses covering all vegetated things.   710
And as they wove and cut from the looms, in various divisions
Stretching over Europe and Asia from Ireland to Japan,
They divided into many lovely daughters to be counterparts
To those they wove; for when they wove a male, they divided
Into a female to the woven male. In opaque hardness
They cut the fibres from the rocks; groaning in pain they weave,
Calling the rocks Atomic Origins of Existence, denying Eternity
By the atheistical Epicurean philosophy of Albion's tree.°
Such are the feminine and masculine when separated from Man.
They call the rocks Parents of Men and adore the frowning   720
    chaos,°
Dancing around in howling pain clothed in the bloody veil,
Hiding Albion's sons within the veil, closing Jerusalem's
Sons without, to feed with their souls the Spectres of Albion:
Ashamed to give love openly to the piteous and merciful man,
Counting him an imbecile mockery. But the warrior
They adore and his revenge cherish with the blood of the
    innocent.
They drink up Dan and Gad to feed with milk Skofield and
    Kotope;
They strip off Joseph's coat and dip it in the blood of battle.°

Tirzah sits weeping to hear the shrieks of the dying. Her knife
Of flint is in her hand; she passes it over the howling victim.   730
The daughters weave their work in loud cries over the rock
Of Horeb, still eyeing Albion's cliffs, eagerly seizing and twisting
The threads of Vala and Jerusalem running from mountain to
    mountain

Over the whole earth. Loud the warriors rage in Beth Peor
Beneath the iron whips of their captains and consecrated
    banners.
Loud the sun and moon rage in the conflict; loud the stars
Shout in the night of battle, and their spears grow to their hands
With blood, weaving the deaths of the mighty into a tabernacle
For Rahab and Tirzah, till the great polypus of generation
    covered the earth.

In Verulam the polypus's head, winding around his bulk   740
Through Rochester and Chichester, and Exeter and Salisbury,
To Bristol; and his heart beat strong on Salisbury Plain,
Shooting out fibres round the earth, through Gaul and Italy
And Greece, and along the Sea of Rephaim into Judaea
To Sodom and Gomorrha, thence to India, China and Japan.

The twelve daughters in Rahab and Tirzah have circumscribed
    the brain
Beneath, and pierced it through the midst with a golden pin.
Blood hath stained her fair side beneath her bosom.

'O thou poor human form!' said she: 'O thou poor child of woe!
Why wilt thou wander away from Tirzah; why me compel to   750
    bind thee?
If thou dost go away from me I shall consume upon these rocks.
These fibres of thine eyes that used to beam in distant heavens
Away from me, I have bound down with a hot iron;
These nostrils, that expanded with delight in morning skies,
I have bent downward with lead melted in my roaring furnaces
Of affliction, of love, of sweet despair, of torment unendurable.
My soul is seven furnaces; incessant roars the bellows
Upon my terribly flaming heart; the molten metal runs
In channels through my fiery limbs. Oh love, oh pity, oh fear,
Oh pain! Oh the pangs, the bitter pangs of love forsaken.   760
Ephraim was a wilderness of joy where all my wild beasts ran;
The River Kanah wandered by my sweet Manasseh's side°
To see the boy spring into heavens sounding from my sight!
Go, Noah, fetch the girdle of strong brass; heat it red-hot;
Press it around the loins of this ever-expanding cruelty.
Shriek not so, my only love. I refuse thy joys, I drink
Thy shrieks, because Hand and Hyle are cruel and obdurate to
    me.

'Skofield, why art thou cruel? Lo, Joseph is thine! To make
You one, to weave you both in the same mantle of skin.
Bind him down, sisters, bind him down on Ebal, mount of   770
    cursing.°
Mahlah, come forth from Lebanon, and Hoglah from Mount
    Sinai;
Come, circumscribe this tongue of sweets, and with a screw of
    iron
Fasten this ear into the rock. Milcah, the task is thine.
Weep not so, sisters; weep not so. Our life depends on this;
Or mercy and truth are fled away from Shechem and Mount
    Gilead,°
Unless my beloved is bound upon the stems of vegetation.

And thus the warriors cry, in the hot day of victory, in songs:

'Look, the beautiful daughter of Albion sits naked upon the stone,
Her panting victim beside her; her heart is drunk with blood
Though her brain is not drunk with wine. She goes forth from   780
    Albion
In pride of beauty, in cruelty of holiness, in the brightness
Of her tabernacle, and her ark, and secret place. The beautiful
    daughter
Of Albion delights the eyes of the kings; their hearts and the
Hearts of their warriors glow hot before Thor and Friga. O
    Molech!°
O Chemosh! O Bacchus! O Venus! O double god of generation.
The heavens are cut like a mantle around from the cliffs of Albion
Across Europe, across Africa. In howling and deadly wars
A sheet and veil and curtain of blood is let down from heaven,
Across the hills of Ephraim and down Mount Olivet to
The valley of the Jebusite. Molech rejoices in Heaven;°   790
He sees the twelve daughters naked upon the twelve stones,
Themselves condensing to rocks and into the ribs of a man.
Lo, they shoot forth in tender nerves across Europe and Asia;
Lo, they rest upon the tribes, where their panting victims lie.
Molech rushes into the kings, in love to the beautiful daughters,
But they frown and delight in cruelty, refusing all other joy.
Bring your offerings, you first-begotten, pampered with milk and
    blood,
Your first-born of seven years old, be they males or females,

To the beautiful daughters of Albion. They sport before the
    kings
Clothed in the skin of the victim! Blood, human blood, is the life   800
And delightful food of the warrior! The well-fed warrior's flesh
Of him who is slain in war fills the valleys of Ephraim with
Breeding women, walking in pride and bringing forth under
    green trees
With pleasure, without pain, for their food is blood of the
    captive.
Molech rejoices through the land from Havilah to Shur; he
    rejoices°
In moral law and its severe penalties. Loud Shaddai and Jehovah
Thunder above, when they see the twelve panting victims
On the twelve stones of power, and the beautiful daughters of
    Albion:
"If you dare rend their veil with your spear you are healed of
    love."
From the hills of Camberwell and Wimbledon, from the valleys   810
Of Walton and Esher, from Stonehenge and from Maldon's cove,
Jerusalem's pillars fall in the rendings of fierce war
Over France and Germany, upon the Rhine and Danube.
Reuben and Benjamin flee; they hide in the valley of Rephaim.
Why trembles the warrior's limbs when he beholds thy beauty
Spotted with victim's blood, by the fires of thy secret tabernacle
And thy ark and holy place? At thy frowns, at thy dire revenge,
Smitten as Uzzah of old, his armour is softened; his spear°
And sword faint in his hand, from Albion across Great Tartary.
O beautiful daughter of Albion, cruelty is thy delight.   820
O virgin of terrible eyes, who dwellest by valleys of springs
Beneath the mountains of Lebanon, in the city of Rehob in
    Hamath,°
Taught to touch the harp, to dance in the circle of warriors
Before the kings of Canaan, to cut the flesh from the victim,
To roast the flesh in fire, to examine the infant's limbs
In cruelties of holiness, to refuse the joys of love, to bring
The spies from Egypt, to raise jealousy in the bosoms of the
    twelve
Kings of Canaan, then to let the spies depart to Meribah Kadesh°
To the place of the Amalekite. I am drunk with unsatiated love;
I must rush again to war, for the virgin has frowned and refused.   830
Sometimes I curse, and sometimes bless thy fascinating beauty.

Once Man was occupied in intellectual pleasures and energies,
But now my soul is harrowed with grief and fear and love and
    desire.
And now I hate and now I love, and intellect is no more;
There is no time for anything but the torments of love and desire.
The feminine and masculine shadows soft, mild and ever-
    varying
In beauty, are shadows now no more, but rocks in Horeb.'

Then all the males combined into one male, and every one
Became a ravening eating cancer growing in the female,
A polypus of roots of reasoning, doubt, despair and death,   840
Going forth and returning from Albion's rocks to Canaan,
Devouring Jerusalem from every nation of the earth.

Envying stood the enormous form, at variance with itself
In all its members, in eternal torment of love and jealousy,
Driven forth by Los time after time from Albion's cliffy shore,
Drawing the free loves of Jerusalem into infernal bondage;
That they might be born in contentions of chastity, and in
Deadly hate between Leah and Rachel, daughters of deceit and
    fraud,°
Bearing the images of various species of contention,
And jealousy, and abhorrence, and revenge, and deadly murder,   850
Till they refuse liberty to the male. And not like Beulah,
Where every female delights to give her maiden to her husband;
The female searches sea and land for gratifications to the
Male genius, who in return clothes her in gems and gold,
And feeds her with the food of Eden. Hence all her beauty beams.
She creates at her will a little moony night and silence,
With spaces of sweet gardens and a tent of elegant beauty,
Closed in by a sandy desert and a night of stars shining,
And a little tender moon and hovering angels on the wing.
And the male gives a time and revolution to her space   860
Till the time of love is passed in ever-varying delights.
For all things exist in the human imagination,
And thence in Beulah they are stolen by secret amorous theft,
Till they have had punishment enough to make them commit
    crimes.
Hence rose the tabernacle in the wilderness and all its offerings—
From male and female loves in Beulah and their jealousies.

But no one can consummate female bliss in Los's world without
Becoming a generated mortal, a vegetating death.

And now the Spectres of the dead awake in Beulah. All
The jealousies become murderous, uniting together in Rahab,   870
A religion of chastity, forming a commerce to sell loves,
With moral law, an equal balance, not going down with decision.
Therefore the male, severe and cruel, filled with stern revenge;
Mutual hate returns, and mutual deceit and mutual fear.

Hence the infernal veil grows in the disobedient female,
Which Jesus rends and the whole Druid law removes away
From the inner sanctuary: a false holiness hid within the centre.
For the sanctuary of Eden is in the camp, in the outline,
In the circumference, and every minute particular is holy.°
Embraces are comminglings from the head even to the feet,   880
And not a pompous High Priest entering by a secret place.

Jerusalem pined in her inmost soul over wandering Reuben,
As she slept in Beulah's night hid by the daughters of Beulah.

And this the form of mighty Hand sitting on Albion's cliffs
Before the face of Albion, a mighty threat'ning form.

His bosom wide and shoulders huge, overspreading, wondrous,
Bear three strong sinewy necks and three awful and terrible
    heads,
Three brains in contradictory council brooding incessantly,
Neither daring to put in act its councils, fearing each other:
Therefore rejecting ideas as nothing, and holding all wisdom   890
To consist in the agreements and disagreements of ideas,
Plotting to devour Albion's body of humanity and love.

Such form the aggregate of the twelve sons of Albion took, and
    such
Their appearance when combined; but often by birth-pangs and
    loud groans
They divide to twelve. The key-bones and the chest dividing in
    pain,°
Disclose a hideous orifice, thence issuing, the giant-brood
Arise as the smoke of the furnace, shaking the rocks from sea to
    sea,

And there they combine into three forms, named Bacon and
   Newton and Locke,
In the oak groves of Albion which overspread all the earth.

Imputing sin and righteousness to individuals, Rahab   900
Sat, deep within him hid: his feminine power unrevealed,
Brooding abstract philosophy (to destroy imagination, the Divine
Humanity), a threefold wonder—feminine, most beautiful,
   threefold
Each within other. On her white marble and even neck, her heart
Inorbed and bonified, with locks of shadowing modesty, shining
Over her beautiful female features, soft flourishing in beauty,
Beams mild, all love and all perfection; that when the lips
Receive a kiss from gods or men, a threefold kiss returns
From the pressed loveliness. So her whole immortal form
   threefold
Threefold embrace returns, consuming lives of gods and men,   910
In fires of beauty melting them as gold and silver in the furnace.
Her brain enlabyrinths the whole heaven of her bosom and loins
To put in act what her heart wills. Oh, who can withstand her
   power?
Her name is Vala in Eternity; in time, her name is Rahab.

The starry heavens all were fled from the mighty limbs of Albion,
And above Albion's land was seen the heavenly Canaan
As the substance is to the shadow; and above Albion's twelve
   sons
Were seen Jerusalem's sons, and all the twelve tribes spreading
Over Albion. As the soul is to the body, so Jerusalem's sons
Are to the sons of Albion, and Jerusalem is Albion's Emanation.   920

What is above is within, for everything in Eternity is translucent.
The circumference is within; without, is formed the selfish
   centre.
And the circumference still expands, going forward to Eternity,
And the centre has eternal states. These states we now explore.

And these the names of Albion's twelve sons, and of his twelve
   daughters,
With their districts. Hand dwelt in Selsey, and had Sussex and
   Surrey

And Kent and Middlesex. All their rivers and their hills of flocks
   and herds,
Their villages, towns, cities, sea-ports, temples, sublime
   cathedrals—
All were his friends, and their sons and daughters intermarry in
   Beulah.

For all are men in Eternity; rivers, mountains, cities, villages,   930
All are human. And when you enter into their bosoms you walk
In heavens and earths, as in your own bosom you bear your
   heaven
And earth. And all you behold, though it appears without, it is
   within,
In your imagination of which this world of mortality is but a
   shadow.

Hyle dwelt in Winchester, comprehending Hampshire, Dorset,
   Devon, Cornwall:
Their villages, cities, sea-ports, their cornfields and gardens,
   spacious
Palaces, rivers and mountains. And between Hand and Hyle
   arose
Gwendolen, and Cambel who is Boadicea; they go abroad and
   return
Like lovely beams of light from the mingled affections of the
   brothers.
The inhabitants of the whole earth rejoice in their beautiful light.   940

Coban dwelt in Bath; Somerset, Wiltshire, Gloucestershire
Obeyed his awful voice. Ignoge is his lovely Emanation;
She adjoined with Gwantok's children. Soon lovely Cordella
   arose.
Gwantok forgave and joyed over South Wales and all its
   mountains.

Peachey had North Wales, Shropshire, Cheshire and the Isle of
   Man.
His Emanation is Mehetabel, terrible and lovely upon the
   mountains.

Brereton had Yorkshire, Durham, Westmorland, and his
   Emanation
Is Ragan; she adjoined to Slayd, and produced Gonorill far-
   beaming.

Slayd had Lincoln, Stafford, Derby, Nottingham, and his lovely
Emanation Gonorill rejoices over hills and rocks and woods and   950
   rivers.

Huttn had Warwick, Northampton, Bedford, Buckingham,
Leicester and Berkshire, and his Emanation is Gwinefred
   beautiful.

Skofield had Ely, Rutland, Cambridge, Huntingdon, Norfolk,
Suffolk, Hertford and Essex, and his Emanation is Gwineverra
Beautiful. She beams towards the east, all kinds of precious stones
And pearl, with instruments of music in holy Jerusalem.

Kox had Oxford, Warwick, Wilts. His Emanation is Estrild;
Joined with Cordella she shines southward over the Atlantic.

Kotope had Hereford, Stafford, Worcester, and his Emanation
Is Sabrina; joined with Mehetabel she shines west over America.   960

Bowen had all Scotland, the Isles, Northumberland and Cum-
   berland.
His Emanation is Conwenna; she shines a triple form
Over the north with pearly beams gorgeous and terrible;
Jerusalem and Vala rejoice in Bowen and Conwenna.

But the four sons of Jerusalem that never were generated
Are Rintrah and Palamabron and Theotormon and Bromion.
   They
Dwell over the four provinces of Ireland in heavenly light,
The four universities of Scotland, and in Oxford and Cambridge
   and Winchester.°

But now Albion is darkened and Jerusalem lies in ruins,
Above the mountains of Albion, above the head of Los.   970

And Los shouted with ceaseless shoutings and his tears poured
    down
His immortal cheeks, rearing his hands to heaven for aid divine!
But he spoke not to Albion, fearing lest Albion should turn his
    back
Against the Divine Vision and fall over the precipice of eternal
    death.
But he receded before Albion and before Vala, weaving the veil
With the iron shuttle of war among the rooted oaks of Albion,
Weeping and shouting to the Lord day and night; and his
    children
Wept round him as a flock, silent seven days of eternity.

And the thirty-two counties of the four provinces of Ireland
Are thus divided. The four counties are in the four camps:°   980
Munster south in Reuben's gate, Connaught west in Joseph's
    gate,
Ulster north in Dan's gate, Leinster east in Judah's gate.

For Albion in Eternity has sixteen gates among his pillars,
But the four towards the west were walled up, and the twelve
That front the three other points were turned four-square
By Los, for Jerusalem's sake, and called the Gates of Jerusalem,°
Because twelve sons of Jerusalem fled successive through the
    gates.
But the four sons of Jerusalem who fled not but remained
Are Rintrah and Palamabron and Theotormon and Bromion,
The four that remain with Los to guard the western wall.   990
And these four remain to guard the four walls of Jerusalem°
Whose foundations remain in the thirty-two counties of Ireland,
And in twelve counties of Wales, and in the forty counties
Of England, and in the thirty-six counties of Scotland.

And the names of the thirty-two counties of Ireland are these.
Under Judah and Issachar and Zebulun are Louth, Longford,
East Meath, West Meath, Dublin, Kildare, King's County,
Queen's County, Wicklow, Catherlow, Wexford, Kilkenny;
And those under Reuben and Simeon and Levi are these:
Waterford, Tipperary, Cork, Limerick, Kerry, Clare;   1000
And those under Ephraim, Manasseh and Benjamin are these:
Galway, Roscommon, Mayo, Sligo, Leitrim;

And those under Dan, Asher and Napthali are these:
Donegal, Antrim, Tyrone, Fermanagh, Armagh, Londonderry,
Down, Monaghan, Cavan. These are the land of Erin.

All these centre in London and in Golgonooza, from whence
They are created continually, east and west and north and south,
And from them are created all the nations of the earth,
Europe and Asia and Africa and America, in fury fourfold.

And thirty-two the nations to dwell in Jerusalem's gates—   1010
Oh come, ye nations; come, ye people; come up to Jerusalem!
Return, Jerusalem, and dwell together as of old, return,
Return! O Albion, let Jerusalem overspread all nations
As in the times of old; O Albion, awake! Reuben wanders.
The nations wait for Jerusalem; they look up for the bride.

France, Spain, Italy, Germany, Poland, Russia, Sweden,
    Turkey,
Arabia, Palestine, Persia, Hindustan, China, Tartary, Siberia,
Egypt, Lybia, Ethiopia, Guinea, Caffraria, Negroland,
    Morocco,°
Congo, Zaara, Canada, Greenland, Carolina, Mexico,°
Peru, Patagonia, Amazonia, Brazil, thirty-two nations—   1020
And under these thirty-two classes of islands in the ocean:
All the nations, peoples and tongues throughout all the earth.

And the four gates of Los surround the universe within and
Without, and whatever is visible in the vegetable earth, the same
Is visible in the Mundane Shell, reversed in mountain and vale.
And a son of Eden was set over each daughter of Beulah to guard
In Albion's tomb the wondrous creation. And the fourfold gate
Towards Beulah is to the south; Fénelon, Guyon, Teresa,
Whitefield and Hervey guard that gate, with all the gentle souls°
Who guide the great winepress of love. Four precious stones that   1030
    gate.

Such are Cathedron's golden halls in the city of Golgonooza.

And Los's furnaces howl loud, living, self-moving, lamenting
With fury and despair; and they stretch from south to north
Through all the four points. Lo, the labourers at the furnaces!
Rintrah and Palamabron, Theotormon and Bromion, loud
    lab'ring

With the innumerable multitudes of Golgonooza, round the
    anvils
Of death. But how they came forth from the furnaces, and how
    long,
Vast and severe the anguish ere they knew their father, were
Long to tell, and of the iron rollers, golden axle-trees and yokes
Of brass, iron chains and braces, and the gold, silver and brass   1040
Mingled or separate: for swords, arrows, cannons, mortars,
The terrible ball, the wedge, the loud-sounding hammer of
    destruction,
The sounding flail to thresh, the winnow to winnow kingdoms,
The water-wheel, and mill of many innumerable wheels resist-
    less
Over the fourfold monarchy from earth to the Mundane Shell.

Perusing Albion's tomb in the starry characters of Og and Anak,
To create the lion and wolf, the bear, the tiger and ounce;
To create the woolly lamb and downy fowl and scaly serpent,
The summer and winter, day and night, the sun and moon and
    stars,
The tree, the plant, the flower, the rock, the stone, the metal   1050
Of vegetative nature, by their hard restricting condensations.

Where Luvah's world of opaqueness grew to a period, it
Became a limit, a rocky hardness without form and void,
Accumulating without end. Here Los, who is of the Elohim,
Opens the furnaces of affliction in the Emanation,
Fixing the sexual into an ever-prolific generation,
Naming the limit of opaqueness Satan, and the limit of contraction
Adam, who is Peleg and Joktan, and Esau and Jacob, and Saul
    and David.°

Voltaire insinuates that these limits are the cruel work of God,°
Mocking the remover of limits and the resurrection of the dead,   1060
Setting up kings in wrath: in holiness of natural religion,
Which Los with his mighty hammer demolishes time and time
In miracles and wonders in the fourfold desert of Albion,
Permanently creating, to be in time revealed and demolished.
Satan, Cain, Tubal, Nimrod, Pharaoh, Priam, Bladud, Belin,°
Arthur, Alfred, the Norman Conqueror, Richard, John,
And all the kings and nobles of the earth and all their glories:

These are created by Rahab and Tirzah in Ulro, but around
These, to preserve them from eternal death, Los creates
Adam, Noah, Abraham, Moses, Samuel, David, Ezekiel,   1070
Dissipating the rocky forms of death by his thunderous hammer.
As the pilgrim passes while the country permanent remains,
So men pass on, but states remain permanent for ever.

The Spectres of the dead howl round the porches of Los
In the terrible family feuds of Albion's cities and villages,
To devour the body of Albion hung'ring and thirsting and
     rav'ning.
The sons of Los clothe them and feed, and provide houses and
     gardens;
And every human vegetated form in its inward recesses
Is a house of pleasantness and a garden of delight, built by the
Sons and daughters of Los in Bowlahoola and in Cathedron.   1080

From London to York and Edinburgh the furnaces rage terrible;
Primrose Hill is the mouth of the furnace and the iron door.
The four Zoas clouded rage; Urizen stood by Albion,
With Rintrah and Palamabron and Theotormon and Bromion.
These four are Verulam and London and York and Edinburgh,
And the four Zoas are Urizen and Luvah and Tharmas and
     Urthona,
In opposition deadly, and their wheels in poisonous
And deadly stupor turned against each other loud and fierce.
Entering into the reasoning power, forsaking imagination,
They became Spectres, and their human bodies were reposed   1090
In Beulah by the daughters of Beulah, with tears and lamen-
     tations.

The Spectre is the reasoning power in Man, and when separated
From imagination, and closing itself as in steel, in a ratio
Of the things of memory, it thence frames laws and moralities
To destroy imagination, the divine body, by martyrdoms and
     wars!

Teach me, O Holy Spirit, the testimony of Jesus! Let me°
Comprehend wondrous things out of the divine law.
I behold Babylon in the opening streets of London. I behold
Jerusalem in ruins wandering about from house to house.
This I behold; the shudderings of death attend my steps.   1100

I walk up and down in six thousand years; their events are
    present before me,
To tell how Los in grief and anger, whirling round his hammer
    on high,
Drove the sons and daughters of Albion from their ancient
    mountains.
They became the twelve gods of Asia opposing the Divine
    Vision.

The sons of Albion are twelve; the sons of Jerusalem sixteen.
I tell how Albion's sons by harmonies of concords and discords
Opposed to melody, and by lights and shades opposed to outline,
And by abstraction opposed to the visions of imagination,
By cruel laws, divided sixteen into twelve divisions;
How Hyle roofed Los in Albion's cliffs, by the affections rent   1110
Asunder and opposed to thought, to draw Jerusalem's sons
Into the vortex of his wheels. Therefore Hyle is called Gog,
Age after age drawing them away towards Babylon:
Babylon, the rational morality deluding to death the little ones
In strong temptations of stolen beauty. I tell how Reuben slept
On London Stone, and the daughters of Albion ran around
    admiring
His awful beauty; with moral virtue the fair deceiver, offspring
Of good and evil, they divided him in love upon the Thames and
    sent
Him over Europe in streams of gore out of Cathedron's looms;
How Los drave them from Albion and they became daughters of   1120
    Canaan.
Hence Albion was called the Canaanite, and all his giant sons.
Hence is my theme. O Lord my Saviour! open thou the gates
And I will lead forth thy words, telling how the daughters
Cut the fibres of Reuben, how he rolled apart and took root
In Bashan. Terror-struck Albion's sons look toward Bashan.
They have divided Simeon; he also rolled apart in blood
Over the nations till he took root beneath the shining looms
Of Albion's daughters, in Philistia by the side of Amalek.
They have divided Levi; he hath shot out into forty-eight roots
Over the land of Canaan. They have divided Judah;   1130
He hath took root in Hebon, in the land of Hand and Hyle.
Dan, Naphtali, Gad, Asher, Issacher, Zebulun roll apart
From all the nations of the earth to dissipate into non-entity.

I see a feminine form arise from the four terrible Zoas,
Beautiful but terrible, struggling to take a form of beauty,
Rooted in Shechem. This is Dinah, the youthful form of Erin.°
The wound I see in South Molton Street and Stratford Place,
Whence Joseph and Benjamin rolled apart away from the
    nations.
In vain they rolled apart; they are fixed into the land of Cabul.°

And Rahab, Babylon the Great, hath destroyed Jerusalem.   1140
Bath stood upon the Severn with Merlin and Bladud and Arthur,
The cup of Rahab in his hand, her poisons twenty-sevenfold.

And all her twenty-seven heavens, now hid and now revealed,
Appear in strong delusive light of time and space, drawn out
In shadowy pomp, by the Eternal Prophet created evermore.
For Los in six thousand years walks up and down continually,
That not one moment of time be lost, and every revolution
Of space he makes permanent in Bowlahoola and Cathedron.

And these the names of the twenty-seven heavens and their
    churches.
Adam, Seth, Enos, Cainan, Mahalaleel, Jared, Enoch,   1150
Methuselah, Lamech; these are the giants mighty, hermaphrodi-
    tic.
Noah, Shem, Arphaxad, Cainan the second, Salah, Heber,
Peleg, Reu, Serug, Nahor, Terah; these are the female-males,
A male within a female hid as in an ark and curtains.
Abraham, Moses, Solomon, Paul, Constantine, Charlemagne,
Luther; these seven are the male-females; the dragon forms,
The female hid within a male. Thus Rahab is revealed:
Mystery, Babylon the Great, the abomination of desolation,
Religion hid in war, a dragon red and hidden harlot.
But Jesus, breaking through the central zones of death and Hell,   1160
Opens Eternity in time and space, triumphant in mercy.

Thus are the heavens formed by Los within the Mundane Shell,
And where Luther ends Adam begins again in eternal circle:
To awake the prisoners of death, to bring Albion again
With Luvah into light eternal, in his eternal day.

But now the starry heavens are fled from the mighty limbs of
    Albion.

CHAPTER IV

TO THE CHRISTIANS

[*see above, p. 33*]

The Spectres of Albion's twelve sons revolve mightily
Over the tomb and over the body, rav'ning to devour
The sleeping humanity. Los with his mace of iron
Walks round; loud his threats, loud his blows fall
On the rocky Spectres, as the potter breaks the potsherds:°
Dashing in pieces self-righteousnesses, driving them from Albion's
Cliffs, dividing them into male and female forms in his furnaces
And on his anvils. Lest they destroy the feminine affections
They are broken. Loud howl the Spectres in his iron furnace.

While Los laments at his dire labours, viewing Jerusalem,   10
Sitting before his furnaces clothed in sackcloth of hair,
Albion's twelve sons surround the forty-two gates of Erin
In terrible armour, raging against the Lamb and against Jerusalem,
Surrounding them with armies to destroy the Lamb of God.
They took their mother Vala, and they crowned her with gold;
They named her Rahab, and gave her power over the earth,
The concave earth round Golgonooza in Entuthon Benython,
Even to the stars exalting her throne: to build beyond the throne
Of God and the Lamb, to destroy the Lamb and usurp the throne of God,
Drawing their Ulro voidness round the fourfold humanity.   20

Naked Jerusalem lay before the gates upon Mount Zion,
The hill of giants, all her foundations levelled with the dust.

Her twelve gates thrown down, her children carried into captivity,
Herself in chains: this from within was seen in a dismal night
Outside, unknown before in Beulah, and the twelve gates were filled
Wih blood, from Japan eastward to the Giants' Causeway, west
Into Erin's continent. And Jerusalem wept upon Euphrates' banks
Disorganised—an evanescent shade, scarce seen or heard among
Her children's Druid temples, dropping with blood wandered weeping;

And thus her voice went forth in the darkness of Philistia: 30
'My brother and my father are no more! God hath forsaken me.
The arrows of the Almighty pour upon me and my children.°
I have sinned and am an outcast from the divine presence!
My tents are fall'n! My pillars are in ruins! My children dashed
Upon Egypt's iron floors and the marble pavements of Assyria.
I melt my soul in reasonings among the towers of Heshbon;
Mount Zion is become a cruel rock; and no more dew
Nor rain, no more the spring of the rock appears, but cold,
Hard and obdurate are the furrows of the mountain of wine and
    oil.
The mountain of blessing is itself a curse and an astonishment;°   40
The hills of Judaea are fallen with me into the deepest hell,
Away from the nations of the earth, and from the cities of the
    nations.
I walk to Ephraim; I seek for Shiloh; I walk like a lost sheep
Among the precipices of despair. In Goshen I seek for light°
In vain, and in Gilead for a physician and a comforter.
Goshen hath followed Philistia, Gilead hath joined with Og!
They are become narrow places in a little and dark land;
How distant far from Albion! His hills and his valleys no more
Receive the feet of Jerusalem; they have cast me quite away,
And Albion is himself shrunk to a narrow rock in the midst of the   50
    sea!
The plains of Sussex and Surrey, their hills of flocks and herds,
No more seek to Jerusalem nor to the sound of my holy ones.
The fifty-two counties of England are hardened against me
As if I was not their mother; they despise me and cast me out.
London covered the whole earth; England encompassed the
    nations;
And all the nations of the earth were seen in the cities of Albion.
My pillars reached from sea to sea; London beheld me come
From my east and from my west. He blessed me and gave
His children to my breasts, his sons and daughters to my knees.
His aged parents sought me out in every city and village.   60
They discerned my countenance with joy; they showed me to
    their sons,
Saying: "Lo, Jerusalem is here! She sitteth in our secret
    chambers.
Levi and Judah and Issachar, Ephraim, Manasseh, Gad and Dan
Are seen in our hills and valleys. They keep our flocks and herds;

They watch them in the night, and the Lamb of God appears
    among us."
The river Severn stayed his course at my command;
Thames poured his waters into my basins and baths;
Medway mingled with Kishon; Thames received the heavenly
    Jordan.°
Albion gave me to the whole earth to walk up and down, to pour°
Joy upon every mountain, to teach songs to the shepherd and   70
    ploughman.
I taught the ships of the sea to sing the songs of Zion.
Italy saw me, in sublime astonishment. France was wholly mine,
As my garden and as my secret bath. Spain was my heavenly
    couch;
I slept in his golden hills; the Lamb of God met me there;
There we walked as in our secret chamber among our little ones.
They looked upon our loves with joy; they beheld our secret joys,
With holy raptures of adoration rapt sublime in the visions of
    God.
Germany, Poland and the north wooed my footsteps; they found
My gates in all their mountains and my curtains in all their vales;
The furniture of their houses was the furniture of my chamber.   80
Turkey and Grecia saw my instruments of music; they arose,
They seized the harp, the flute, the mellow horn of Jerusalem's
    joy;
They sounded thanksgivings in my courts. Egypt and Lybia
    heard.
The swarthy sons of Ethiopia stood round the Lamb of God
Enquiring for Jerusalem; he led them up my steps to my altar.
And thou, America! I once beheld thee, but now behold no more
Thy golden mountains, where my cherubim and seraphim
    rejoiced
Together among my little ones. But now my altars run with
    blood!
My fires are corrupt! My incense is a cloudy pestilence
Of seven diseases! Once a continual cloud of salvation rose   90
From all my myriads; once the fourfold world rejoiced among
The pillars of Jerusalem, between my winged cherubim.
But now I am closed out from them in the narrow passages
Of the valleys of destruction, into a dark land of pitch and
    bitumen,
From Albion's tomb afar and from the fourfold wonders of God

Shrunk to a narrow doleful form in the dark land of Cabul.
There is Reuben and Gad and Joseph and Judah and Levi, closed
    up
In narrow vales. I walk and count the bones of my beloveds
Along the valley of destruction, among these Druid temples
Which overspread all the earth in patriarchal pomp and cruel   100
    pride.
Tell me, O Vala, thy purposes. Tell me wherefore thy shuttles
Drop with the gore of the slain; why Euphrates is red with blood;
Wherefore in dreadful majesty and beauty outside appears
Thy masculine from thy feminine, hardening against the heavens
To devour the human! Why dost thou weep upon the wind
    among
These cruel Druid temples? O Vala! Humanity is far above
Sexual organisation and the visions of the night of Beulah,
Where sexes wander in dreams of bliss among the Emanations,
Where the masculine and feminine are nursed into youth and
    maiden
By the tears and smiles of Beulah's daughters, till the time of   110
    sleep is past.
Wherefore then do you realize these nets of beauty and delusion,
In open day to draw the souls of the dead into the light,
Till Albion is shut out from every nation under heaven?
Encompassed by the frozen net and by the rooted tree,
I walk weeping in pangs of a mother's torment for her children.
I walk in affliction; I am a worm, and no living soul!°
A worm going to eternal torment, raised up in a night
To an eternal night of pain, lost, lost, lost, for ever!'

Beside her Vala howled upon the winds in pride and beauty,
Lamenting among the timbrels of the warriors, among the   120
    captives
In cruel holiness; and her lamenting songs were from Arnon
And Jordan to Euphrates. Jerusalem followed, trembling,
Her children in captivity, listening to Vala's lamentation
In the thick cloud and darkness. And the voice went forth from
The cloud: 'Oh rent in sunder from Jerusalem the harlot
    daughter,
In an eternal condemnation, in fierce burning flames
Of torment unendurable! And if once a delusion be found
Woman must perish, and the heavens of heavens remain no more.

'My father gave to me command to murder Albion
In unreviving death; my love, my Luvah, ordered me in night　130
To murder Albion, the king of men. He fought in battles fierce;
He conquered Luvah my beloved. He took me and my father;
He slew them. I revived them to life in my warm bosom.
He saw them issue from my bosom; dark in jealousy
He burned before me. Luvah framed the knife, and Luvah gave
The knife into his daughter's hand. Such thing was never known
Before in Albion's land: that one should die a death never to be
　revived.
For in our battles we the slain men view with pity and love;
We soon revive them in the secret of our tabernacles.
But I, Vala, Luvah's daughter, keep his body embalmed in moral　140
　laws
With spices of sweet odours of lovely jealous stupefaction
Within my bosom, lest he arise to life and slay my Luvah.
Pity me then, O Lamb of God! O Jesus, pity me!
Come into Luvah's tents, and seek not to revive the dead!'

So sang she, and the spindle turned furious as she sang.
The children of Jerusalem, the souls of those who sleep,
Were caught into the flax of her distaff and in her cloud,
To weave Jerusalem a body according to her will,
A dragon form on Zion hill's most ancient promontory.

The spindle turned in blood and fire. Loud sound the trumpets　150
Of war; the cymbals play loud before the captains,
With Cambel and Gwendolen in dance and solemn song,
The cloud of Rahab vibrating with the daughters of Albion.
Los saw terrified; melted with pity and divided in wrath
He sent them over the narrow seas in pity and love,
Among the four forests of Albion which overspread all the earth.
They go forth and return, swift as a flash of lightning.
Among the tribes of warriors, among the stones of power,
Against Jerusalem they rage through all the nations of Europe,
Through Italy and Grecia, to Lebanon and Persia and India.　160

The serpent temples through the earth, from the wide plain of
　Salisbury,
Resound with cries of victims, shouts and songs and dying groans
And flames of dusky fire, to Amalek, Canaan and Moab.

And Rahab like a dismal and indefinite hovering cloud
Refused to take a definite form. She hovered over all the earth,
Calling the definite 'sin', defacing every definite form,
Invisible or visible, stretched out in length or spread in breadth:
Over the temples drinking groans of victims, weeping in pity,
And joying in the pity, howling over Jerusalem's walls.

Hand slept on Skiddaw's top, drawn by the love of beautiful   170
Cambel; his bright-beaming counterpart divided from him,
And her delusive light beamed fierce above the mountain,
Soft, invisible, drinking his sighs in sweet intoxication,
Drawing out fibre by fibre. Returning to Albion's tree
At night and in the morning to Skiddaw, she sent him over
Mountainous Wales into the loom of Cathedron, fibre by fibre.
He ran in tender nerves across Europe to Jerusalem's shade
To weave Jerusalem a body repugnant to the Lamb.

Hyle on East Moor in rocky Derbyshire raved to the moon
For Gwendolen; she took up in bitter tears his anguished heart   180
That, apparent to all in Eternity, glows like the sun in the breast.
She hid it in his ribs and back; she hid his tongue with teeth,
In terrible convulsions pitying and gratified, drunk with pity,
Glowing with loveliness before him, becoming apparent
According to his changes. She rolled his kidneys round
Into two irregular forms, and looking on Albion's dread tree
She wove two vessels of seed, beautiful as Skiddaw's snow,
Giving them bends of self-interest and selfish natural virtue.
She hid them in his loins; raving he ran among the rocks,
Compelled into a shape of moral virtue against the Lamb,   190
The invisible lovely one giving him a form according to
His law: a form against the Lamb of God, opposed to mercy
And playing in the thunderous loom in sweet intoxication,
Filling cups of silver and crystal with shrieks and cries, with
    groans
And dolorous sobs, the wine of lovers in the winepress of Luvah.

'O sister Cambel,' said Gwendolen, as their long beaming light
Mingled above the mountain, 'What shall we do to keep
These awful forms in our soft bands? Distracted with trembling,
I have mocked those who refused cruelty and I have admired

The cruel warrior. I have refused to give love to Merlin the   200
    piteous;
He brings to me the images of his love, and I reject in chastity
And turn them out into the streets for harlots, to be food
To the stern warrior. I am become perfect in beauty over my
    warrior.
For men are caught by love, woman is caught by pride:
That love may only be obtained in the passages of death.
Let us look, let us examine. Is the cruel become an infant,
Or is he still a cruel warrior? Look sisters, look! Oh, piteous!
I have destroyed wand'ring Reuben who strove to bind my will;
I have stripped off Joseph's beautiful integument for my beloved,
The cruel one of Albion, to clothe him in gems of my zone.°   210
I have named him Jehovah of Hosts. Humanity is become
A weeping infant in ruined lovely Jerusalem's folding cloud.

'In Heaven love begets love, but fear is the parent of earthly love!
And he who will not bend to love must be subdued by fear.
I have heard Jerusalem's groans; from Vala's cries and lamen-
    tations
I gather our eternal fate! Outcasts from life and love!
Unless we find a way to bind these awful forms to our
Embrace, we shall perish annihilate, discovered our delusions.
Look, I have wrought without delusion. Look! I have wept!
And given soft milk mingled together with the spirits of flocks,   220
Of lambs and doves, mingled together in cups and dishes
Of painted clay. The mighty Hyle is become a weeping infant;
Soon shall the Spectres of the dead follow my weaving threads.'

The twelve daughters of Albion attentive listen in secret shades,
On Cambridge and Oxford beaming, soft uniting with Rahab's
    cloud,
While Gwendolen spoke to Cambel, turning soft the spinning
    reel,
Or throwing the winged shuttle, or drawing the cords with
    softest songs.
The golden cords of the looms animate beneath their touches soft
Along the island white, among the Druid temples, while
    Gwendolen
Spoke to the daughters of Albion standing on Skiddaw's top.   230

So saying, she took a falsehood and hid it in her left hand,
To entice her sisters away to Babylon on Euphrates.
And thus she closed her left hand and uttered her falsehood;
Forgetting that falsehood is prophetic, she hid her hand behind
   her,
Upon her back behind her loins, and thus uttered her deceit.

'I heard Enitharmon say to Los: "Let the daughters of Albion
Be scattered abroad and let the name of Albion be forgotten.
Divide them into three! Name them Amalek, Canaan and Moab.
Let Albion remain a desolation without an inhabitant,
And let the looms of Enitharmon and the furnaces of Los   240
Create Jerusalem and Babylon and Egypt and Moab and Amalek,
And Helle and Hesperia and Hindustan and China and Japan.°
But hide America, for a curse, an altar of victims and a holy
   place."
See, sisters: Canaan is pleasant; Egypt is as the Garden of Eden;
Babylon is our chief desire, Moab our bath in summer.
Let us lead the stems of this tree; let us plant it before Jerusalem
To judge the friend of sinners to death without the veil,
To cut her off from America, to close up her secret ark,
And the fury of Man exhaust in war. Woman permanent remain.
See how the fires of our loins point eastward to Babylon.   250
Look, Hyle is become an infant love. Look! Behold! See him lie
Upon my bosom! Look! Here is the lovely wayward form
That gave me sweet delight by his torments beneath my veil.
By the fruit of Albion's tree I have fed him with sweet milk,
By contentions of the mighty for sacrifice of captives.
Humanity, the great delusion, is changed to war and sacrifice;
I have nailed his hands on Bath Rabbim and his feet on
   Heshbon's wall.°
Oh, that I could live in his sight. Oh, that I could bind him to my
   arm.'

So saying, she drew aside her veil from Mam-Tor to Dovedale,
Discovering her own perfect beauty to the daughters of Albion   260
And Hyle a winding worm beneath, and not a weeping infant.
Trembling and pitying she screamed and fled upon the wind;
Hyle was a winding worm and herself perfect in beauty.
The deserts tremble at his wrath; they shrink themselves in fear.

Cambel trembled with jealousy. She trembled! She envied!
The envy ran through Cathedron's looms into the heart
Of mild Jerusalem, to destroy the Lamb of God. Jerusalem
Languished upon Mount Olivet, east of mild Zion's hill.

Los saw the envious blight above his seventh furnace
On London's tower on the Thames. He drew Cambel in wrath   270
Into his thundering bellows, heaving it for a loud blast,
And with the blast of his furnace upon fishy Billingsgate,
Beneath Albion's fatal tree, before the gate of Los,
Showed her the fibres of her beloved to ameliorate
The envy. Loud she laboured in the furnace of fire
To form the mighty form of Hand according to her will,
In the furnaces of Los and in the winepress treading day and
    night.
Naked among the human clusters, bringing wine of anguish
To feed the afflicted in the furnaces, she minded not
The raging flames, though she returned instead of beauty   280
Deformity. She gave her beauty to another, bearing abroad
Her struggling torment in her iron arms, and like a chain
Binding his wrists and ankles with the iron arms of love.

Gwendolen saw the infant in her sister's arms. She howled
Over the forests with bitter tears, and over the winding worm
Repentant, and she also in the eddying wind of Los's bellows
Began her dolorous task of love, in the winepress of Luvah
To form the worm into a form of love by tears and pain.
The sisters saw; trembling ran through their looms, softening
    mild
Towards London. Then they saw the furnaces opened, and in   290
    tears
Began to give their souls away in the furnaces of affliction.

Los saw and was comforted at his furnaces, uttering thus his
    voice:

'I know I am Urthona, keeper of the gates of Heaven,
And that I can at will expatiate in the gardens of bliss.
But pangs of love draw me down to my loins, which are
Become a fountain of veiny pipes. O Albion! My brother!
Corruptibility appears upon thy limbs, and never more

Can I arise and leave thy side, but labour here incessant
Till thy awaking! Yet, alas, I shall forget Eternity!
Against the patriarchal pomp and cruelty labouring incessant,   300
I shall become an infant horror. Enion! Tharmas! Friends,
Absorb me not in such dire grief. O Albion, my brother,
Jerusalem hungers in the desert! Affection to her children!
The scorned and contemned youthful girl, where shall she fly?
Sussex shuts up her villages. Hants, Devon and Wilts,
Surrounded with masses of stone in ordered forms, determine
    then
A form for Vala and a form for Luvah, here on the Thames,
Where the victim nightly howls beneath the Druid's knife;
A form of vegetation nail them down on the stems of mystery.
Oh, when shall the Saxon return with the English, his redeemed   310
    brother?
Oh, when shall the Lamb of God descend among the Reprobate?
I woo to Amalek to protect my fugitives; Amalek trembles.
I call to Canaan and Moab in my night watches; they mourn,
They listen not to my cry, they rejoice among their warriors.
Woden and Thor and Friga wholly consume my Saxons
On their enormous altars built in the terrible north,
From Ireland's rocks to Scandinavia, Persia and Tartary,
From the Atlantic sea to the universal Erythrean.
Found ye London! Enormous city! Weeps thy river?
Upon his parent bosom lay thy little ones, O land   320
Forsaken. Surrey and Sussex are Enitharmon's chamber,
Where I will build her a couch of repose, and my pillars
Shall surround her in beautiful labyrinths. Oothoon?
Where hides my child? In Oxford hidest thou with Antamon?
In graceful hidings of error, in merciful deceit,
Lest Hand the terrible destroy his affection, thou hidest her
In chaste appearances for sweet deceits of love and modesty,
Immingled, interwoven, glistening to the sickening sight.
Let Cambel and her sisters sit within the Mundane Shell,
Forming the fluctuating globe according to their will.   330
According as they weave the little embryon nerves and veins,
The eye, the little nostrils, and the delicate tongue and ears
Of labyrinthine intricacy, so shall they fold the world:
That whatever is seen upon the Mundane Shell, the same
Be seen upon the fluctuating earth woven by the sisters.
And sometimes the earth shall roll in the abyss, and sometimes

Stand in the centre, and sometimes stretch flat in the expanse,
According to the will of the lovely daughters of Albion.
Sometimes it shall assimilate with mighty Golgonooza,
Touching its summits, and sometimes divided roll apart.   340
As a beautiful veil, so these females shall fold and unfold
According to their will the outside surface of the earth
(An outside shadowy surface superadded to the real surface,
Which is unchangeable for ever and ever. Amen; so be it!),
Separate Albion's sons gently from their Emanations,
Weaving bowers of delight on the current of infant Thames.
Where the old parent still retains his youth, as I, alas!
Retain my youth eight thousand and five hundred years,
The labourer of ages in the valleys of despair.
The land is marked for desolation, and unless we plant   350
The seeds of cities and of villages in the human bosom
Albion must be a rock of blood. Mark ye the points
Where cities shall remain and where villages; for the rest,
It must lie in confusion till Albion's time of awaking.
Place the tribes of Llewellyn in America for a hiding place,
Till sweet Jerusalem emanates again into Eternity.
The night falls thick; I go upon my watch; be attentive.
The sons of Albion go forth; I follow from my furnaces—
That they return no more, that a place be prepared on Euphrates.
Listen to your watchman's voice; sleep not before the furnaces.   360
Eternal death stands at the door. O God, pity our labours.'

So Los spoke to the daughters of Beulah, while his Emanation
Like a faint rainbow waved before him in the awful gloom
Of London City on the Thames, from Surrey hills to Highgate.
Swift turn the silver spindles, and the golden weights play soft
And lulling harmonies beneath the looms, from Caithness in the
     north
To Lizard Point and Dover in the south. His Emanation
Joyed in the many weaving threads in bright Cathedron's dome,
Weaving the web of life for Jerusalem; the web of life
Down flowing into Entuthon's vales glistens with soft affections   370

While Los arose upon his watch and down from Golgonooza—
Putting on his golden sandals to walk from mountain to
     mountain—
He takes his way, girding himself with gold, and in his hand

Holding his iron mace. The Spectre remains attentive;
Alternate they watch in night, alternate labour in day,
Before the furnaces labouring, while Los all night watches
The stars rising and setting, and the meteors and terrors of night.
With him went down the dogs of Leutha; at his feet
They lap the water of the trembling Thames, then follow swift.
And thus he heard the voice of Albion's daughters on Euphrates: 380

'Our father Albion's land, oh, it was a lovely land! And the
    daughters of Beulah
Walked up and down in its green mountains. But Hand is fled
Away, and mighty Hyle, and after them Jerusalem is gone.
    Awake,
Highgate's heights and Hampstead's, to Poplar, Hackney and
    Bow,
To Islington and Paddington and the brook of Albion's river.
We builded Jerusalem as a city and a temple; from Lambeth
We began our foundations, lovely Lambeth! O lovely hills
Of Camberwell, we shall behold you no more in glory and pride,
For Jerusalem lies in ruins and the furnaces of Los are builded
    there;
You are now shrunk up to a narrow rock in the midst of the sea. 390
But here we build Babylon on Euphrates, compelled to build
And to inhabit, our little ones to clothe in armour of the gold
Of Jerusalem's cherubims, and to forge them swords of her
    altars.
I see London blind and age-bent begging through the streets
Of Babylon, led by a child; his tears run down his beard.
The voice of wandering Reuben echoes from street to street
In all the cities of the nations: Paris, Madrid, Amsterdam.
The corner of Broad Street weeps; Poland Street languishes;
To Great Queen Street and Lincoln's Inn all is distress and woe.

'The night falls thick. Hand comes from Albion in his strength; 400
He combines into a mighty one, the double Molech and
    Chemosh,
Marching through Egypt in his fury. The east is pale at his
    course,
The nations of India; the wild Tartar that never knew Man
Starts from his lofty places and casts down his tents and flees
    away.

But we woo him all the night in songs. O Los, come forth; O Los,
Divide us from these terrors and give us power them to subdue.
Arise upon thy watches; let us see thy globe of fire
On Albion's rocks, and let thy voice be heard upon Euphrates.'

Thus sang the daughters in lamentation, uniting into one
With Rahab as she turned the iron spindle of destruction.   410
Terrified at the sons of Albion, they took the falsehood which
Gwendolen hid in her left hand; it grew and grew till it
Became a space and an allegory around the winding worm.
They named it Canaan, and built for it a tender moon.
Los smiled with joy, thinking on Enitharmon, and he brought
Reuben from his twelvefold wand'rings and led him into it,
Planting the seeds of the twelve tribes and Moses and David,
And gave a time and revolution to the space, six thousand years.
He called it Divine Analogy, for in Beulah the feminine
Emanations create space, the masculine create time and plant   420
The seeds of beauty in the space. List'ning to their lamentation
Los walks upon his ancient mountains in the deadly darkness,
Among his furnaces, directing his laborious myriads watchful,
Looking to the east; and his voice is heard over the whole earth,
As he watches the furnaces by night and directs the labourers.

And thus Los replies upon his watch. The valleys listen silent;
The stars stand still to hear. Jerusalem and Vala cease to mourn;
His voice is heard from Albion. The Alps and Apennines
Listen; Hermon and Lebanon bow their crowned heads.
Babel and Shinar look toward the western gate; they sit down   430
Silent at his voice; they view the red globe of fire in Los's hand
As he walks from furnace to furnace, directing the labourers.
And this is the song of Los, the song that he sings on his watch:

'O lovely mild Jerusalem! O Shiloh of Mount Ephraim!
I see thy gates of precious stones, thy walls of gold and silver.
Thou art the soft reflected image of the sleeping man
Who, stretched on Albion's rocks, reposes amidst his twenty-
     eight
Cities, where Beulah lovely terminates in the hills and valleys of
     Albion:
Cities not yet embodied in time and space. Plant ye
The seeds, O sisters, in the bosom of time and space's womb,   440

To spring up for Jerusalem. Lovely shadow of sleeping Albion,
Why wilt thou rend thyself apart and build an earthly kingdom,
To reign in pride, and to oppress, and to mix the cup of delusion?
O thou that dwellest with Babylon, come forth, O lovely one!

'I see thy form, O lovely mild Jerusalem, winged with six wings
In the opacous bosom of the sleeper, lovely threefold
In head and heart and reins, three universes of love and beauty.
Thy forehead bright, 'Holiness to the Lord' with gates of pearl,°
Reflects Eternity beneath thy azure wings of feathery down,
Ribbed delicate and clothed with feathered gold and azure and   450
   purple
From thy white shoulders shadowing, purity in holiness!
Thence feathered with soft crimson of the ruby bright as fire,
Spreading into the azure wings, which like a canopy
Bends over thy immortal head, in which Eternity dwells.
Albion, beloved land, I see thy mountains, and thy hills,
And valleys, and thy pleasant cities: 'Holiness to the Lord'.
I see the Spectres of thy dead, O Emanation of Albion.

'Thy bosom white, translucent, covered with immortal gems,
A sublime ornament not obscuring the outlines of beauty,
Terrible to behold for thy extreme beauty and perfection: 460
Twelvefold here all the tribes of Israel I behold
Upon the holy land. I see the River of Life and Tree of Life.
I see the new Jerusalem descending out of Heaven°
Between thy wings of gold and silver, feathered immortal,
Clear as the rainbow, as the cloud of the sun's tabernacle.

'Thy reins covered with wings translucent, sometimes covering
And sometimes spread abroad, reveal the flames of holiness
Which like a robe covers, and like a veil of seraphim
In flaming fire unceasing burns from eternity to eternity.
Twelvefold I there behold Israel in her tents;   470
A pillar of cloud by day, a pillar of fire by night
Guides them. There I behold Moab and Ammon and Amalek.
There bells of silver round thy knees, living, articulate
Comforting sounds of love and harmony; and on thy feet
Sandals of gold and pearl; and Egypt and Assyria before me,
The isles of Javan, Philistia, Tyre and Lebanon.'°

Thus Los sings upon his watch, walking from furnace to furnace.
He seizes his hammer every hour; flames surround him as
He beats. Seas roll beneath his feet; tempests muster
Around his head; the thick hailstones stand ready to obey   480
His voice in the black cloud. His sons labour in thunders
At his furnaces; his daughters at their looms sing woes.
His Emanation separates in milky fibres, agonising
Among the golden looms of Cathedron, sending fibres of love
From Golgonooza with sweet visions for Jerusalem, wanderer.

Nor can any consummate bliss without being generated
On earth, of those whose Emanations weave the loves
Of Beulah for Jerusalem and Shiloh in immortal Golgonooza:
Concentering in the majestic form of Erin in eternal tears,
Viewing the winding worm on the deserts of Great Tartary,   490
Viewing Los in his shudderings, pouring balm on his sorrows.
So dread is Los's fury, that none dare him to approach
Without becoming his children in the furnaces of affliction.

And Enitharmon like a faint rainbow waved before him,
Filling with fibres from his loins, which reddened with desire,
Into a globe of blood beneath his bosom, trembling in darkness
Of Albion's clouds. He fed it with his tears and bitter groans,
Hiding his Spectre in invisibility from the timorous shade;
Till it became a separated cloud of beauty, grace and love
Among the darkness of his furnaces dividing asunder; till   500
She separated stood before him, a lovely female, weeping.
Even Enitharmon separated outside, and his loins closed
And healed after the separation; his pains he soon forgot,
Lured by her beauty outside of himself in shadowy grief.
Two wills they had, two intellects, and not as in times of old.

Silent they wandered hand in hand, like two infants wand'ring,
From Enion in the deserts, terrified at each other's beauty,
Envying each other yet desiring, in all-devouring love,
Repelling weeping Enion, blind and age-bent, into the fourfold
Deserts. Los first broke silence and began to utter his love: 510

'O lovely Enitharmon, I behold thy graceful forms
Moving beside me, till intoxicated with the woven labyrinth
Of beauty and perfection my wild fibres shoot in veins

Of blood through all my nervous limbs. Soon overgrown in roots
I shall be closed from thy sight. Seize therefore in thy hand
The small fibres as they shoot around me, draw out in pity,
And let them run on the winds of thy bosom. I will fix them
With pulsations; we will divide them into sons and daughters,
To live in thy bosom's translucence as in an eternal morning.'

Enitharmon answered: 'No! I will seize thy fibres and weave   520
Them: not as thou wilt but as I will. For I will create°
A round womb beneath my bosom, lest I also be overwoven
With love. Be thou assured I never will be thy slave.
Let Man's delight be love, but Woman's delight be pride.
In Eden our loves were the same; here they are opposite.
I have loves of my own; I will weave them in Albion's Spectre.
Cast thou in Jerusalem's shadows thy loves: silk of liquid
Rubies, jacinths, crysolites, issuing from thy furnaces. While
Jerusalem divides thy care, while thou carest for Jerusalem,
Know that I never will be thine. Also thou hidest Vala;   530
From her these fibres shoot to shut me in a grave.
You are Albion's victim; he has set his daughter in your path.'

Los answered, sighing like the bellows of his furnaces:

'I care not! The swing of my hammer shall measure the starry
    round.
When in Eternity man converses with man they enter
Into each other's bosom (which are universes of delight)
In mutual interchange, and first their Emanations meet,
Surrounded by their children. If they embrace and commingle
The human fourfold forms mingle also in thunders of intellect,
But, if the Emanations mingle not, with storms and agitations   540
Of earthquakes and consuming fires they roll apart in fear.
For man cannot unite with man but by their Emanations,
Which stand, both male and female, at the gates of each
    humanity.
How then can I ever again be united as man with man
While thou, my Emanation, refusest my fibres of dominion?
When souls mingle and join through all the fibres of brother-
    hood,
Can there be any secret joy on earth greater than this?'

Enitharmon answered: 'This is Woman's world, nor need she any
Spectre to defend her from Man. I will create secret places,
And the masculine names of the places Merlin and Arthur.   550
A triple female tabernacle for moral law I weave,
That he who loves Jesus may loathe, terrified, female love,
Till God himself become a male subservient to the female.'

She spoke in scorn and jealousy, alternate torments, and
So speaking she sat down on Sussex shore, singing lulling
Cadences and playing in sweet intoxication among the glistening
Fibres of Los, sending them over the ocean eastward into
The realms of dark death. O perverse to thyself; contrarious
To thy own purposes. For when she began to weave,
Shooting out in sweet pleasure, her bosom in milky love   560
Flowed into the aching fibres of Los—yet contending against
    him
In pride, sending his fibres over to her objects of jealousy
In the little lovely allegoric night of Albion's daughters,
Which stretched abroad, expanding east and west and north and
    south
Through all the world of Erin and of Los and all their children.

A sullen smile broke from the Spectre in mockery and scorn,
Knowing himself the author of their divisions and shrinkings.
    Gratified
At their contentions, he wiped his tears, he washed his visage.

'The man who respects woman shall be despised by Woman,
And deadly cunning and mean abjectness only shall enjoy them.   570
For I will make their places of joy and love excrementitious,
Continually building, continually destroying in family feuds.
While you are under the dominion of a jealous female,
Unpermanent for ever because of love and jealousy,
You shall want all the minute particulars of life.'

Thus joyed the Spectre in the dusky fires of Los's forge, eyeing
Enitharmon, who at her shining looms sings lulling cadences;
While Los stood at his anvil in wrath, the victim of their love
And hate, dividing the space of love with brazen compasses
In Golgonooza, and in Udan-Adan, and in Entuthon of Urizen.   580

The blow of his hammer is justice; the swing of his hammer
    mercy;
The force of Los's hammer is eternal forgiveness. But
His rage or his mildness were vain; she scattered his love on the
    wind
Eastward into her own centre, creating the female womb
In mild Jerusalem around the Lamb of God. Loud howl
The furnaces of Los! Loud roll the wheels of Enitharmon.
The four Zoas in all their faded majesty burst out in fury
And fire. Jerusalem took the cup which foamed in Vala's hand
Like the red sun upon the mountains in the bloody day,
Upon the hermaphroditic winepresses of love and wrath.   590

Though divided by the cross and nails and thorns and spear
In cruelties of Rahab and Tirzah, permanent endure
A terrible indefinite hermaphroditic form,
A winepress of love and wrath, double, hermaphroditic,
Twelvefold in allegoric pomp, in selfish holiness:
The Pharisaion, the Grammateis, the Presbyterion,°
The Archiereus, the Iereus, the Saddusaion, double
Each withoutside of the other, covering eastern heaven.

Thus was the covering cherub revealed, majestic image
Of selfhood, body put off, the Antichrist accursed,   600
Covered with precious stones. A human dragon terrible
And bright stretched over Europe and Asia gorgeous,
In three nights he devoured the rejected corse of death.

His head dark, deadly, in its brain encloses a reflection
Of Eden all-perverted, Egypt on the Gihon many-tongued
And many-mouthed, Ethiopia, Lybia, the Sea of Rephaim.
Minute particulars in slavery I behold among the brick-kilns
Disorganized, and there is Pharaoh in his iron court,
And the dragon of the river, and the furnaces of iron.°

Outwoven from Thames and Tweed and Severn, awful streams,   610
Twelve ridges of stone frown over all the earth in tyrant pride,
Frown over each river, stupendous works of Albion's Druid sons.
And Albion's forests of oaks covered the earth from pole to pole.

His bosom wide reflects Moab and Ammon, on the River
Pison, since called Arnon. There is Heshbon beautiful,
The rocks of Rabbath on the Arnon, and the fish-pools of
    Heshbon°
Whose currents flow into the Dead Sea by Sodom and
    Gomorrah.
Above his head high-arching wings, black, filled with eyes,
Spring upon iron sinews from the *scapulae* and *os humeri*;°
There Israel in bondage to his generalizing gods,   620
Molech and Chemosh. And in his left breast is Philistia,
In Druid temples over the whole earth with victim's sacrifice,
From Gaza to Damascus, Tyre and Sidon, and the gods
Of Javan, through the isles of Grecia and all Europe's kings.
Where Hiddekel pursues his course among the rocks
Two wings spring from his ribs of brass, starry, black as night,
But translucent their blackness as the dazzling of gems.

His loins enclose Babylon on Euphrates beautiful,
And Rome in sweet Hesperia; there Israel scattered abroad
In martyrdoms and slavery I behold. Ah, vision of sorrow!   630
Enclosed by eyeless wings, glowing with fire as the iron
Heated in the smith's forge, but cold the wind of their dread fury.

But in the midst of a devouring stomach, Jerusalem
Hidden within the covering cherub as in a tabernacle
Of threefold workmanship, in allegoric delusion and woe.
There the seven Kings of Canaan and five Baalim of Philistia,
Sihon and Og, the Anakim and Emim, Nephilim and Gibborim,°
From Babylon to Rome. And the wings spread from Japan,
Where the Red Sea terminates the world of generation and death,
To Ireland's farthest rocks where giants builded their causeway   640
Into the Sea of Rephaim; but the sea o'erwhelmed them all.

A double female now appeared within the tabernacle:
Religion hid in war, a dragon red and hidden harlot—
Each within other, but, without, a warlike mighty one
Of dreadful power, sitting upon Horeb pondering dire
And mighty preparations, mustering multitudes innumerable
Of warlike sons among the sands of Midian and Aram.
For multitudes of those who sleep in Alla descend,
Lured by his warlike symphonies of tabret, pipe and harp,
Burst the bottoms of the graves and funeral arks of Beulah.   650

Wandering in that unknown night beyond the silent grave,
They become one with the Antichrist and are absorbed in him.
The feminine separates from the masculine and both from Man,
Ceasing to be his Emanations, life to themselves assuming.
And while they circumscribe his brain, and while they circum-
  scribe
His heart, and while they circumscribe his loins, a veil and net
Of veins of red blood grows around them like a scarlet robe,
Covering them from the sight of Man like the woven veil of sleep;
Such as the flowers of Beulah weave to be their funeral mantles,
But dark, opaque, tender to touch, and painful, and agonizing   660
To the embrace of love and to the mingling of soft fibres
Of tender affection—that no more the masculine mingles
With the feminine, but the sublime is shut out from the pathos
In howling torment, to build stone walls of separation, compel-
  ling
The pathos to weave curtains of hiding secrecy from the torment.

Bowen and Conwenna stood on Skiddaw, cutting the fibres
Of Benjamin from Chester's river. Loud the river, loud the
  Mersey
And the Ribble thunder into the Irish sea, as the twelve sons
Of Albion drank and imbibed the life and eternal form of Luvah.
Cheshire and Lancashire and Westmorland groan in anguish;   670
As they cut the fibres from the rivers he sears them with hot
Iron of his forge, and fixes them into bones of chalk and rock.
Conwenna sat above; with solemn cadences she drew
Fibres of life out from the bones into her golden loom.
Hand had his furnace on Highgate's heights, and it reached
To Brockley Hills across the Thames; he with double Boadicea
In cruel pride cut Reuben apart from the hills of Surrey,
Commingling with Luvah and with the sepulchre of Luvah.
For the male is a furnace of beryl; the female is a golden loom.

Los cries: 'No individual ought to appropriate to himself,   680
Or to his Emanation, any of the universal characteristics
Of David or of Eve, of the woman or of the Lord,
Of Reuben or of Benjamin, of Joseph or Judah or Levi.
Those who dare appropriate to themselves universal attributes
Are the blasphemous selfhoods, and must be broken asunder.
A vegetated Christ and a virgin Eve are the hermaphroditic

Blasphemy; by his maternal birth he is that evil one,
And his maternal humanity must be put off eternally,
Lest the sexual generation swallow up regeneration.
Come, Lord Jesus; take on thee the Satanic body of holiness.'   690

So Los cried in the valleys of Middlesex in the spirit of prophecy,
While in selfhood Hand and Hyle and Bowen and Skofield
   appropriate
The divine names, seeking to vegetate the Divine Vision
In a corporeal and ever-dying vegetation and corruption.
Mingling with Luvah in one, they become one great Satan.

Loud scream the daughters of Albion beneath the tongs and
   hammer;
Dolorous are their lamentations in the burning forge.
They drink Reuben and Benjamin as the iron drinks the fire;
They are red hot with cruelty, raving along the banks of Thames
And on Tyburn's brook among the howling victims, in loveli-   700
   ness.
While Hand and Hyle condense the little ones and erect them
   into
A mighty temple even to the stars; but they vegetate
Beneath Los's hammer, that life may not be blotted out.

For Los said: 'When the individual appropriates universality
He divides into male and female; and when the male and female
Appropriate individuality, they become an eternal death.
Hermaphroditic worshippers of a god of cruelty and law!
Your slaves and captives you compel to worship a god of mercy.
These are the demonstrations of Los, and the blows of my
   mighty hammer.'

So Los spoke. And the giants of Albion, terrified and ashamed   710
With Los's thunderous words, began to build trembling rocking-
   stones,
For his words roll in thunders and lightnings among the temples;
Terrified, rocking to and fro upon the earth, and sometimes
Resting in a circle in Maldon or in Strathness or Jura,
Plotting to devour Albion and Los the friend of Albion,
Denying in private, mocking God and eternal life, and in public
Collusion calling themselves Deists, worshipping the maternal
Humanity, calling it Nature, and Natural Religion.

But still the thunder of Los peals loud, and thus the thunder's
    cry:
'These beautiful witchcrafts of Albion are gratified by cruelty.   720
It is easier to forgive an enemy than to forgive a friend.
The man who permits you to injure him deserves your vengeance;
He also will receive it. Go, Spectre, obey my most secret desire,
Which thou knowest without my speaking. Go to these fiends of
    righteousness,
Tell them to obey their humanities, and not pretend holiness
When they are murderers. As far as my hammer and anvil permit
Go, tell them that the worship of God is honouring his gifts
In other men, and loving the greatest men best, each according
To his genius, which is the Holy Ghost in Man. There is no
    other
God than that God who is the intellectual fountain of humanity.   730
He who envies or calumniates, which is murder and cruelty,
Murders the Holy One. Go, tell them this and overthrow their
    cup,
Their bread, their altar table, their incense and their oath,
Their marriage and their baptism, their burial and consecration.
I have tried to make friends by corporeal gifts but have only
Made enemies; I never made friends but by spiritual gifts,
By severe contentions of friendship and the burning fire of
    thought.
He who would see the Divinity must see him in his children:
One first, in friendship and love; then a divine family; and in the
    midst
Jesus will appear. So he who wishes to see a vision, a perfect   740
    whole,
Must see it in its minute particulars, organised: and not as thou,
O fiend of righteousness, pretendest. Thine is a disorganised
And snowy cloud, brooder of tempest and destructive war.
You smile with pomp and rigour; you talk of benevolence and
    virtue.
I act with benevolence and virtue and get murdered time after
    time.
You accumulate particulars, and murder by analysing, that you
May take the aggregate. And you call the aggregate Moral Law;
And you call that swelled and bloated form, a Minute Particular.
But general forms have their vitality in particulars; and every
Particular is a man, a divine member of the divine Jesus.'   750

So Los cried at his anvil, in the horrible darkness weeping.

The Spectre builded stupendous works, taking the starry heavens°
Like to a curtain and folding them according to his will,
Repeating the Smaragdine Table of Hermes to draw Los down°
Into the indefinite, refusing to believe without demonstration.
Los reads the stars of Albion; the Spectre reads the voids
Between the stars, among the arches of Albion's tomb sublime:
Rolling the sea in rocky paths, forming Leviathan
And Behemoth, the war by sea enormous and the war
By land astounding, erecting pillars in the deepest hell   760
To reach the heavenly arches. Los beheld undaunted; furious
His heaved hammer. He swung it round and at one blow,
In unpitying ruin driving down the pyramids of pride,
Smiting the Spectre on his anvil, and the integuments of his eye
And ear unbinding in dire pain, with many blows
Of strict severity self-subduing, and with many tears labouring.

Then he sent forth the Spectre; all his pyramids were grains
Of sand, and his pillars dust on the fly's wing, and his starry
Heavens a moth of gold and silver mocking his anxious grasp.
Thus Los altered his Spectre, and every ratio of his reason   770
He altered time after time, with dire pain and many tears,
Till he had completely divided him into a separate space.

Terrified Los sat to behold, trembling and weeping and howling.
'I care not whether a man is good or evil; all that I care
Is whether he is a wise man or a fool. Go! Put off holiness
And put on intellect, or my thundrous hammer shall drive thee
To wrath which thou condemnest, till thou obey my voice.'

So Los terrified cries, trembling and weeping and howling!
   'Beholding,
What do I see? The Briton, Saxon, Roman, Norman amalgamat-
   ing
In my furnaces into one nation, the English, and taking refuge   780
In the loins of Albion; the Canaanite united with the fugitive
Hebrew, whom she divided into twelve and sold into Egypt,
Then scattered the Egyptian and Hebrew to the four winds.
This sinful nation created in our furnaces and looms is Albion.'

So Los spoke. Enitharmon answered in great terror in Lambeth's
vale:
'The poet's song draws to its period, and Enitharmon is no more.
For if he be that Albion I can never weave him in my looms;
But when he touches the first fibrous thread, like filmy dew
My looms will be no more and I, annihilate, vanish for ever.
Then thou wilt create another female according to thy will.'   790

Los answered swift as the shuttle of gold: 'Sexes must vanish and
cease
To be, when Albion arises from his dread repose, O lovely
Enitharmon.
When all their crimes, their punishments, their accusations of
sin,
All their jealousies, revenges, murders, hidings of cruelty in
deceit
Appear only in the outward spheres of visionary space and time,
In the shadows of possibility by mutual forgiveness for evermore,
And in the vision and in the prophecy. That we may foresee and
avoid
The terrors of creation and redemption and judgement, behold-
ing them
Displayed in the emanative visions of Canaan in Jerusalem and in
Shiloh,
And in the shadows of remembrance, and in the chaos of the   800
spectre,
Amalek, Edom, Egypt, Moab, Ammon, Asshur, Philistia,
around Jerusalem:
Where the Druids reared their rocky circles to make permanent
remembrance
Of sin, and the Tree of Good and Evil sprang from the rocky
circle and snake
Of the Druid, along the Valley of Rephaim from Camberwell to
Golgotha,
And framed the Mundane Shell, cavernous in length, breadth
and height.'

Enitharmon heard. She raised her head like the mild moon:

'O Rintrah! O Palamabron! What are your dire and awful
purposes?
Enitharmon's name is nothing before you; you forget all my love!

The mother's love of obedience is forgotten, and you seek a love
Of the pride of dominion that will divorce Ocalythron and   810
     Elynittria
Upon East Moor in Derbyshire and along the valleys of Cheviot.
Could you love me, Rintrah, if you pride not in my love,
As Reuben found mandrakes in the field and gave them to his
     mother?°
Pride meets with pride upon the mountains in the stormy day,
In that terrible day of Rintrah's plough and of Satan's driving the
     team.
Ah! then I heard my little ones weeping along the valley.
Ah! then I saw my beloved ones fleeing from my tent.
Merlin was like thee, Rintrah, among the giants of Albion;
Judah was like Palamabron. O Simeon! O Levi! ye fled away.
How can I hear my little ones weeping along the valley,   820
Or how upon the distant hills see my beloved's tents?'

Then Los again took up his speech as Enitharmon ceased:

'Fear not, my sons, this waking death; he is become one with me.
Behold him here! We shall not die! We shall be united in Jesus.
Will you suffer this Satan, this body of doubt that seems but is
     not,
To occupy the very threshold of eternal life? If Bacon, Newton,
     Locke
Deny a conscience in Man, and the communion of saints and
     angels,°
Contemning the Divine Vision and fruition, worshipping the
     Deus
Of the heathen, the god of this world, and the goddess Nature,
Mystery, Babylon the Great, the Druid dragon and hidden   830
     harlot,
Is it not that signal of the morning which was told us in the
     beginning?'

Thus they converse upon Mam-Tor; the graves thunder under
     their feet.

Albion cold lays on his rock; storms and snows beat round him,
Beneath the furnaces and the starry wheels and the immortal
     tomb.

Howling winds cover him; roaring seas dash furious against him.
In the deep darkness broad lightnings glare, long thunders roll.

The weeds of death enwrap his hands and feet, blown incessant
And washed incessant by the for-ever restless sea-waves foaming
     abroad
Upon the white rock. England, a female shadow, as deadly
     damps°
Of the mines of Cornwall and Derbyshire, lays upon his bosom   840
     heavy,
Moved by the wind in volumes of thick cloud, returning, folding
     round
His loins and bosom, unremovable by swelling storms and loud
     rending
Of enraged thunders. Around them the starry wheels of their
     giant sons
Revolve; and over them the furnaces of Los, and the immortal
     tomb around,
Erin sitting in the tomb to watch them unceasing night and day.
And the body of Albion was closed apart from all nations.

Over them the famished eagle screams on bony wings, and
     around
Them howls the wolf of famine; deep heaves the ocean black,
     thundering
Around the wormy garments of Albion, then pausing in deathlike
     silence.

Time was finished! The breath divine breathed over Albion   850
Beneath the furnaces and starry wheels and in the immortal
     tomb;
And England, who is Britannia, awoke from death on Albion's
     bosom.
She awoke pale and cold; she fainted seven times on the body of
     Albion.

'Oh piteous sleep, oh piteous dream! O God, O God, awake. I
     have slain
In dreams of chastity and moral law; I have murdered Albion!
     Ah!
In Stonehenge and on London Stone and in the oak groves of
     Maldon

I have slain him in my sleep with the knife of the Druid. Oh
    England,
O all ye nations of the earth, behold ye the jealous wife.
The eagle and the wolf and monkey and owl and the king and
    priest were there.'

Her voice pierced Albion's clay-cold ear; he moved upon the   860
    rock.
The breath divine went forth upon the morning hills. Albion
    moved
Upon the rock; he opened his eyelids in pain. In pain he moved
His stony members; he saw England. Ah! shall the dead live
    again?

The breath divine went forth over the morning hills. Albion rose
In anger, the wrath of God breaking bright flaming on all sides
    around
His awful limbs. Into the heavens he walked, clothed in flames,
Loud thund'ring, with broad flashes of flaming lightning and
    pillars
Of fire, speaking the words of Eternity in human forms, in direful
Revolutions of action and passion, through the four elements on
    all sides
Surrounding his awful members. Thou seest the sun in heavy   870
    clouds
Struggling to rise above the mountains. In his burning hand
He takes his bow, then chooses out his arrows of flaming gold;
Murmuring the bowstring breathes with ardour! Clouds roll
    round the
Horns of the wide bow; loud sounding winds sport on the
    mountain brows
Compelling Urizen to his furrow, and Tharmas to his sheepfold,
And Luvah to his loom. Urthona he beheld mighty labouring at
His anvil, in the great Spectre Los unwearied labouring and
    weeping.
Therefore the sons of Eden praise Urthona's Spectre in songs:
Because he kept the Divine Vision in time of trouble.

As the sun and moon lead forward the visions of Heaven and   880
    Earth,
England, who is Britannia, entered Albion's bosom rejoicing,

Rejoicing in his indignation, adoring his wrathful rebuke.
She who adores not your frowns will only loathe your smiles.

Then Jesus appeared standing by Albion, as the good shepherd
By the lost sheep that he hath found; and Albion knew that it
Was the Lord, the universal humanity. And Albion saw his form,
A man, and they conversed as man with man, in ages of eternity.
And the divine appearance was the likeness and similitude of
    Los.

Albion said: 'O Lord, what can I do? My selfhood cruel
Marches against thee deceitful from Sinai and from Edom   890
Into the wilderness of Judah to meet thee in his pride.
I behold the visions of my deadly sleep of six thousand years
Dazzling around thy skirts like a serpent of precious stones and
    gold.
I know it is my self, O my divine creator and redeemer.'

Jesus replied: 'Fear not, Albion; unless I die thou canst not live.
But if I die I shall arise again and thou with me.
This is friendship and brotherhood; without it Man is not.'

So Jesus spoke. The covering cherub coming on in darkness
Overshadowed them and Jesus said: 'Thus do men in Eternity,
One for another to put off by forgiveness every sin.'   900

Albion replied: 'Cannot Man exist without mysterious
Offering of self for another? Is this friendship and brotherhood?
I see thee in the likeness and similitude of Los my friend.'

Jesus said: 'Wouldest thou love one who never died
For thee, or ever die for one who had not died for thee?
And if God dieth not for Man and giveth not himself
Eternally for Man, Man could not exist. For Man is love,
As God is love. Every kindness to another is a little death
In the divine image, nor can Man exist but by brotherhood.'

So saying the cloud overshadowing divided them asunder.   910
Albion stood in terror, not for himself but for his friend
Divine; and self was lost in the contemplation of faith,
And wonder at the divine mercy and at Los's sublime honour.

'Do I sleep amidst danger to friends? O my cities and counties!
Do you sleep? Rouse up, rouse up! Eternal death is abroad.'

So Albion spoke and threw himself into the furnaces of affliction.
All was a vision, all a dream. The furnaces became
Fountains of living waters flowing from the Humanity Divine.
And all the cities of Albion rose from their slumbers, and all
The sons and daughters of Albion on soft clouds waking from   920
   sleep.
Soon all around remote the heavens burnt with flaming fires,
And Urizen and Luvah and Tharmas and Urthona arose into
Albion's bosom. Then Albion stood before Jesus in the clouds
Of Heaven, fourfold among the visions of God in Eternity.

'Awake! Awake, Jerusalem! O lovely Emanation of Albion,
Awake and overspread all nations as in ancient time.
For lo! the night of death is past and the eternal day
Appears upon our hills. Awake Jerusalem, and come away.'

So spake the vision of Albion, and in him so spake in my hearing
The universal father. Then Albion stretched his hand into   930
   infinitude
And took his bow. Fourfold the vision: for bright beaming
   Urizen
Laid his hand on the south and took a breathing bow of carved
   gold;
Luvah his hand stretched to the east and bore a silver bow bright
   shining;
Tharmas westward a bow of brass pure flaming, richly wrought;
Urthona northward in thick storms a bow of iron terrible
   thundering.

And the bow is a male and female, and the quiver of the arrows of
   love
Are the children of this bow, a bow of mercy and loving-
   kindness, laying
Open the hidden heart in wars of mutual benevolence, wars of
   love;
And the hand of Man grasps firm between the male and female
   loves.
And he clothed himself in bow and arrows in awful state, fourfold 940
In the midst of his twenty-eight cities, each with his bow
   breathing.

Then each an arrow flaming from his quiver fitted carefully.

They drew fourfold the unreprovable string, bending through
the wide heavens

The horned bow fourfold. Loud sounding flew the flaming arrow
fourfold.

Murmuring the bow-string breathes with ardour. Clouds roll
round the horns

Of the wide bow; loud sounding winds sport on the mountain's
brows.

The Druid Spectre was annihilate, loud thund'ring, rejoicing,
terrific vanishing,

Fourfold annihilation. And at the clangour of the arrows of
intellect

The innumerable chariots of the Almighty appeared in Heaven,

And Bacon and Newton and Locke, and Milton and Shakespeare   950
and Chaucer,

A sun of blood-red wrath surrounding Heaven on all sides
around,

Glorious, incomprehensible by mortal man; and each chariot was
sexual threefold.

And every man stood fourfold. Each four faces had: one to the
west,°

One toward the east, one to the south, one to the north. The
horses fourfold.

And the dim chaos brightened beneath, above, around! Eyed as
the peacock,

According to the human nerves of sensation, the four rivers of the
water of life.

South stood the nerves of the eye. East in rivers of bliss the nerves
of the

Expansive nostrils. West flowed the parent sense, the tongue.
North stood

The labyrinthine ear. Circumscribing and circumcising, the
excrementitious

Husk and covering into vacuum evaporating, revealing the   960
lineaments of Man,

Driving outward the body of death in an eternal death and
resurrection,

Awaking it to life among the flowers of Beulah, rejoicing in unity
In the four senses, in the outline, the circumference and form, for
    ever
In forgiveness of sins which is self-annihilation. It is the
    covenant of Jehovah.

The four living creatures, chariots of Humanity Divine incom-
    prehensible,
In beautiful paradises expand. These are the four rivers of
    paradise
And the four faces of humanity fronting the four cardinal points
Of Heaven, going forward, forward, irresistible from eternity to
    eternity.

And they conversed together in visionary forms dramatic, which
    bright
Redounded from their tongues in thunderous majesty, in visions,   970
In new expanses, creating exemplars of memory and of intellect,
Creating space, creating time according to the wonders divine
Of human imagination, throughout all the three regions immense
Of childhood, manhood and old age; and the all-tremendous
    unfathomable non-ens°
Of death was seen in regenerations terrific or complacent,
    varying
According to the subject of discourse. And every word and every
    character
Was human, according to the expansion or contraction, the
    translucence or
Opaquenes of nervous fibres. Such was the variation of time and
    space,
Which vary according as the organs of perception vary, and they
    walked
To and fro in Eternity as one man, reflecting each in each and   980
    clearly seen
And seeing, according to fitness and order. And I heard Jehovah
    speak
Terrific from his holy place and saw the words of the mutual
    covenant divine
On chariots of gold and jewels, with living creatures starry and
    flaming
With every colour: lion, tiger, horse, elephant, eagle, dove, fly,
    worm,

And the all-wondrous serpent clothed in gems and rich array,
    humanize

In the forgiveness of sins according to the covenant of Jehovah.
    They cry:

'Where is the covenant of Priam, the moral virtues of the
    heathen?

Where is the Tree of Good and Evil that rooted beneath the cruel
    heel

Of Albion's Spectre, the patriarch Druid? Where are all his
    human sacrifices

For sin, in war and in the Druid temples of the accuser of sin,   990
    beneath

The oak groves of Albion that covered the whole earth beneath
    his Spectre?

Where are the kingdoms of the world and all their glory that grew
    on desolation,°

The fruit of Albion's poverty-tree, when the triple-headed Gog-
    Magog giant

Of Albion taxed the nations into desolation, and then gave the
    spectrous oath?'

Such is the cry from all the Earth, from the living creatures of the
    Earth

And from the great city of Golgonooza in the shadowy
    generation,

And from the thirty-two nations of the Earth among the living
    creatures,

All human forms identified, even tree, metal, earth and stone. All

Human forms identified, living, going forth, and returning
    wearied

Into the planetary lives of years, months, days and hours,   1000
    reposing

And then awaking into his bosom in the life of immortality.

And I heard the name of their Emanations; they are named
    Jerusalem.

The End of the Song of Jerusalem

NOTES

For reasons mentioned in the Introduction Blake's mythology is not systematically glossed in these notes. His invented names and special motifs are often best investigated by reference to the whole range of contexts in which they appear, and these are listed in the Index of Names and Motifs. Where a Blakean name or motif is glossed in these notes the gloss is usually attached to the first occurrence of the item, which can be located by means of the Index of Names and Motifs.

BLAKE ON RELIGION AND KNOWLEDGE

3 *From the annotations to Lavater's Aphorisms on Man.* In 1788 the Swiss-born artist Henry Fuseli published a translation of this minor work by his compatriot and long-standing friend, and probably gave Blake a set of unbound sheets. Lavater's invitation to the reader to indicate approval or disapproval of the various aphorisms by remarks on the page and to 'show your copy to whom you please', combined with Fuseli's availability for this role, gave rise to an idiosyncratically Blakean kind of text: marginal annotations intented for reading by others. Lavater was a Zwinglian pastor whose religious beliefs had an optimistic and pantheistic tendency. His aphorisms are orthodox in their moral views, but are distinguished from the conventional sequence of prudential maxims by their celebration of strong instinctual feeling and creativity. Blake was delighted by their spirit, but Lavater does not seem to have become permanently important to him.

exuberant: a term that became important for Blake at this period, and enters his vocabulary from Fuseli's translation (German *üppig*).

You are my brother . . . God in him: Mark 3: 35, and John 4: 16.

4 *amulet*: probably in the sense of a spiritual or occult medicine.

stamina: probably a plural: the intrinsic or original elements of a thing.

causes and consequences. Causality was certainly an important element in recent fashions in thought (in the laws of Newtonian physics and the epistemology of Hume, for example) but it was not closely connected with contemporary treatments of the accident versus essence/substance issue. That Hume was a historian added to the circumstantial association in Blake's eyes (see p. 96).

5 *From the annotations to Swedenborg's Divine Love and Wisdom.* Allusions to Emanuel Swedenborg (1688–1772), the Swedish Christian mystic and founder of a millenarian sect, are scattered throughout Blake's work. He seems to have known at least five of Swedenborg's books, of which he annotated two, and he attended the first conference of the Swedenborgian Church in London in 1789, though he seems not to have become a

member. The best accounts of the subject, with suggested reconstructions of Blake's changing views of Swedenborg, are John Howard, 'An audience for *The Marriage of Heaven and Hell*', *Blake Studies*, 3 (1970), 19–52, and Morton D. Paley, '"A New Heaven is Born": William Blake and Swedenborgianism', *Blake*, 13 (1979), 64–90.

All Religions are One. Each of the paragraphs here and in *There is No Natural Religion* was printed by Blake on a separate sheet and decorated with ornament and illustration, making this venture almost certainly the prototype for the publication of most of Blake's verse. Blake never bound up these sheets and, while the order followed here is that favoured by modern scholarship, many alternative orders are possible, limited chiefly by Blake's numbering of the majority of his propositions.

The voice ... wilderness. The phrase is applied in all the gospels to John the Baptist, and derives from Isa. 40: 3.

6 *Spirit of Prophecy*: Rev. 19: 10.
   There is No Natural Religion (1 and 2). 'Natural religion' is the standard 18th-c. term for any religious belief which relies on reason and our knowledge of the physical world, and not on revelation. Always allowing for the problem of ordering Blake's plates, in these two sequences he seems, characteristically, to be attacking such a position both from within and from without.

7 *ratio.* Blake's complicated use of this term includes a play on Latin *ratio* (reason) and here perhaps also the sense, current in the 18th century, of 'ration'.

8 *The Marriage of Heaven and Hell.* Blake's title enshrines an impossible and absurd notion, which is nevertheless the kind of event which becomes possible in apocalypses. This tension informs the whole work. *The Marriage* attacks errors about Man's moral and spiritual life (especially those which interchange and confound the heavenly and hellish in him), but at another level does not admit the possibility of error, celebrating as it does the whole of existence in all its plurality and diversity (and in particular the diverse inner worlds of all living things). This is matched in the unusual tone of the work, at once vehement and affable (indeed comic), expressing both 'opposition' and 'friendship'. Much of it consists of 'Memorable Fancies', which are parodies of Swedenborg, yet also tributes to his visionary writings. For all its plainness of utterance, *The Marriage* is full of obliqueness and ambiguity.

The Argument. According to Alicia Ostriker, the leading modern authority on Blake's prosody, the first example of free verse in English.

l. 2. *swag*: droop down heavily.

l. 5. *vale of death.* The valley and its dangerous path are a reminiscence of Bunyan's Valley of the Shadow of Death, which in turn derives from Ps. 23: 4 and Jer. 2: 6.

l. 12. *bleached bones.* For this image see Ezek. 37.

9 l. 17. *serpent walks*. Many biblical commentators held that before the Fall the serpent had an upright posture.

thirty-three years. The time elapsed since 1757 (Blake's birth-year and Swedenborg's date for a Last Judgement), and also Christ's age at his crucifixion and resurrection.

folded up. John 20: 5–7 emphasizes the disposition of Christ's empty grave-clothes.

Edom: alludes to the eventual triumph of the dispossessed Esau over Jacob (see Gen. 27).

Isaiah xxxiv and xxxv. The imagery of the work's 'Argument' draws on the latter chapter.

Attraction . . . existence. There may be an echo here of a remark by the French free thinker Baron d'Holbach (for this and other parallels see T. A. Hoagwood's 'Holbach and Blake's philosophical statement in "The Voice of the Devil"', *English Language Notes*, 15 (1978), 81–6).

10 *called Satan*. In Job Satan is one of the 'sons of God' (1: 6) and the tormentor of Job on God's instructions.

the comforter: John 14: 16.

cuts . . . way. Faerie Queene II. viii. 5. For a full discussion of the phrase in context, see Robert F. Gleckner, 'Blake's Miltonizing of Chatterton', *Blake*, 7 (1977), 27–9.

11 *Excess of joy*. For Milton's Satan (*Paradise Lost* i. 123) God the tyrant reigns 'in the excess of joy'.

12 *Sublime*: a noun, and an important aesthetic positive for Blake.

13 *all gods . . . ours*: a thought more in the spirit of enlightenment religious historiography than of the Old Testament.

King David: traditionally the composer of the Psalms.

14 *three years*: Isa. 20: 3.

Diogenes the Grecian: the Greek philosopher (*c*.400–*c*.325 BC) whose main tenet, as expressed in his own practice, was that humans should live in as simple and unartificial a manner as possible.

perception of the infinite: Ezek. 4. Despite the allusion to North America Blake may also see Ezekiel's fakir-like action (he lay 390 days on his left side, and 40 on his right) as an example of how 'the philosophy of the east taught the first principles of perception'.

six thousand years: an interval often mentioned by Blake and, according to a long-standing tradition in Christian theology, the age of the world. There is also an even older tradition to the effect that this is the life-span of the world. Blake's remark amounts to a declaration that the apocalypse is imminent.

Tree of Life: Gen. 3: 24, Rev. 22: 2.

by corrosives: with etched plates rather than with conventional letter-press.

15 *The giants*. Blake, like some ancient authors, seems to conflate the Giants and the Titans, both of whom rebelled against the Gods and were defeated and imprisoned. A second-generation Titan, Prometheus, made mankind out of clay (cf. 'The Argument' above) and was chained to a rock.

a sword: Matt. 10: 34, 25: 31–3.

Tempter: the Satan of Job.

he took me. The five stages of this trip mirror the sequence in the 'printing-house in Hell'.

17 *fixed stars*. Swedenborg held that spirits and angels populated all the bodies in the solar system, becoming progressively less spiritual beyond the orbit of Mars. Moreover, with the discovery of Uranus in 1781 there was no longer a 'void' beyond Saturn.

Analytics. The two treatises of the *Analytics* are concerned with the science of logic.

18 *Paracelsus or Jacob Behmen*. Paracelsus was an extremely influential exponent of occult and alchemical thought in the early 16th century. Behmen, or Boehme, born in 1575, was a German mystic whose vision, in its own fashion, centred on progression through contraries.

blue . . . pink. The sequence is accurate for the light emitted by a cooling body as its wavelength increases.

Bray a fool . . . out of him: Prov. 27: 22.

ten commandments. Blake's references to Christ's violations of the sixth and eighth commandments (forbidding murder and theft) are obscure, but may be clarified by reference to *Everlasting Gospel* 93–8 (below, p. 47). The other transgressions are mentioned at Mark 2: 27–8, John 8: 2–11, Matt. 27: 13–14, 10: 14.

Elijah: 2 Kgs. 2: 11.

19 *A Song of Liberty*. While this section may have been composed later than the rest of *Marriage*, and evidently emphasizes social and political questions not previously mentioned in the work, it was consistently employed by Blake as its conclusion (and as a structural device is anticipated in his *Edward III*). Such a blending of the moral/epistemological/religious and the social/political is indeed the whole tendency of Blake's writing in the years of the French Revolution. All these strands are present in the closing sentence.

Female groaned. Another point of close connection with the rest of *Marriage*, since this is the revival of 'eternal Hell' announced at the outset. The event continues to be paired with the story of Christ, and here starts with a momentous birth. This whole section has connections with Milton's 'Ode on the Morning of Christ's Nativity'. For the 'Eternal Female' see 'Vision of the last Judgement' (below p. 38).

The Eternal . . . the earth. This verse should be compared with Rev. 12: 1–2; for further possible parallels see Randell Helms, 'Blake's use of the Bible in "A Song of Liberty"', *English Language Notes* 16 (1979), 287–91.

thy dungeon. The Bastille was sacked in 1789.

keys, O Rome: emblem of the Pope's spiritual power.

Flagged: covered.

20 *stony law.* The ten commandments were written on stone tablets.

loosing . . . dens of night. In Blake's story of the 'son of fire' there is generally a reminiscence of Phaeton, who disobediently drove the chariot of the sun and scorched the earth.

Empire. Probably combines the idea of monarchy and England's recently lost American colonies.

From the annotations to Watson's Apology for the Bible. In this work Richard Watson, Bishop of Llandaff, mounted a famous attack on Thomas Paine's even more famous *The Age of Reason* (1791–2). Paine was a major figure in the American Revolution and a minor one in the French. He believed in God, but utterly denied the specific theology of Christianity, and *The Age of Reason* ridicules the theological claims of Old and New Testament alike. In his contribution to this hierarchy of texts Blake counter-attacks on Paine's behalf with considerable loyalty, but his position is complicated by the loyalty he also felt to the Bible.

whole duty of man: Eccles. 12: 13.

Matthew. The chapter consists of Christ's furious attack on Scribes and Pharisees, with its proverbial 'whited sepulchres' and 'generation of vipers'.

examiners. Watson had queried whether Paine had 'examined calmly' the arguments for revealed religion.

21 *so many thousands.* Watson had maintained, against Paine, that a divine instruction to the Israelites to slaughter the Canaanites could be consistent with God's providence.

Holy Ghost. The sin against the Holy Ghost is the unforgivable sin.

Phidias, Glycon: both Greek sculptors.

Songs of Fingal. Blake seems to have accepted the 'Ossian' poems published by James Macpherson in the 1760s as genuine translations of Gaelic heroic verse.

22 *where unbelief hindered*: Matt. 13: 28.

The manner . . . impossibility. To 20th-c. readers Blake's view of curative miracles (essentially, that they are psychosomatic) will sound more 'modern' than the claim that they are purely supernatural phenomena, not dependent on the wishes of the patient.

a small pamphlet: probably Paine's *Common Sense* (1776), which articulated a revolutionary philosophy for the American colonies' uprising.

want not to be confounded: do not need to be defeated.

23 *abomination . . . desolate*: Matt. 24: 15, Mark 13: 14.

everlasting gospel. See headnote to Blake's poem of this name (below, p. 524).

'*To my friend Butts*'. This poem is contained in a letter Blake wrote to his friend and loyal patron Thomas Butts just two weeks after moving to the Sussex seaside village of Felpham.

l. 15. *particles bright*. In Blake's day the corpuscular theory of light (as sanctioned by Newton) had yet to be displaced by the 19th-c. wave theory.

24 l. 41. *friend*: William Hayley (1745–1820), the man-of-letters and connoisseur who had invited Blake to Felpham so that he could employ him and generally assist his career. Over the next two years Blake became increasingly distressed in his situation, and antagonistic towards Hayley.

25 '*With happiness stretched across the hills*': also contained in a letter from Felpham to Butts, written in November 1802.

l. 7. *Hayley*. See note to l. 41 of preceding poem.

26 l. 39. *Fuseli*. The Swiss-born Henry Fuseli (1741–1825) was a dynamic personality and a highly original painter and illustrator, in particular of frightening, erotic, and dream-like subjects. He and Blake were intimate for about ten years from 1787, and he probably introduced the latter to some of the radicals of the day. Thereafter their friendship, though it continued, became more fraught.

l. 47. *Flaxman*. John Flaxman (1755–1826), the neo-classical sculptor and illustrator, was a friend of Blake's from their earliest manhood, and throughout his life he exerted himself on behalf of Blake's career, though Blake later entertained suspicions of his loyalty.

l. 54. *weltering*. Blake's 'double vision' expresses itself in a play on words. 'Weltering' means withering, as applied to the thistle, and staggering, as applied to the old man.

l. 59. *Los*. If 'sun' and 'Los' are also outward and inward respectively in spelling, the outward form of 'Los' would be 'Sol'.

27 l. 86. *Beulah*. See note to *Milton* i 1 (below, p. 567).

Auguries of Innocence

28 l. 17. *gamecock . . . fight*. Fighting cocks have their feathers and combs heavily cut, and their natural spurs sheathed in artificial ones.

l. 33. *boy . . . fly*: *King Lear* IV. i. 38–9.

l. 42. *polar bar*. See note to *Thel* 108 (below, p. 548).

29 l. 65. *hands*: farmworkers.

30 ll. 97–8. *poison . . . crown*. The fact of the poison in laurel being prussic acid (recently isolated chemically; see note to *Jerusalem III*. 551, p. 583 below) enables Blake to attack the militarism of the Roman Empire with peculiar intensity and wit.

l. 105. *emmet*: an ant.

31   l. 127 *born . . . night*: Jonah 4: 10.

Prefaces to chs. II, III and IV of Jerusalem. Blake's way of couching the three prefaces that follow may owe something to Rev. Richard Clarke's *Jesus the Nazarene: Addressed to Jews, Deists, and Believers* (1795). For the first preface see below, p. 72.

patriarchal religion: the early religion of the Old Testament after the Fall, there claimed to be exclusively Jewish.

Huzza: Hurray! (thoroughly English, though Hebrew-looking).

Selah: a musical direction used in the Psalms, often mistaken for a pious ejaculation.

Heber . . . Noah. An ascending genealogy: Heber is the son of Shem, who is the son of Noah.

Druids. The Druids were a priestly caste in pre-Roman Britain whose ceremonies, according to the few sources, were conducted in oak groves. Theories linking the Druids and the Jewish patriarchs were fashionable in Blake's day. He is unusual in proposing that the Druids are prior to the patriarchs.

Druid temples . . . earth. Blake combines Druidic oak groves with the megalithic monuments, such as Stonehenge, conjectured to be Druidic. The most conspicuous megalithic remains outside Britain are in northern France.

a tradition. Blake may be referring to Hebrew scriptural embellishments and commentary (Kabbalah) springing from twin references in Genesis to the first man: as divine (1: 27) and earthly (2: 7). The distinction finds its way into Gnosticism, and the double Adam is discussed by St Paul (1 Cor. 15: 45–8). The outstanding review of the possibility, and possible elements, of Blake's knowledge of Kabbalah is Sheila Spector 'Kabbalistic sources—Blake and his critics', *Blake*, 17 (1983), 84–99.

Elohim: the ordinary Hebrew word for God, and used for God the creator of the world in Genesis rather than Jehovah. Blake is reported as 'triumphantly' pointing out this fact in conversation with Henry Crabb Robinson, and expounding the Gnostic doctrine that the world is not the creation of the supreme deity.

32   *Rahab*: a Gentile prostitute who aids the Israelites in Josh. 2. In *Jerusalem* Blake gives her name also to the Whore of Babylon of Rev. 17, probably because of Ps. 87: 4 (with the encouragement of the apocalyptic overtones of Isa. 51: 9).

Deists. Deism is the common term for a religious attitude which spread in Christian culture in the enlightenment: it accepted the existence of a God (above all, as the creator and designer of the universe) but was sceptical of the more specifically Christian doctrines and institutions.

synagogue of Satan: Rev. 3: 9.

god of this world: 2 Cor. 4: 4.

natural philosophy: science.

33   *Methodist*. Methodism was the vigorous movement within the Church of England—eventually becoming a separate church—founded in the 18th century by Wesley and Whitefield, which stressed a personal, non-institutional relationship to God, and the experience of personal conversion.

player. Blake seems to attach a special but enigmatic meaning to this term (see also 'Vision of the Last Judgement', p. 43), which may show him to be impressively at ease in Greek, since the English word 'hypocrite' is derived from the Greek for actor.

Foote. Samuel Foote (1720–77) wrote a comedy satirizing Methodism.

Louises and Fredericks: generic names for militaristic leaders of modern European states, such as France and Prussia.

persecutest thou me: Acts 9: 4.

34   *curse to hide*: Matt. 25: 14–29.

for ourselves: Matt. 6: 20.

science: a major term for Blake, but deployed in several senses and with both pejorative and affirmative force (see Mark L. Greenberg, 'Blake's "Science"', *Studies in 18th-Century Culture*, 12 (1983), 115–30). Here the meaning is something like 'learning'.

more than raiment: Matt. 6: 25.

35   l. 12. *a watcher and a holy one*: Dan. 4: 13.

'Vision of the Last Judgement'. These extracts are drawn from a manuscript text, dispersed through Blake's notebook and probably written at various times, which provides a commentary on a painting by Blake of the Last Judgement (now lost, or never completed, and not to be identified with any of the surviving works on this theme). Its religious subject links with other topics. Those interested in the detail of Blake's reading of human history (of especial relevance to *Milton* and *Jerusalem*) should consult the full text. The extracts here are instructive as to the status, and semantic character, of the figures in Blake's own narratives. The 'Vision' is an informal document, a place where Blake was testing formulations, often to reject them; but it contains many arresting utterances.

36   *Daughters of Memory*. Conventionally the phrase would refer to the nine muses, daughters of Mnemosyne.

the fable: roughly, 'plot' (whereas 'fable' in the first half of this sentence means something like 'fiction'). For Blake, the former is much more veridical.

Socrates say: in his *Apology of Socrates*.

37   *true vine*: John 15: 1.

churches . . . extinguished. Abel prepares a sacrificial offering for God, and there is great emphasis on burnt offerings in the Old Testament well after Abraham (e.g. the first ten chapters of Leviticus).

Apuleius's . . . Metamorphoses. Apuleius' *Golden Ass* is a reworking of a Greek original, embellished with many traditional tales. Ovid's *Metamorphoses* brings together scores of mythical and legendary narratives of the classical world.

Lot's wife: Gen. 19: 26.

Changing . . . Moses: John 2, Exod. 7: 19–20.

38 *as . . . multitude*. The phrase is used several times in the Old Testament in reference to Adam's offspring and the Israelites. The version of his picture which Blake may be describing is recorded as containing more than a thousand figures.

Between the figures. As Blake indicates, many elements in this section derive from Revelation. The relevant chapters are 9, 11, 12, 13, 17, and 20. The passage, and the biblical chapters it draws on, are a necessary background to the reading of *Milton* and *Jerusalem*.

Gog and Magog: Rev. 20: 8 is more relevant than Ezekiel (where Magog is the land ruled by the Israelite enemy Gog at 38: 2; Blake's use of verse 8 of this chapter is characteristically ingenious). Blake would also have known of the giant figures in the Guildhall called Gog and Magog.

39 *Britannica.* It is characteristic of the fluid nature of Blake's mythology that this figure is introduced in a correction to the manuscript. Blake had originally written of Albion: 'his Emanation or wife is Jerusalem'. This is partly consistent with *Jerusalem*, for in that poem Jerusalem is Albion's Emanation, but not his wife. Blake also has a 'Britannia' in *Jerusalem*, who appears briefly at the end as a kind of additional consort for Albion, but 'Britannica' is not found elsewhere in Blake's work.

mould: broken, fertile covering of soil.

Lord in the air: 1 Thess. 4: 17.

40 *bread . . . Supper.* The sacrament of Baptism involves water, that of the Eucharist bread and wine. Both sacraments have been regarded favourably by Protestant churches.

the ark. The elements in this description associated with the ark of the covenant—cherubim, curtains, candlestick, table of the shewbread—derive from Exod. 25, but they are all present more compactly at Heb. 9: 2–5, and Blake's perspective is evidently that of the New Testament. The rending of the veil at the time of Christ's death is recorded in the first three gospels.

41 *Tree of Life*: Rev. 22: 1–2.

Satan's labyrinth. Cf. *Paradise Lost* ii. 555–65.

the glory: a zone of light surrounding a holy figure.

nursing . . . mothers: Isa. 49: 23.

42 *by their . . . them.* Cf. Matt. 7: 20.

Satan the accuser: Rev. 12:9–10. Blakes' phrasing suggests that he was familiar with the passage in the original Greek.

woe to you hypocrites! Imitates a repeated formula in Matt. 23.

groans to be delivered: combines 2 Rom. 8: 20–3 and Rev. 12: 2.

43 *Seth*: the third son of Adam.

Og: a giant-king and opponent of the Israelites.

Molech: an Old Testament deity associated particularly with human sacrifice (also called Moloch).

44 *Holy, holy, holy*: from the canticle *Te Deum*.

45 *Annotation to Spurzheim's Observations on . . . Insanity*. The context of Blake's note is the assertion by Spurzheim, a founder of phrenology, that 'primitive feelings of religion', such as Methodism embodies, can 'produce insanity'.

Cowper: William Cowper (1731–1800), the intensely religious poet, of Calvinist persuasion, whose depressive episodes assumed psychiatric proportions.

The Everlasting Gospel. Blake's drafts for a couplet poem, at least part of which he entitled 'The Everlasting Gospel', are dispersed through his manuscript remains. Further sections may have been lost. It is not certain which pieces of text, in which order, might have appeared in a completed form of the work. Here a version is offered in which Blake's repetitions and false starts have been suppressed; the order chosen seems cogent and consistent with Blake's own indications. For the title of the work see Rev. 16: 6.

l. 3. *Caiaphas and Pilate*: the Jewish High Priest and the Roman governor of Judea, between them responsible for the arrest, condemnation, and execution of Christ.

l. 11. *Antichrist*: 1 John 2: 22.

l. 14. *Rhadamanthus*: in classical lore, one of the judges in Hades and ruler of Elysium. Hence, in Blake's version, he is the agent of moral condemnation in the afterlife as understood by false Christianity.

l. 18. *Plato and Cicero*: apparently chosen simply as two exponents of a pre-Christian world-view.

46 l. 62. *seven devils*: Mark 16: 9.   *pen*: perhaps from the dialect sense of a sow's vagina.

47 ll. 63–4. The mother of Jesus could have been a harlot and still a virgin if she had consorted with 'sucking devils'.

l. 73. *mocked the Sabbath*: e.g. Mark 2: 27–8.

ll. 74–6. *unlocked . . . divines*: refers to the casting out of devils from the insane, and the enlisting of Peter, Andrew, and John as disciples.

l. 83. Mark 5: 1–19, Luke 8: 27–37.

l. 87. *Obey your parents*: as instructed by the fifth commandment.

l. 88. John 2: 4.

l. 91. *seventy disciples*: Luke 10: 1.

ll. 101–2. *adulteress . . . law*: John 8: 3–11.

48   l. 108. *Sinai's trumpet*. In Exod. 19 this announced the delivery of the ten commandments.

l. 110. Luke 2: 42–9.

l. 119. Matt. 14: 3–10.

ll. 120–6. The various temptations put to Jesus by the devil are recounted in the fourth chapters of Matthew and Luke.

l. 128. *thunder's sound*: also part of the sound effects in Exod. 19.

l. 134. *bound old Satan*: Rev. 20: 2.

l. 139. See Matt. 33.

ll. 142–3. *gate's . . . bar*: Ps. 107: 15–18.

49   ll. 151–2. *scourged . . . temple*: adapts John 2: 13–17.

l. 159. *Church of Rome*. Catholic religious imagery frequently depicts the dead Christ.

l. 167. Matt. 7: 9, Luke 11: 11.

ll. 168–71. John 3: 1–8.

l. 173. Matt. 7: 29, Mark 1: 22.

l. 175. Matt. 11: 28–9.

50   l. 198. *Priestley . . . Bacon . . . Newton*. The common thread between these three is natural science (Priestley did important work on gases; Bacon was an early theorist of science). Priestley, being a Unitarian, was also an enemy of Christianity as Blake saw it. They do not literally resemble Caesar; the connection is that of militaristic values and false religion, so stressed in *The Everlasting Gospel*.

51   l. 231. Isa. 57: 15.

l. 240. Luke 23: 34.

l. 243. John 17: 9.

l. 244. *in a garden*: in the garden of Gethsemane, just before his arrest.

l. 246. *of woman born*: Job 25: 4. The Book of Job is generally important for this difficult section of the poem.

l. 252. *fiction*: apparently in the sense (generally obsolete in Blake's day) of making, creation.

52   l. 267. *Mary*. Blake, as is traditional, identifies the woman taken in adultery (see John 8) with the Mary Magdalen of Mark 16: 1–9. The rest of this section is Blake's version of the episode.

l. 268. Recalls *Paradise Lost* ix. 1000–3, the moment at which Adam tastes the forbidden fruit.

ll. 275–7. *heavens . . . roll*: Rev. 6: 14.

ll. 284–310. The speaker is Jesus.

l. 286. According to Exod. 31: 18 the ten commandments were 'written with the finger of God'.

l. 287. Job 15: 15.

l. 291. *devil*. Blake at first wrote 'God', thus clearly echoing Luke 18: 18–19.

l. 292. The angel of the presence, frequently mentioned in Blake's later writing, is his development of a solitary reference in the Bible (Isa. 63: 9). He evidently interpreted several other biblical passages as references to this figure, citing (in a section of 'A Vision of the Last Judgement' not reproduced here) as one of them Exod. 14: 19, which mentions the 'angel of God' leading the Israelites over the Red Sea. God's guiding agency is called his 'presence' later in Exodus (33: 14–15).

ll. 301–3. Derives from Milton's picture of the punishment of the defeated angels.

l. 304. *breath divine*. Blake's anglicization of Gen. 1: 2 is a more literal translation of the Hebrew than the Authorized Version's.

53   l. 317. Gen. 3: 14.

l. 319. Luke 16: 20.

54   l. 370. *Melitus*: one of the accusers of Socrates (whose teachings were alleged to corrupt the young).

From the annotations to Berkeley's Siris. The thought of Bishop Berkeley (1685–1753) has affinities with that of Blake, in so far as he asserted the essential spirituality of the material world in a post-Lockean philosophical environment (indeed it was essential for Berkeley's metaphysics that Locke should be correct); these annotations are the only allusions to Berkeley in Blake's surviving works, though Blake's first biographer Alexander Gilchrist records that Berkeley 'was one on the list of Blake's favourite authors'. *Siris* is Berkeley's last philosophical utterance, and an oddity in his output: an eccentrically structured work in which arguments for immaterialism are couched in a more Platonic idiom than hitherto.

I will not leave . . . to you: John 13: 33, 14: 18–21.

spiritual body . . . heavenly Father: Matt. 18: 10.

imaginative form. Berkeley had denied that harmony and proportion are 'objects of sense': in empiricist terms, they are secondary qualities. In response, Blake in this paragraph simultaneously claims more for sense-perception than empiricism (harmony and proportion are primary qualities), and less (the 'reality' of a thing is 'imaginative').

55   *perception or sense*. Blake is again, in a sense, ultra-empirical.

From the annotations to Thornton's Lord's Prayer. R. J. Thornton (1768–1837), physician, botanical writer and illustrator, and part-time classicist, seems to have intended his new version of the Lord's Prayer (which seeks to be more faithful to the original Greek than the Authorized Version) for missionary work at home and abroad. Blake does not treat him as a learned man in his own right, but as a pawn of the learned. Thornton's translation is as follows:

> O Father of Mankind, Thou, who dwellest in the highest of the Heavens, reverenc'd be Thy name!
>
> May Thy reign be, everywhere, proclaim'd so that Thy will may be done upon the Earth, as it is in the Mansions of Heaven:
>
> Grant unto me, and the whole world, day by day an abundant supply of spiritual and corporeal food:
>
> Forgive us our transgressions against Thee, as we extend our kindness, and forgiveness to all:
>
> O God! Abandon us not, when surrounded by trials;
>
> But preserve us from the dominion of Satan: for Thine only, is the Sovereignty, the Power, and the Glory, throughout Eternity!!!
>
> Amen.

56 *Scotch*: so called because of Scotland's reputation for prowess in learning.

allegory of kings. Augustus Caesar, like many of the Roman emperors, was deified.

not kingdom but kingship. Gk. *basileia* (Authorized Version *kingdom*) occurs twice in the Lord's Prayer (see Mat. 6: 10, 13), and is rendered by Thornton as 'reign' and then 'sovereignty'; so the speaker of these words is Thornton, or the spirit of his text. Interestingly, Thornton in his commentary defends his phrasing in the first case on the grounds that the traditional *Thy kingdom come* 'does not sufficiently express, it is a spiritual kingdom only'. Blake appears to side with the traditional translation, although he knows how the original reads, and hence should know that *basileia* indeed means 'kingship' rather than 'kingdom'.

BLAKE ON ART AND LITERATURE

Letter of 16 August 1799

59 *Cumberland*: George Cumberland (1754–184?), occasional writer on art and amateur artist, a long-standing friend of Blake's who remained loyal to him as an artist, although he could not agree with his metaphysics.

lost art of the Greeks: Greek pictorial art, which has not survived like Greek sculpture and architecture.

your dictate. Dr Trusler, a religious writer, had evidently commissioned Blake to provide an illustration for a book, which has not been identified.

when he says: Paradise Lost. vii. 28–30.

I cannot . . . bad: Num. 24: 13.

cabinet pictures: small or medium-scale pictures. *Malevolence* measured about 12 in. by 9 in.

Letter of 23 August 1799

60 *your ideas . . . differ so much*. Trusler's letter has not survived; but see the next item.

61 *mediately*: indirectly, secondarily.

Letter of 26 August 1799

62 *those of Christ*: John 18: 36.

your outlines: probably Cumberland's *Thoughts on Outline* (1796), for which Blake had engraved a number of plates.

Night Thoughts. An edition of Edward Young's *Night Thoughts* was issued in 1797 with 43 plates by Blake.

Johnson: Joseph Johnson (1738–1809), radical bookseller and publisher, who knew Blake well and commissioned work from him in the 1780s and 1790s.

my employer: Thomas Butts.

ocean of business. Blake completed his apprenticeship in 1779.

Letter of 2 July 1800

63 *your plan . . . execution*. Cumberland's proposal came to nothing, and it was to be a quarter of a century before anything of the sort was established.

take in originals. Cumberland's plan was simply for plaster casts of sculptures and reliefs.

as well as France. The Louvre was first used as a public museum of art in 1793.

An Essay . . . John Flaxman: published by Hayley in 1800.

To my Dearest Friend, John Flaxman

64 *taken to Italy*. Flaxman was in Italy between 1787 and 1794.

65 *From the annotations to Boyd's Dante*. Blake annotated Boyd's preface to his translation, consistently objecting to the view that Dante, or any great poetry, has a moral aim and effect.

Capaneus: one of the 'Seven against Thebes' of Greek legend and literature, struck down by a thunderbolt after defying Jove.

wine-bibber: Matt. 11: 19, Luke 7: 34.

Letter of 11 September 1801

in London. Blake is writing from Felpham.

66 *distress*. There were bread riots in 1801 when bread-prices rose as a result of the failure of the previous year's harvest and the impact of the

Napoleonic War on the cost of imported grain. Soon after this letter Pitt's administration was under fierce attack for having failed to take the opportunity of concluding a peace with the French.

Mr Johnson's: John Johnson, a friend of Hayley's.

Historical designing. 'Historical' or 'history' as classificatory terms for art (Blake uses both) had an important function in Blake's day of distinguishing depictions of narrative subjects. These might be actual or fictitious events, but in either case the artist's activity would be one of imaginative recreation. There was also an assumption that the treatment would be impressive, even idealized.

From a letter of 10 January 1802

67 *soldier of Christ*: from the first words of Charles Wesley's well-known hymn.

68 *crowned with . . . honour*: Ps. 8: 5.

Letter of 22 November 1802

My brother: James Blake.

in a letter: Printed in William Gilpin's *Three Essays* (1792). Gilpin's account of the picturesque, and of its associated visual qualities of angular, broken form and contrasts of tone and colour, was extremely influential.

69 *Carracci*. Of the three Carracci brothers, Bolognese Baroque painters, the most celebrated, and the one probably referred to here, is Annibale (1560–1609).

70 *Michelangelo's opinion*. This does not correspond to any known remark of Michelangelo's, but Blake may have in mind several statements by Michelangelo recorded in Francisco de Hollanda's *Four Dialogues on Painting* (1548) to the effect that Nature should not be slavishly copied, but approached as a repository of ideal forms.

express image: Heb. 1: 3.

not unlike a champion: In the manner of the knightly hero of a chivalric romance.

stars of God. Isa. 14: 13.

enthusiasm. With overtones of religious zeal and inspiration, as also in Blake's preface to ch. I of *Jerusalem*.

From a letter of 25 April 1803

71 *against me:* Matt. 12: 30.

three years . . . banks of the ocean: Blake's sojurn at Felpham.

long poem: probably refers to *The Four Zoas*, a narrative poem of over 4,000 lines that Blake never brought to a finished form.

machinery: the supernatural agencies in the poem.

without premeditation: recalls Milton's 'unpremeditated verse' (*Paradise Lost* ix. 24).

Preface to Milton

stolen and perverted: probably because Blake believed them to be derivative from more ancient sources.

72 *delight is in destroying*: not a specific allusion.

Preface to Ch. I of Jerusalem. For unknown reasons Blake erased from his etched plate several parts of this preface, most extensively in the second paragraph.

consolidated: uniting the hitherto scattered.

God of fire: perhaps an allusion to Zoroastrianism.

73 *types*: printed lettering.

'Εδόθη ... γῆς: 'All power is given me in Heaven and Earth' (Matt. 28: 18).

dictated to me. As often, Blake echoes Milton on his inspiration in writing *Paradise Lost* (here recalling ix. 23).

the modern bondage of rhyming: a quotation from Milton's prefatory remarks to the second edition of *Paradise Lost*, justifying his use of blank verse. Even if the phrase did not stem from such a context Blake's 'derived' just before it would be odd; it is almost certainly an error for 'delivered'.

terrific: fierce, awe-inspiring.

numbers: metrical writing.

From the annotations to vol. 1 of Reynolds's Works. Reynolds's fifteen occasional presidential addresses to the Royal Academy were delivered over a period of more than twenty years between 1769 and 1790. As far as we know, Blake only owned and annotated the first eight, but he is likely to have read the whole sequence. If so, he would have found that Reynolds' views changed (above all, Reynolds is very positive on the imagination as a 'divine' faculty in Discourse XIII). Even without this, Blake's confrontation with Reynolds is complex. He is generally extremely hostile, but has to cope with the fact that he shares Reynolds's enthusiasm for the High Renaissance Roman art of Rafael and Michelangelo as opposed to the Venetians, and that even the detested notion of 'general form' has affinities with a central doctrine of Blake's aesthetics (see Blake on Chaucer in *A Descriptive Catalogue*, p. 84 below).

74 *the reader*: clear evidence that Blake expected his marginalia to be seen by others.

Barry: James Barry (1741–1806), history painter, whose works indicate, and whose writings assert, ambitions for this branch of art as grand as Blake's.

Mortimer. John Hamilton Mortimer (1741–79), another flamboyant history painter, was more notorious for the disorderliness of his life than of his mind, but his subjectmatter was often macabre, and Blake may have heard the anecdote that Reynolds was alarmed by a set of etchings of dream-monsters that Mortimer dedicated to him.

empire: non-pejoratively, as generally in Blake's later writing: government, political rule.

on peut dire . . . trop savans: 'One could say that Leo X, by encouraging learning, furnished weapons against himself. I have it on hearsay from an English nobleman that he had seen a letter from de la Pole, later cardinal, to this Pope, in which, while congratulating him on his dissemination of learning in Europe, he warned him that it was dangerous to make men too wise.' See Voltaire, *Essai sur l'Histoire Générale, et sur les Mœurs et l'Esprit des Nations* (1761), iii 326. I do not know which edition Blake is citing.

Léon Xme: Pope Leo X (1475–1521), patron of Rafael and Michelangelo, and instigator of many of the finest achievements of the High Renaissance in Rome.

depuis Cardinal. Reginald de la Pole (1500–58) lived in Rome in the early 1520s and subsequently became a cardinal and the most powerful English Catholic in the first years of the Reformation.

for nothing. Barry decorated the premises of the Society in the Adelphi with a vast mural scheme entitled 'Human Culture', 12 ft. high and 140 ft. long. In due course he was amply rewarded, both directly by the Society and from exhibition and reproduction revenues, but Blake is thinking of the indigence that Barry was said to have endured during his six years' work on the project. There were several anecdotes current about this, but Blake seems to have heard a version from Barry's own lips (this is the only evidence that Blake knew Barry).

75 *Fuseli's Milton*. Fuseli mounted a exhibition of more than thirty paintings entitled the 'Milton Gallery' in 1799 and again in 1800. It had a certain number of illustrious admirers, but on the whole was a failure.

intellectuals: powers and activities of the intellect.

Moser: George Michael Moser (1704–83), prime mover in the foundation of the Royal Academy and its first Keeper.

Le Brun: Charles Le Brun (1619–90), French painter who, due to the favour of Louis XIV, was an extremely influential figure in French art.

galleries: volumes reproducing mural schemes by these artists.

candour: approving, friendly disposition.

praises Michelangelo. Reynolds emphasizes the 'sublime' in Michelangelo, and the qualities of 'grandeur', 'vehemence', 'imagination', 'energy', and 'poetical inspiration'.

blames Rafael. Reynolds finds fault with certain features of Rafael's oil paintings (not frescos): their 'dryness' and 'littleness of manner'.

76 *a society*: The British Institution for Promoting the Fine Arts in the United Kingdom, founded in 1805.

Minute discrimination. Reynolds had written of the 'dry, Gothic' manner of Rafael before he fell under Michelangelo's influence, which 'attends to the minute and accidental discriminations of particular and individual objects'.

77 *Mere enthusiasm*. Blake picks up a derisive phrase from Reynolds.

Epicurus: cited as an exponent of irreligious or sceptical philosophy.

con- or innate science: to be expanded, as Blake does below, 'conscience or innate science'. Blake means that innate wisdom knows these principles, with no assistance from experience.

following discourse: Discourse III.

Innate . . . in every man. Blake attacks the fundamental postulate of Locke's epistemology, which Reynolds accepted and tried to square with his aesthetics.

78 *particular not general*. It was recognized in 18th-c. Lockean philosophy that there was a difficulty about how a precise form could be possessed by a general concept that had been derived from many differing particular instances of the concept—which concepts had to be, if there were no innate ideas (in aesthetics there is an equivalent difficulty about visual representations of 'ideal' subjects). Blake claims that our general concepts can be precise because they are innate or generated by the imagination.

bolstered: padded.

79 *lights*: where Rubens depicts a lighted surface.

outrageous: violent, furious.

Oppression . . . mad: Eccles. 7: 7.

following discourse: Discourse VII.

80 *Reynolds pretends*. Reynolds calls the traditional language of artistic inspiration 'metaphors'.

Sublime and Beautiful. Edmund Burke's *Philosophical Enquiry into . . . the Sublime and Beautiful* (1756) is particularly important for locating aesthetic 'sublimity' in characteristics such as obscurity and indefiniteness—Baroque visual qualities that were anathema to Blake's neoclassical enterprise.

81 *From A Descriptive Catalogue of Pictures*. In May 1809 Blake took the unusual step of mounting an exhibition of paintings in his brother's shop, for which he wrote and had printed a polemical catalogue. The exhibition received one abusive review and some scattered unofficial sympathy. It seems to have been scarcely visited, and Blake's professional fortunes, far from being improved as he had hoped, declined very seriously in the next few years.

four years after. Blake's chronology is extremely inaccurate. Correggio and Titian were both born at least fifteen years after Michelangelo.

the Venetian: Titian.

unmudded by oil. The exhibition consisted of seven watercolours and nine tempera paintings, though Blake tends to refer to all as watercolours or 'frescos' (the latter term because they all have some kind of plaster ground). In this paragraph and the next Blake is referring particularly to his use of tempera: that is, of a stiff medium for pigment made not from oil, but from egg-white or from a colloidal solution in water (for example, with starch). Six of the watercolours survive (*The Body of Abel found by Adam and Eve* at the Fogg; *The Soldiers Casting Lots for Christ's Garments* at the Fitzwilliam; *Ruth* at the Southampton Art Gallery; *Jacob's Ladder* at the British Museum; *The Angels Hovering over the Body of Jesus in the Sepulchre* at the Victoria and Albert; and *The Penance of Jane Shore* at the Tate), and five of the tempera paintings (*The Spiritual Form of Nelson Guiding Leviathan*, *The Spiritual Form of Pitt Guiding Behemoth*, and *The Bard* at the Tate; *Sir Jeffery Chaucer and the nine and twenty Pilgrims* at Pollok House, Glasgow; and *Satan Calling up his Legions* at the Victoria and Albert).

This picture: *The Spiritual Form of Pitt Guiding Behemoth*.

brought oil . . . practice. Blake's dating of this development to the 17th century is at least 150 years too late; he is obviously aware that his opinion is heterodox.

82 *work on painting*: not known, but there is evidence that it was written, and perhaps even published.

John of Bruges. Jan van Eyck (?1390–1440). Blake almost certainly learned of the claim about his invention of oil-painting from Vasari, who calls him 'John of Bruges'.

whiting grounds. Historically tempera has been used on surfaces first painted white. Blake connected this in his mind with the plaster base of fresco, and used such a base for the tempera paintings in this exhibition.

apotheoses: representations of the mortal as divine.

Cherubim. Blake was not alone in his day in conjecturing that a vanished Eastern tradition had been ancestral to the surviving art of the ancient world. The leading evidence for such a tradition among the Jews was the Bible's account of Solomon's temple, with its cherubim-adorned walls and its two wooden figures of cherubim in the sanctuary, each with a height and wingspan of ten cubits (about eight feet) (1 K. 6: 23–8). Later in the *Descriptive Catalogue* Blake proposes that a neighbouring culture, the Phoenicians, also created cherubim, almost certainly because Hiram, builder of the temple, is said to be from the Phoenician city of Tyre (1 K. 7: 13–14).

Moab, Edom, Aram: non-Jewish kingdoms in the Old Testament whose founders (Lot, Esau, and Shem respectively) are all sons of Jewish patriarchs.

the Torso: the Belvedere Torso, a famous sculptural fragment in the Vatican.

83 *journey to Canterbury*. The remarkable passage that follows can still stand up as a discussion of the General Prologue and of Chaucer's artistic stance. Though Blake's views are thoroughly contrary to modern orthodoxy he is too astutely at home with the detail of Chaucer's text to be ignored.

early morning. Chaucer's pilgrims rise early for their journey (General Prologue 33).

youthful: Blake's sensible deduction from Chaucer's description.

84 *in his glossary*. This might seem good evidence of which edition of Chaucer Blake used, but there is no such thing as a glossary by either William or Francis Thynne, early Chaucer scholars. It is, in fact, surprisingly hard to identify the edition(s) that Blake knew (his quotations from Chaucer seem to be virtually Blake's own reconstruction of the text in wording and spelling). His main source may have been Thomas Tyrwhitt's edition (various issues from 1775), and he probably also knew those by Thomas Speght (the 1687 edition especially) and John Urry (1721) (see Karl Kiralis' discussion in *Blake Studies*, 1 (1969), 167–74 and Claire Pace, 'Blake and Chaucer', *Art History*, 3 (1980), 388–409).

The characters . . . existence. Dryden makes a comparable claim, in a more restricted way, about the pilgrims: 'their general characters are still remaining in mankind, and even in England . . . for mankind is ever the same, and nothing lost out of Nature, though everything is altered' (Preface to *Fables*, 1700). Though Blake was writing at a time of rising enthusiasm for Chaucer (see Pace op. cit.) there had been no discussion, except Dryden's, to match Blake's for scale and seriousness.

85 *now lost*. The so-called Epilogue to 'The Nun's Priest's Tale', which does give a glimpse of one of the companions, is a cancelled passage not known to Blake.

of London: Blake deduces, again sensibly, that Chaucer's Haberdasher-Webber-Dyer-Tapiser group are Londoners. See below for his exclusion of the Carpenter.

expensively attended. Presumably by virtue of the companions Blake has given him.

86 *housekeeper*: person in charge of an establishment.

Goddesses of Destiny. Blake thinks of them as the three Fates of Greek mythology. Shakespeare does classicize the witches to the extent of associating them with Hecate and Acheron.

87 *so are Chaucer's*: Shakespeare's fairies as in *Midsummer Night's Dream* and Chaucer's as in 'The Merchant's Tale'. This aspect of the connection between ancient and modern mythologies (which Blake is so keen to

demonstrate in this discussion) may be due to Tyrwhitt's Introductory Discourse to his edition, where he associates Oberon and Titania with Pluto and Proserpine.

most holy sanctuary: Ezek. 45: 3–4.

messenger of Heaven. The world 'apostle' derives from the Greek for 'messenger.'

greatest of his age: Luke 22: 24–7.

citizens of London: the Haberdasher and the Dyer.

88 *the gods of Priam*. Blake's common, but puzzling, way of referring to the classical deities in his later writings (and at *Milton* i. 678 even the religion of Plato is said to be 'Trojan'). Priam was the king of Troy, and in Homer the Greeks and Trojans follow the same polytheistic religion. Blake may have it in mind that the Greeks, like the Israelites, should have sought to extirpate their enemy's religion. Priam, father of many and famous offspring, is the classical equivalent of a Jewish patriarch.

made to number four citizens. That is, commentators have identified four guildsmen (excluding the Carpenter). As Blake explains in the next paragraph, the four can be reduced to three by either of two readings— but the problem of Chaucer's 'nine and twenty' remains if he counts himself as one of the 'company'. In the picture's title Blake excludes Chaucer but includes the Host as a pilgrim, squaring the arithmetic by silently omitting the Carpenter from the depicted procession. Blake's second suggested reading of 'Webbe Dyer' is original to him, and has gained some modern scholarly support.

89 *competitors*. Blake refers to the engraver of his Blair designs of 1808 (Louis Schiavonetti) and to Thomas Stothard, who, at the same time as Blake, produced his own picture of the Canterbury pilgrims, to be issued as a large commercial engraving. The idea of such an engraving was probably borrowed from Blake.

90 *shall know them*: modulates Matt. 7: 20.

91 *dumb dollies*: heavily swathed figures (such as featured in a traditional Twelfth Night game).

his friend: probably Robert Cromek, the engraver-publisher who set up the Stothard project.

H——: the artist John Hoppner, in a published letter.

92 *Mr. S——*: Thomas Stothard (1755–1834), painter and book-illustrator, until this point a long-standing friend of Blake's. For his rival Chaucer project see the note to p. 89 above.

touch: A term that naturally tends to get used for painterly, rather than graphic, skill.

a Goldsmith. This corruption of Chaucer's text did have the authority of the 1598 folio edition of his works.

93 *from Gray*. Thomas Gray's poem *The Bard* (1757) uses a tradition to the effect that Edward I put Welsh bards to death, and imagines a survivor defiantly prophesying to Edward and his sons the misfortunes of their descendants, and the continuing mission of poetry to expose vice and tyranny.

94 *the Ugliest Man*. Blake is drawing on a fragmentary Welsh legend.

95 *Arcturus, or Boötes*. Arcturus is the brightest star in (and sometimes used as the name of) the constellation Boötes, near the Great Bear. In Greek Arcturus signifies 'bear-keeper'.

Watling Street. The purpose and original course of Watling Street in London was a controversial subject in Blake's day; William Stukeley, the Druidist, denied that it started near London Stone.

London Stone: the monolith in Cannon Street, which features importantly in Blake's epics as a sacrificial altar. Various ancient purposes have been conjectured for it.

written it: probably refers to *Jerusalem*.

British prince: Arthur. Presumably the true era of ancient British civilization is thought to have preceded the Roman conquest in the 1st century.

ancient British history. Blake is thinking of Milton's *History of Britain* (1670). Book 3 discusses Arthur, but with considerable scepticism.

96 *Jacob Bryant*. Bryant (1715–1804) is chiefly known for his *A New System, or an Analysis of Ancient Philosophy* (1774), an extraordinary exercise in the collapsing of the known ancient myths and religions into one another, with the ultimate purpose of showing that they are all distortions of historical facts most accurately recorded in the Old Testament.

Hume: as author of the *History of Great Britain*.

Echard: Laurence Echard (?1670–1730), English and clerical historian.

Rapin: Paul Rapin de Thoyras (1661–1725), author of a history of England.

97 *frontals*: the bony components of the forehead (here, presumably, when prominent).

like . . . whirlwind: Isa. 17: 13.

98 *Universal Theology, No. 623*. The vision is described by Swedenborg in *True Christian Religion* (1781).

99 *more perfect picture*. This also survives, at Petworth House, Sussex.

Chiaroscuro: that approach to painting which treats a picture as an arrangement of areas of different tonal value: bright, dark, and gradations between. It therefore de-emphasizes colour and outline, and, historically, the painters thought of as masters of chiaroscuro favour a dark picture-surface. Blake would also have disliked the way in which this 'machine', or purely formal approach to a picture, is antagonistic to narrative and iconic aspects of art.

100 *altogether false*: the legend of Correggio's poverty, due to Vasari, is probably incorrect, and Blake's account nearer the truth.

ideal design: imaginative rather than literal in treatment.

by Mr. Wilkin. Charles Wilkins' pioneering translation of the *Bhagvad-geeta* was published in 1785.

naked: in the celebrated sculptural group depicting the death of the Trojan priest of Apollo who warned against the Trojan Horse.

four drawings. The Abel and Jacob watercolours and the two depicting Christ's passion (see above).

101 *England ... duty*: alludes to Nelson's order-of-the-day before Trafalgar (1805).

Whither ... thee and me: Ruth 1: 16–17.

102 *Protogones and Apelles*: two celebrated painters of ancient Greece. Apelles supposedly excelled Protogones only in ease of execution.

plagiary: plagiarist.

inflexions: curves.

the line of the Almighty: suggested by *Paradise Lost* vii. 225–31.

Jane Shore: mistress of Edward IV and other 15th-c. English potentates; forced by Richard III to do public penance as a harlot.

he has ... shoe: Blake's metaphoric application of the belief that fairies will leave money in a shoe during the night.

From 'Public Address'. Materials, sometimes slightly repetitious, for a manifesto on art, following up on Blake's exhibition and catalogue, are distributed through his notebook. It may be part or all of the otherwise lost 'work on painting' mentioned in the *Descriptive Catalogue*. It is the most personal and circumstantial of all Blake's writings on his profession, but it still contains much that is directly accessible to a modern reader. Blake's attack on the contemporary pre-eminence of landscape painting in England is a new theme, and one still interesting to an age that values Turner and Constable very highly. Even more remarkable is his pioneering assault on representationalism.

103 *slobbering*: slovenly execution.

Chalcographic Society. This organization (chalcography is copper-engraving) was established by Blake's enemy Cromek in 1807, but despite favourable notice in the *Examiner* (another of Blake's dislikes), and a number of members and illustrious patrons, it soon collapsed. Blake's 'Public Address' is primarily directed at this body, though having deplored the ineptitude and commercialism of current English print-making he turns towards the Society of Arts with an appeal for greater support for the visual arts in general (see below).

my print: of the *Canterbury Pilgrims* picture.

Bartolozzi, Woollett, Strange: Francesco Bartolozzi (1727–1815), William Woollett (1735–85), and Sir Robert Strange (1721–92), the three most celebrated engravers of the late 18th century in Britain. Blake inherited much of his own technique from the latter two; he objects chiefly to the aesthetic and commercial uses to which they put their skills, and may have been aware that Cromek had invoked their names in his Chalcographic Society publicity.

Heath: the engraver James Heath (1757–1834).

Durer's Histories: Dürer's illustrations of narrative subjects.

104 *years ago*. In the early 1780s Blake had engraved some 29 book-illustrations from paintings by Stothard.

Examiner. In August 1808 and September 1809 Leigh and Robert Hunt published derisive reviews of Blake's work in their magazine, calling him among other things a 'lunatic'.

a poem: probably *Milton*.

Macklin: Thomas Macklin, engraver-publisher, for whom Blake did plates in the early 1780s.

to Italy. Flaxman designed and engraved his *Iliad*, *Odyssey*, and Dante illustrations in Rome in 1793.

wound . . . my shoulder. The Hunts had also abused Fuseli for his support of Blake.

West: Benjamin West (1738–1820), the influential American-born history painter, at this period President of the Royal Academy.

Basire's: James Basire (1730–1802), the engraver under whom Blake served his apprenticeship.

105 *fribbles' toilettes*: their workplaces were like the dressing-tables or dressing-rooms of frivolous, idle men.

Marc Antonio: Marc Antonio Raimondi (?1488–?1533), the greatest Italian engraver, whose engravings after Rafael, Michelangelo, and Julio Romano were important for Blake's acquaintance with these artists.

Browne: John Browne (1741–1801), engraver.

Aliamet: François Germain Aliamet (1734–90).

Cook: Thomas Cook (?1744–1818).

106 *Goltzius*: Hendrik Goltzius (1558–1617), an important Dutch engraver.

Sadeler: any of several members of the illustrious Dutch engraving family, contemporaries of Goltzius.

Edelinck: Gerard Edelinck (1649–1707), French engraver.

Hall: John Hall (1739–97), English engraver of middling reputation.

ruled sky: one represented by parallel horizontal lines in an etching.

out of conceit: dissatisfied.

line of beauty. In Hogarth's influential formulation the 'line of beauty', or essential element of all beautiful forms, was specifically an elongated S-shape.

107 *pugilists*. Blake lived in the heyday of English bare-fisted prize-fighting.

Gravelot: Hubert François Gravelot (1699–1773), French book-illustrator who worked mainly in England from 1733 to 1754.

108 *poco piud*: From Italian *poco più*, a little more: hence, needlessly embellished, spoiled by fussy reworking.

Bartelloze: Bartolozzi.

Dryden . . . only planned. Blake probably means that an intention to improve on Milton was attributed to Dryden in rhyme by Nathaniel Lee (see below, p. 113; there, however, Blake purports to doubt Lee's authorship of the passage, and he may be proposing Dryden instead).

toilette: dressing-table.

Pope: an allusion to Pope's translations of the *Iliad* and *Odyssey*.

poor Schiavonetti. Cromek had employed Luigi Schiavonetti, rather than Blake, to engrave the latter's illustrations for Blair's *The Grave*. Schiavonetti died in 1810 and received an obituary from Cromek in the *Examiner*.

metaphysical: excessively intellectual or witty.

jargon: jangling, medley of sound.

109 *like a drayman*: in a heavily-muscled Baroque manner.

amateurs: art-lovers.

tints: tonal rather than linear effects.

generous contention. In the *Iliad* 'generous' is a common epithet for the Greeks and Greek valour.

combination: illegal confederacy.

premiums: prizes.

110 *Vasari tells us*. There are remarks in Vasari's life of Michelangelo that could be taken to attribute this view to the latter, but in his lives of both artists Vasari also stresses their careful study of Nature.

intellects: mental faculties.

fresco originals. in Blake's usage, watercolour or tempera paintings on a whiting or plaster ground.

111 *Anytus, Melitus and Lycon*: the three accusers of Socrates (here probably denoting the three Hunt brothers).

name . . . rot: Prov. 10: 7.

aquafortis: nitric acid.

Americus Vesputius: Columbus made the first landfall on America, but it was the later Amerigo Vespucci who first navigated long stretches of its coastline.

relief: tonal contrast.

112 *Rembrandt's hundred guilders*. Rembrandt's etching of *Christ Healing the Sick* was generally known in Blake's day as the 'Hundred Guilders Piece'.

Bacon says. Not a direct citation of Bacon. The original wording in the manuscript indicates that Blake is developing rather than quoting a remark of Bacon's, hitherto unidentified.

Vanloo: probably the French painter Jean Baptiste Vanloo (1684–1745).

Luxembourg Gallery: Rubens' mural decoration in the Luxembourg Palace, Paris.

rattletraps: odds and ends.

113 *Milton's . . . finished it*. Lee's preface to Dryden's opera *The State of Innocence, and Fall of Man* has remarks to this curious effect.

114 *exceeding great reward*: Gen. 15: 1.

Society for Encouragement of Arts. Blake uses the full name of the organization usually called the Society of Arts. This is his usual practice, but he may be mindful of the Society for the Encouragement of the Art of Engraving, a daughter organization of the Chalcographic Society set up by Cromek expressly to attract private investment in engraving, and hence a good example of the commercial, as opposed to national and public, funding of the arts so disliked by Blake.

Anglus: the average Englishman.

115 *dots . . . clean strokes*: discussion of engraving devices, rather than of underlying design.

On Homer's Poetry. This and the following *On Virgil* are both engraved on the same elaborately decorated plate. They express the anti-classicism of Blake's last years very directly, and anticipate early Victorian developments in English taste.

Homer's subject. Homer makes no allusion to the Judgement of Paris (by which Paris won the right to abduct Helen), whereas the story of Bellerophon (who is simply grandfather to two of the Trojan combatants) is given in *Iliad* vi.

116 *From the annotations to Wordsworth's Poems*. The writer Henry Crabb Robinson (in whose copy of Wordsworth Blake made these notes) recorded that 'his delight in Wordsworth's poetry was intense, nor did it seem less notwithstanding by the reproaches he continually cast on Wordsworth for his imputed worship of nature'.

natural piety. Blake is glossing the phrase in l. 3 of Wordsworth's Immortality Ode.

This: Wordsworth's 'To H.C., Six Years Old'.

greatest in the Kingdom of Heaven: Matt. 18: 1–4.

Natural objects . . . imagination. Blake is inverting the wording of Words-worth's heading to the 'Wisdom and spirit of the universe' section of *The Prelude*, as published separately from 1809 onwards.

Michelangelo's sonnet: as translated by Wordsworth 'No mortal object . . .'.

117   *prefaces*: Wordsworth's 1815 preface, and his 1802 *Lyrical Ballads* preface with its appendix and supplementary essay.

Ossian . . . Rowley: imaginary authors of the purportedly archaic poetry of William Macpherson and Thomas Chatterton respectively.

EARLY VISIONARY AND NARRATIVE WRITINGS

121   *To Spring*. The extremely free blank verse of this sonnet (and of parts of its companion poems) tends to fall into septenaries, or seven-stress lines, such as Blake used extensively in the prophetic writings.

ll. 2–3. *turn . . . western isle*: Dan. 11: 18.

122   *To Autumn*. The ending imitates that of Milton's *Lycidas*, and both poems run through the annual cycle of vegetation.

To Winter

123   l. 16. *Mount Hecla*: either Hekla in Iceland or Hecla in the Western Isles.

To Morning

124   l. 8. *buskined*. Buskins (or half-boots) were sometimes associated with the goddess Diana, who is perhaps also alluded to in the motifs of virginity and hunting.

Fair Elenor. Though set out to look like a ballad the poem is unrhymed, and free in metre. The action starts abruptly, as in many of Blake's short prophetic writings, as if this were a fragment of an old text. In the ballads collected in Percy's *Reliques* there are several 'Eleanors' (or similar names), perhaps most notably in 'Lord Thomas and Fair Ellinor', a poem which ends with a beheading.

125   ll. 34–5. *arrows . . . darkness*. Ps. 91: 5–6.

126   ll. 62–3: There is a reminiscence of the Ghost's speech in *Hamlet*.

To the Muses

l. 1. *Ida*: either the Phrygian or the Cretan mountain that bore this ancient Greek name.

l. 4. *ceased*: presumably with the departure of Apollo, god of music and the sun.

127   *Gwin, King of Norway*. Historically the setting of Blake's poem would seem to be the Viking invasions of Celtic Britain, though Gwin is a Celtic name and the general effect is more like the 'Ossian' tales of warring Celtic

nations. But the persons and places are all invented by Blake, and the poem may be thought of as the first surviving Blakean myth. The alternate rhyming of the ballad stanza means that it can be read as septenary couplets.

ll. 5–8. *feed . . . door*: Chatterton, 'Bristowe Tragedie' 162–4: 'To feede the hungrie poore,/ Ne lett mye servants dryve awaie / The hungrie fromme my doore'.

131 *An Imitation of Spenser*. The stanzas are usually of the same length as Spenser's in the *Faerie Queene* and Blake uses some of Spenser's archaisms.

ll. 10–15. Midas was given asses' ears by Apollo for adjudging the latter's music inferior to Pan's.

ll. 11. *nervous*: energetic and strongly knit.

l. 13. *leesing*: lying, from Spenser's noun 'lesing', and from the psalter (e.g. Ps. 5: 6).

133 *King Edward the Third*. This impressive short play, or commencement of a longer play, concerns the build-up to an ambiguous event in 14th-c. history, the Battle of Crécy (1346), which may be thought of as a great national triumph, or an appalling slaughter by a marauding invader. Blake has modelled his drama on the Tudor, particularly Shakespearian, chronicle play, and his source is the French chronicler Froissart.

Scene I

134 ll. 43–5. *revenge . . . Paris*. Alludes to the execution of Olivier de Clisson and others by the French king in 1343.

l. 46. *here in Brittany*. Blake has confused Brittany (scene of Edward's 1343 campaign) with Normandy, which Edward chose for his 1346 excursion on Harcourt's advice, and which his armies thoroughly looted and terrorized.

ll. 47–9. In the context (of taking 'a just revenge') Edward means that Brittany too will now be devastated, but Blake is too sophisticated to allow an explicit reference to this to disturb the stately high-flown rhetoric of the king.

Scene III

137 l. 8. *Philip*: Philip of Valois, the French king.

138 ll. 56–7. *distemper . . . running away*. Edward was making his withdrawal eastwards across northern France when obliged to confront the French at Crécy.

139 l. 99. *gilt . . . blood*: *King John* II. i. 316.

140 l. 106. *not . . . parted with*: reminiscent of *Hamlet* II. ii. 216.

141 ll. 158–9. *a candle half burned out*: *2 Henry IV* I. ii. 156.

ll. 159–60. *pig . . . pattle*: supposedly Welsh pronunciations of 'big' and 'battle': a consonantal device employed also by Shakespeare for Fluellen in *Henry V*.

l. 175. *ribs of death*: *Comus* 562.

l. 176. *mortal dart*: as possessed by Milton's Death (*Paradise Lost* ii. 729).

ll. 184–5. *more . . . servants*: from the French epigram to the effect that no man is a hero to his valet.

Scene IV

145 *natural philosopher*: *As You Like It* III. ii. 33.

Scene V

146 l. 6. *wons*: dwells.

147 l. 43. The allegedly glorious occasion of Crécy left some 6,000 French dead; and England was soon to be ravaged by the Black Death. An anonymous account of Blake towards the end of his life has him enjoying a conversation with the spirit of Edward III, in which Blake challenged the latter on his 'butcheries'; the king is supposed to have replied 'what you and I call carnage is a trifle unworthy of notice . . . destroying five thousand men is doing them no real injury . . . their important parts being immortal, it is merely removing them from one state of existence to another' (G. E. Bentley, Jr., *Blake Records* (Oxford 1969), 299).

148 *Scene VI*. Blake's short play or dramatic fragment ends with a song on British liberty, anticipating the equally unusual conclusion to *Marriage of Heaven and Hell*. Blake's historical imagination ran readily to conceiving bardic, monitory comment on historical situations. The two 'prologues' that follow are in this vein.

l. 1. *Trojan Brutus*: draws on the legend that Britain was founded by survivors from Troy, under the leadership of Brutus.

l. 12. *navies black*: just as the Greek ships in the *Iliad* are consistently black.

ll. 15–16. Though here Albion is addressed as female it need not follow that Blake's later male figure, the giant Albion, undergoes a sex-change. It is orthodox to use feminine pronouns for countries, which is all Albion is seen as at this date (and the female pronouns in the last stanza of the song mainly refer to Liberty).

149 l. 51. *prevented*: anticipated.

Prologue . . . King Edward the Fourth

150 l. 1. *Oh . . . thunder*: in clear imitation of the opening to Shakespeare's *Henry V* Prologue.

ll. 4–13. *who can stand?* This formula is used several times in the Bible with apocalyptic associations, most notably at Rev. 6: 17. In a play on Edward IV Blake perhaps would have conferred an enhanced significance on the Wars of the Roses and the closing stages of the Hundred Years War, much as he does with the French adventures of Edward III.

Prologue to King John. The historical foundation of this semi-metrical prose piece is presumably the civil conflict that followed Magna Carta.

151 *sow itself*: since the grain is not harvested.

A War Song to Englishmen. This is conceived as an exhortation to an unjust war justified (like war in the Old Testament) by alleging God's authority. Such wars cause misgivings in the speaker (who is like the humane Sir Walter Manny of *Edward III*), but are congenial to the Norman dynasty (including the warmongering Edward III).

ll. 1-2. The Homeric practice of drawing lots from a helmet.

l. 11. *fatal scroll*: a record of sins.

152 l. 28. *learned clerk*: Henry I (nicknamed 'Beauclerc' for his learning).

The Couch of Death. A speculation about an ancient phase in the development of human religion. The religious intuitions reported are regrettable in Blake's view: of a remote, moralistic God whose existence is deducible from the fact of physical creation, and of a space-filled universe bordered by Heaven and Hell. The reality of the youth's death gives the lie to these ideas.

153 *sin . . . cloak*. Cf. John 15: 22.

wiped away his tears: Rev. 7: 17, 21: 4.

hung not out her lamp. Cf. Samuel Butler, *Hudibras* Part 2 (1664), Canto I, l. 910.

Contemplation. Its closest affinities are with verse exercises in personification of the 18th century, such as Joseph Warton's 'To Evening' (1746) or his brother Thomas' 'Retirement: an Ode' (1747), themselves indebted to early Milton.

154 *Samson.* Blake's source is the often confusing narrative in Judg. 13-16, though he is influenced by Milton's *Samson Agonistes* (e.g. in the depiction of Dalila and Manoa).

I sing: as if this text were the opening of an epic on Samson's deeds.

Philista's lords: the Philistine leaders.

155 *Destruction . . . suckled thee*: cf. *Aeneid* iv. 365.

156 *Twice . . . heaven*: Cf. *Samson Agonistes* 23-4.

Hail, highly favoured. The angel's words are those used to the Virgin Mary at the annunciation (Luke 1: 28). In Christian theology Samson has a strong traditional association with Christ.

157 *name is wonderful*. Cf. Is. 9: 6.

The Vision of Pride. This enigmatic and highly original prose fragment, left in manuscript, adapts and combines a variety of models: legendary and mythical genealogies; grouped personifications (such as the Seven

Deadly Sins) and procreating personifications (which run from Hesiod to Milton's Sin and Death); traditions of emblematic and allegorical signification (some are noted below; part of 'The Vision of Pride' is used by Blake also in the Wartonesque 'Contemplation' printed above); Gen. 4–11 (and Adam's vision of this history in *Paradise Lost*). These traditional modes give way every now and then to very personal declarations. Blake conceived the passage as metrical, as his occasional elisions in spelling show. It is usually known by its opening words, but here a title is provided.

she: Fear.

kingdoms . . . glory: as the devil shows Christ (Matt. 4: 8).

giants . . . flood: Gen. 6: 4, but influenced by Milton's referring to them as 'giants of mighty bone' (*Paradise Lost* xi. 642).

Cain's city: the city called Enoch (Gen. 4: 17).

Confusion. The Authorized Version's gloss on Gen. 11: 9's 'Babel' is 'that is, Confusion'.

158 *good king's life.* A difficult passage. Blake sees the Roman Empire as temporarily saved from destruction by the transfer of its headquarters to Constantinople, founded by the first Christian Emperor, and likens this to the retarding of the death of King Hezekiah (see 2 Kgs. 20: 8–11). The Empire's eventual fall is depicted as a pagan sacrifice, no doubt because of the persistence of such observances at the time of the sack of Rome, as deplored by St Augustine in the early sections of the *City of God*.

triple crown. Rome is now the seat of the popes, who are often symbolized by their triple crown.

a sun . . . the sea: obscure: perhaps Islam.

serpent's head. The snake is a traditional accompaniment of Envy (though Blake adds a memory of Gen. 3: 14–15). See for example *Faerie Queene* v. xii. 30–1 (where Envy also eats her own entrails).

fabled Hecate. Hecate's wide powers are originally asserted in Hesiod's *Theogony*.

159 *her father*: not named in the fragment.

flies abroad: as does Virgil's celebrated personification of Rumour at *Aeneid* iv. 187.

image rather: inanimate representation.

Pandora. Created by Vulcan and animated by Athena, Pandora was endowed with charms by all the Gods; the intention was that she, with her box full of evils for mankind, would be accepted by Prometheus' brother.

worse ills: than those Pandora brought.

160 *stands between.* These memorable closing sentences read like an unrhymed quatrain.

'"*Woe!*" *cried the Muse*.' Another puzzling text surviving in manuscript. Like *Lycidas*, which is echoed, it deals with a poetic summons to mourn premature death: but, confusingly, there are two deaths. The second may be that of Blake's beloved brother Robert, who died in 1787—which would mean a subsequent date for this and, probably, the associated 'Vision of Pride'; but the personification technique, and its strange blending with personal despair, is in both pieces too close to 'Contemplation' in *Poetical Sketches* (1783) to make this likely.

the Muse: reminiscent of the opening of a celebrated pastoral elegy, Bion's 'Lament for Adonis'.

Elfrid: Blake's invention ('Elfrida' is an Anglo-Saxon woman's name).

SEPTENARY VERSE OF THE FRENCH REVOLUTIONARY PERIOD

163 *Tiriel*. Almost certainly Blake's first long mythological poem, *Tiriel* fulfils more obviously than any other of his works his ideal of dealing with universal, permanent psychological types and states. Tiriel has elements of Lear (dying king evicted by children whom he curses, and dependent on daughter), of Oedipus (blind, deposed king, also laying a curse on his offspring, and, led by daughter, finding a haven where he dies), of Odin (half-blind father of many sons who wanders disguised among men), and of other analogues (Moses, Theseus, etc.) mentioned in the notes. Blake's procedure is more or less to reverse the moral standing of his archetypes, and he depicts Tiriel in the last forty lines as recognizing the wickedness of all his past actions and beliefs. In the manner of several of the sources the narrative concerns a family: all the characters appear to be related to one another.

l. 1. *And*. As with several Blake texts, the effect of a fragment is deliberately sought by the device of incomplete syntax.   *Tiriel*. The name seems to have been invented by the alchemical writer Cornelius Agrippa, and denotes the spiritual agency common to earthly elements (sulphur and mercury) and a heavenly body (Mercury). The name would also have appealed to Blake for its Hebrew/Old Testament feel, and Tiriel is a patriarch in a pastoral society who has inherited a legacy of cruel laws.

l. 2. *Myratana*: Perhaps derived from Jacob Bryant's alleged Amazonian queen of Mauretania, Myrina.

l. 21. *Serpents, not sons*. Cf. *King Lear* IV. ii. 40, 'Tigers, not daughters'.

l. 24. *Heuxos, Yuva*: both probably Blake's coinages. The first is evidently Greek in inspiration. Yuva, given Blake's conventions for naming in this poem, may be a daughter.

164 l. 32. The dispute about burial outside gates and Myrtana's Amazonian connection in a general way recall Theseus (burier of the seven against Thebes, husband of Hippolyta, and a major figure in *Oedipus at Colonus*).

l. 34. *Zazel*. Tiriel's brother also bears a Hebrew-sounding name (cf. Azazel: scapegoat) which is found in Cornelius Agrippa.

l. 55. *Har*: probably attractive to Blake as a Hebrew word rather than for its specific meaning (which is 'mountain').

l. 56. *Heva*: Blake's coinage, but evoking 'Eve'.

l. 57. *Mnetha*. The name has a Greek form, and may be thought to combine 'Mnemosyne' (Memory) and 'Athena' (Wisdom). True to the convention that *Tiriel* is part of a larger body of narrative, Mnetha is introduced as if already known to the reader.

165 l. 88. *Bless thy poor eyes*. Cf. *Lear* IV. i. 54.

167 l. 119. *did . . . thee*: reminiscent of Lear's delusion about Edgar in *Lear* III. iv.

l. 142. *Ijim*: a Hebrew name, found at Isa. 13: 22: but a plural, denoting something like 'jackals'.

169 l. 183. *Lotho*: Blake's Germanic-sounding coinage.

l. 197. Cf. *Paradise Lost* iv. 800 (of Satan).

l. 200. *so I'll keep him*. The second link between Ijim and a veteran of Troy. Like Aeneas carrying Anchises he bears his father on his shoulders; and he subdues a multiform enemy, as Menelaus subdues Proteus while returning to Sparta.

170 l. 215. *Matha*: presumably a mythological region or aspect of the world, rather than a person (in 'Ossian' the name is borne by a man).   *Orcus*: a Latin term for the underworld and its god.

l. 225. *centre*: of the earth.

l. 226. *pestilence . . . lakes*. Cf. Lear's cursing of Regan (*Lear* II. iv. 167–9).

171 l. 242. *Hela*: a Norse name for the goddess of Hell.

l. 250. This climax to the tribulations brought down by Tiriel echoes the last of the punishments of Egypt wrought on behalf of Moses: the slaying of the first-born.

l. 253. *silent of the night*: 2 *Henry VI* I. iv. 19.

174 l. 339. *lack of mother's nourishment*. There was widespread antagonism in England in the late 18th century to the use of wet-nurses.

l. 346. *repugnant*: unwilling, resistant.

l. 350. *Returns my thirsty hissings*. Cf. *Paradise Lost* x. 518.

175 *The Book of Thel*. The poem is, most immediately, a text about death and the purpose of human life (and in this regard it is indebted to Job, even to the number of Thel's interlocutors). It also incorporates, more allusively, four ancient themes: a world that is parallel to ours and contains the elements of the latter in pre-existent forms; a fall from Paradise; sexual initiation; a descent to the underworld. Each has been emphasized at the expense of others by various critics, who disagree particularly over the meaning of the poem's closing events. Any single reading is simplistic.

Thel. In Greek this is the root element in the vocabulary of wishing and willing.

ll. 3–4. *wisdom . . . bowl*: Eccles. 12: 6, Job 28: 12–15 are the most important of several biblical echoes, and golden bowls are part of the traditional altar-furniture of the tabernacle in the Bible.

l. 5. *Mne Seraphim*. Blake has taken a name from Cornelius Agrippa, 'Bne Seraphim', and made the first element look more Greek.

l. 9. *river of Adona*. Milton has a river Adonis (*Paradise Lost* i. 450). This is one of *Thel*'s links with Spenser's Garden of Adonis, but Blake has not wanted it to be emphatic. There is a very broad resemblance between Thel's lament and that of Spenser's Venus over the transience of the garden (*Faerie Queene* III. vi. 40).

l. 12. This imagery may derive from a passage in Swedenborg's *Wisdom of the Angels* which Blake ringed in his copy (for a resulting Swedenborgian reading of *Thel* see John Adlard, *Studia Neophilologica*, 46 (1974), 172–4).

l. 13. *shadows in the water*. Cf. Job 14: 2.

l. 16. *lay me down*: *Paradise Lost* x. 777. There are several parallels between Adam's state of mind just after the fall and Thel's. And Blake may be recalling James Hervey's *Meditations among the Tombs* (1746) i. 83, a work that influences *Thel* at several points (see Dennis M. Read, *Colby Library Quarterly*, 17 (1982), 160–7). For an honorific mention of Hervey some years later see *Jerusalem* iii. 1029.

l. 19. *walketh . . . evening time*: Cf. Gen. 3: 8.

l. 22. *gilded butterfly*: *Lear* v. iii. 13.

ll. 25–9. *Rejoice . . . vales*. The whole encounter with the lily is indebted to Matt. 7: 28–33.

176 l. 36. *meekin*: probably Blake's dialect-like coinage, combining the senses of timid and weak.

l. 55. *Luvah*: Blake's equivalent of the Greek sun-god Helios, who is normally envisaged as a charioteer.

ll. 57–63. Blake's ingenious personification has water-vapour turning to dew and re-evaporating in the morning to descend as refreshing rain.

177 ll. 79–80. The worm, as the associate of death, should be stronger than a little infant (and, perhaps, more repellent).

178 l. 94. *crown . . . away*: Rev. 3: 11.

ll. 106–7. *given . . . return*. The Clod's role is broadly comparable to that of the Sybil in Book 6 of the *Aeneid*, enabling Aeneas' descent to the underworld.

l. 108. *northern*: probably because the north entrance to the cave of the nymphs (as described in Book 13 of the *Odyssey*) is for mortals, and the southern for gods. There was some tradition of interpreting this cave as an image of mortal life. Twin gates, and the 'porter' or gatekeeper, are further tenuous links with Spenser's Garden of Adonis.

l. 113. *list'ning*. The verb could still take a direct object in Blake's day. See also *French Revolution* 19.

180 *The French Revolution*. Blake's subtitle was 'A Poem in Seven Books': he claimed that the rest was already written, but this has generally been doubted. This is Blake's first and last attempt to write a mythological narrative based at all closely on history. Blake identifies as a suitable leading event for this opening book the withdrawal of royal troops from the neighbourhood of Paris, but the way he envisages this to have come about (National Assembly and Royal Council in session; ultimatum from former to latter concerning the army conveyed by Sieyès; Lafayette ordering the withdrawal when this is rebuffed) is almost entirely imaginary, as are most of the attitudes ascribed to the characters. But the atmosphere of tension between the King's circle and the newly formed National Assembly is convincing. In handling the former Blake explores the psychology of tyranny with great seriousness, and on its own terms.

l. 2. *the Prince*: Louis XVI.

l. 7. *Necker*: Jacques Necker (1732–1804), Louis' Director-General of Finance.

l. 14. *regions of twilight*: in the classical underworld.

181 l. 26. *the prisoners*. There were seven prisoners in the Bastille at the time of its fall, but not seven towers to contain them.

ll. 31–2. *mask . . . kings*: a reminiscence of the 17th-c. legend of an imprisoned, masked royal relative, the 'man in the iron mask'.

184 l. 90. *From the Atlantic mountains*: influenced by the example of the American Revolution.

l. 91. *hind of Geneva*. Necker was a Swiss.

185 l. 104. *Versailles*: where the National Assembly was sitting.

ll. 107–10. The King decides that he has already threatened the interests of his party quite enough by his concessions to reforming demands.

186 l. 117. *thunder . . . burial day*. Blake seems to have in mind the belief that church bells will disperse thunder.

ll. 131–5. The vision owes something to traditional depictions of the Last Judgement, with God and Heaven above, and the damned descending to Hell beneath.

188 l. 165. *Henry the Fourth*: France's celebrated popular monarch of the 16th century.

189 ll. 187–8. *Fayette, Mirabeau, Target, Bailly, Clermont*: all major or minor members of the National Assembly, and hence thought of as in the vanguard of reform.

190 l. 209. *blights of the fife*: the damage done by militaristic policies to France's economic well-being.

191 l. 226. *Devour thy labour*: by taking tithes.

193   l. 264. *ardours*: probably from *Paradise Lost* v. 249, where angels are called 'celestial ardours'.

196   *Visions of the Daughters of Albion*. The predominantly dramatic procedure of the work, evidently understood by Blake to be the form taken by the 'visions' of English women (one of many references to visual perception in the poem), is a highly original experiment on his part. Each speaker displays responses to three levels of our existence: philosophical, moral (especially sexual), and political. The sexual theme is the leading one, and is astonishingly emancipated in its treatment. The effect is typically Blakean, in that each of the conflicting ideologies is explored and expressed, rather than judged, though there is clearly an implicit verdict in favour of Oothoon. There is nothing to choose between the other two figures, however, in unsatisfactoriness: they amount to a dismaying, disabling predicament for Oothoon.

l. 1. *Theotormon*: Blake's coinage. The names borne by the three speakers are phonetically of one type, as if all belonged to the same lost language (in which 'Albion' could also be a word). Only 'Oothoon' is not entirely Blake's invention, being a reminiscence of 'Oithona' in the 'Ossian' tales (where she is a hero's wife who is ravished by an enemy).

197   l. 27. *jealous dolphins*. Dolphins have an association with the sea-nymph Galatea, who was wooed by the Cyclops Polyphemus and, in the most famous version of her story, hated his uncouth courtship; she was discovered by him with her lover Acis, whom Polyphemus killed. This version has evident, though not straightforwardly treated, connections with Blake's Oothoon (moreover it is told, in Ovid, *Metamorphoses* xiii, as a lamentation by the female victim to a female audience: Galatea to Scylla).

l. 29. *signet*: the mark or stamp of a name (here as branded on a slave).

198   ll. 71–82. In this important section the falsity of Locke's denial of innate ideas in Man is deduced from the diverse and highly specialized instinctual drives and capacities which animals possess (which evidently cannot be explained by anything they have learned through their sensory apparatus). And if Locke is wrong Man must have 'thoughts . . . hid of old' analogous to those of animals.

199   l. 122. *Urizen*: probably the earliest surviving allusion by Blake to a figure that he was to write about a great deal. Characteristically, he is mentioned as if he needed no introduction, and in fact several of his later traits have evidently already been determined by Blake. The form of his name fits the conventions of the poem, though it must have been a formation from 'horizon', and perhaps Greek *horizein* 'to limit'. The name is stressed on the first syllable.

200   ll. 133–44. Blake associates the expenditure of natural wealth and manpower on war with enclosure and the decline of agriculture. Hostilities between Britain and revolutionary France commenced in February 1793.   *fat-fed . . . drum*: the recruited soldier.

l. 136. *the parson claim the labour of the farmer*: In tithes.

l. 146. *wake her womb*. In the sexual physiology of Blake's day the whole reproductive apparatus of a woman was thought to participate in arousal and orgasm, with the female equivalent of male ejaculation occurring at the ovary.

203 *Africa*. The date given here is the one that appears on the title-page of *The Song of Los*, under which heading Blake issued 'Africa' (as plates 3 and 4) and 'Asia' (plates 6 and 7). Even if it corresponds to the date of composition (and it is known that Blake's titlepage dates in some cases do not) Blake evidently meant 'Africa' to stand at the head of his sequence of four blank-verse texts covering the earth continent by continent (the Australian continent was still an uncertain entity in Blake's day), and it is so positioned here.

l. 2. *four harps*: perhaps a reference to the sequence of four texts which commences with 'Africa'.

l. 4. *Ariston*. Probably Blake's invention; for the Ariston of Greek legend see note to 'America' 149 (below, p. 552).

ll. 8–31. The passage illustrates the idea that all religions pass through a phase of systematization or codification: the Hindu *Bhagvadgeeta* (and perhaps a reputed Brahmin proto-Aristotelian system of logic); the Jewish Pentateuch; the supposedly Egyptian Smaragdine code; the Christian gospel; the Koran; Norse mythology.

l. 16. Gen. 11: 31.

l. 18. *Trismegistus*: the Egyptian god Thoth supposed to have written the *Hermetica*, the foundation of the alchemical tradition.

l. 21. *Orc*. See note to America 1 (below, p. 552).

l. 23. *man of sorrows*. See note to 'On Another's Sorrow' 27 (below, p. 556).

l. 29. *loose Bible*: presumably because of the considerable overlap in events and persons between the Koran and the Bible, which means that the former can be thought of as an approximation to the latter.

204 l. 52. *nightly*: dark as night. The opening line of the main text of *America*.

205 *America, a Prophecy*. In so far as *America* has a naturalistic content it is mainly the same as that of *The French Revolution*: the earliest, triggering phase of a revolution. But the transactions of humans and human institutions are no longer the main subject; there is a brief episode of conflict which stands for the whole War of Independence, and even the more extensive account of preliminaries has little to do with historical fact. *America* and its successor *Europe* are the only works which Blake subtitled 'prophecies' (although it has become conventional to call all his blank-verse narratives 'prophetic'). 'Prophecy' in this context has nothing to do with prediction; Blake is a 'prophet' in relation to the events of his time (and this explains much about the mode of *America* and *Europe*) in the same fashion as the Old Testament prophets—as a man of vision or

imagination who can discern the purposes and agencies behind the history his readers are witnessing.

l. 1. *shadowy daughter of Urthona*: not a named element in Blake's mythology, and an example of its plasticity. She reappears in the Preludium to *Europe* (which seems to be a post-coital continuation of that to *America*) invoking Enitharmon as her mother, who is not a consort of Urthona.   *Orc*. Orc here commences his memorable but brief activity in Blake's myths (he is only mentioned once in *Jerusalem*). His name clearly derives from the Latin word *Orcus* for the underworld and its god.

l. 6. *pestilence . . . heaven*. Comets and meteors were sometimes supposed to carry pestilence.

207 ll. 68–71. In this astronomical myth of Blake's own invention Mars stood at the centre of the solar system, with Mercury, Venus, and the Earth as satellites, and contained the comets and the sun.

l. 75. All the gospels except Matthew mention the spices used for Christ's burial; John 20: 7 refers to the wrapping up of Christ's discarded grave-clothes.

ll. 76–7. A vision that derives from, but modifies, Ezek. 37: 1–10.

l. 77. *inspiring*: breathing in.

l. 79. *grinding at the mill:* Matt. 24: 41.

208 l. 103. *deserts blossom*: Isa. 35: 1.

l. 111. *Amidst . . . he walks*: Dan. 3: 25.

ll. 111–12. *feet . . . like gold*: Dan. 2: 32–3.

209 l. 145. *Atlantean*: as mighty as Atlas.

l. 149. *Ariston*. No Ariston is traditionally associated with Atlantis or with beauty. Ariston king of Sparta did steal a friend's wife, but Blake may be thinking of Plato's story that Poseidon lord of Atlantis captured a mortal bride.

210 l. 166. *fat of lambs*: Deut. 32: 14.

l. 180. *Bernard's*: Sir Francis Bernard, governor of Massachusetts.

211 ll. 198–206. This passage has the closest resemblances to one that influences *America* at several points: Christ's final overwhelming of the rebel angels in *Paradise Lost* viii. 824–65.

212 l. 223. *mitred York*: alludes to York, one of England's two archdioceses, or to its archbishop.

l. 230. *Bard of Albion*: perhaps the poet laureate William Whitehead.

l. 233. *doors of marriage are open*. Blake married Catherine Boucher in 1782.

213 ll. 254–5. Evidently, until the French Revolution. There are a number of events in the two revolutions that are separated by twelve years, and it is not clear which Blake would have thought decisive.

l. 262. *five gates*: the five physical senses.

214 *Europe, a Prophecy. Europe* is in many respects the climax of the blank-verse narratives of these years: the most complex, ambitious, and strange of them all. Though in its title and subtitles *Europe* seems to be a close parallel to *America* it has a very different procedure, which is due to a quite different approach to time and human history, and one that lays the foundation for all Blake's subsequent prophetic writing. Such contemporary events as can be discerned in *Europe*, and other elements from our orthodox idea of human history, are, with startling effect, placed within unfamiliar schemes that deny conventional time: the 'shadowy female's' perpetual procreation, the 1,800 years' sleep of Enitharmon, and the identity of pre-Roman and modern Britain. Blake has abandoned historical time in the spirit of his investigation of permanent states of human nature.

l. 1. *caverned man*: the man equipped only with physical senses, as depicted in Plato's allegory of the cave.

l. 6. For the phrasing see Prov. 9: 17. Blake thinks of the sense of touch, the last of his five senses, above all as the modality of sex.

215 ll. 58–9. *the secret child Descended*: in human history, the birth of Christ.

216 l. 86. *horrent*: bristling; but probably also with the sense of 'frightening'.

l. 92. *lovely Woman.* Blake echoes Goldsmith's famous line 'When lovely woman stoops to folly'.

217 l. 102. *skipping upon the mountains.* Cf. *S. of S.* 2: 8.

l. 112. *Eighteen hundred years*: the time between Christ's birth and Blake's own day.

l. 122. *Angels of Albion*: historically, the government of Pitt, at war with revolutionary France.

ll. 129–30. *ancient temple . . . island white.* Blake generalizes the theory, due to William Stukeley, that the megalithic ruins at Avebury described a serpent-shape.

l. 132. *Verulam*: a major town in Roman times, modern St Albans.

218 l. 136. *colours twelve*: Rev. 21: 19–20.

l. 138. *fluxile*: presumably, here and elsewhere, 'malleable': a usage peculiar to Blake.

l. 140. *spiral ascents*: the ears, in their original form.

219 l. 178. Generally accepted to be a reference to the dismissal of Lord Chancellor Thurlow in June 1792 at Pitt's instigation.

220 l. 204. Cf. *Paradise Lost* i. 301–2.

221 l. 237. *seven churches*: Rev. 1: 20.

222 *Asia.* 'Asia' (plates 6 and 7 of *Los*) is less complex and adventurous than *Europe*, and if it was indeed composed after the latter it perhaps indicates Blake's loss of interest in his project of a series of geographical prophecies.

223 l. 51. *pillar of fire*: the first appearance of a major Blake motif, derived from
     Exod. 13: 21–2.

     l. 53. *sullen earth*: Shakespeare, *Sonnets* xxix. 12.

     ll. 55–6. Ezek. 37: 7–10.

     l. 62. *glandous*: apparently Blake's invention, and only found in this
     context—which hence must be the main guide to the word's meaning.

THE LYRICS

Song ('*I love the jocund dance*').

229 l. 11. *White or brown*: varieties of bread.

230 *Mad Song.* George Saintsbury (*A History of English Prosody* (London,
     1910), iii, p. 11) judged that for 'pure verse-effect' there were 'few pieces
     in English to beat' this lyric. Analysing his texts in terms of traditional feet
     (and coping with irregularities under the rubric of 'substituted feet')
     Saintsbury believed that Blake in the lyrics and shorter prophetic books
     was the greatest English prosodist since Shakespeare.

231 *Song* ('*When early morn*'). To be read (unlike its preceding companion
     poem) as a satire on the kind of love that is happier with sexual deprivation
     than with sexual fulfilment.

232 *Song First by a Shepherd.* The second shepherd's song which was a
     companion is a close variant of 'Laughing Song' in *Innocence*.

     [*The Cynic's First Song*]. Sung by 'Quid the Cynic' in Blake's prose satire
     'An Island in the Moon'; the character represents Blake himself. Quid also
     sings a version of 'The Little Boy Lost' from *Songs of Innocence*.

     l. 2. *yellow vest*. Vest, as in modern American English, means waistcoat. A
     yellow one was probably dandified.

233 l. 18. Presumably a reference to morbid dissection in medical research and
     teaching, a practice much distrusted in Blake's day.

     l. 22. Vitamin-C deficiency and typhus. The ravages of these diseases in
     the navy were finally arrested in the 1780s by naval medical adminis-
     trators.

234 [*Miss Gittipin's First Song*]. Miss Gittipin is also a character in 'An Island
     in the Moon', and has not been definitely identified. Her main trait is her
     hankering after fashionable London pleasures, though here she celebrates
     rustic festivities.

     [*The Cynic's Second Song*].

     l. 7. *fingerfooted*: with feet of finger-like slimness.

235 l. 23. *the pip*: a common disease of birds.

     [*Obtuse Angle's Song*]. Uttered by another unidentified character in 'An

Island in the Moon'. He also sings a version of 'Holy Thursday' from *Songs of Innocence*.

ll. 4–5. *South . . . Sherlock*. Richard South and William Sherlock, two Anglican divines of the latter 17th century, who engaged in a pamphlet controversy. South was celebrated as a preacher, and Sherlock for his *Practical Discourse Concerning Death* (1689).

l. 6. *Sutton*. Thomas Sutton, founder of the Charterhouse, a boys' school and home for the aged.

236 [*The Lawgiver's Song*]. Steelyard the Lawyer in 'An Island in the Moon' is Blake's friend Flaxman (see note to 'With happiness stretched across the hills' 47, above, p. 520).

238 *Songs of Innocence*. (For the significance of the title see the general note to *Songs of Experience* below.) The various orders in which Blake bound up different copies of the *Songs* offer his editor many choices in the matter, though not an unlimited number: there are some groupings that remain very constant. The order adopted here tries to be faithful to as many of these groupings as possible. It is slightly unusual in the three songs placed after 'Introduction' (two of which, moreover, were later transferred to *Experience*); but they are so arranged in the copy which Geoffrey Keynes regarded as personally preferred by Blake, and Keynes used this arrangement in his own editions. This way of starting the collection also makes good literary sense, establishing its predominant motif of the threatened/guarded child in night ('innocent' in the sense of unharmed) versus the free, joyful child in day.

Introduction

l. 18. Refers to the manufacture of ink or watercolour.

239 *The Little Girl Lost*. In all but one of the known copies of the twin volume, *Songs of Innocence and Experience*, this and its sequel 'Little Girl Found' are inserted among the Experience lyrics, and they are omitted from several of the known copies of the Innocence lyrics published separately. Blake's motif is essentially the traditional one of the child nurtured by animals. His affirmative treatment of the child's experience is in keeping with the ancient belief that such children can have superior powers. There was a revived, rationalistic interest in the subject in the 18th century. The 16th-c. 'Babes in the Wood' and 'Children in the Wood' ballads had made the topic an occasion for sentimental fear; Blake would have been aware of the ballads through their appearance in Percy's *Reliques* and in chapbooks.

The Little Girl Found

241 ll. 18–19. *Pressed with feet*. 'Pressed' probably has the sense 'hastened', but the whole phrase is just possibly a Latinism (see *Aenied* vi. 197), and hence means 'halted'.

l. 30. *allay*. The intransitive sense would have been almost obsolete in Blake's day.

The Shepherd

243 l. 4. The phrase recalls many in the Psalms (e.g. 71: 8).

On Another's Sorrow

245 l. 22. Cf. Rev. 7: 17, 21: 4.

l. 27. *man of woe*: Cf. Isa. 53: 3.

246 *The Schoolboy*. In a few copies of the combined lyrics this poem is included among the Experience songs.

The Little Black Boy

248 l. 28. *be like him*: 1 John 3: 2.

249 *The Voice of the Ancient Bard*. In a few copies of the combined lyrics this poem is included among the Experience songs.

250 *Nurse's Song*. A version also appeared in 'An Island in the Moon' sung by Mrs Nannicantipot (a coarsely irreligious character) as her mother's (but the modern sense of 'nanny' existed in Blake's day, so she is perhaps herself a nurse).

Holy Thursday. Blake describes the combined service for all the charity schools in London which was held in St Paul's, generally on the first Thursday in May, from 1782 onwards.

251 l. 5. *multitude*. There may have been 6,000 children present at these services.

l. 6. *in companies they sit*: perhaps in groups according to their schools. But there is also an important echo of the feeding of the five thousand (Mark 6: 39).

l. 9. *mighty wind*. The Pentecostal wind is called a 'rushing mighty wind' (Acts 2: 2). Pentecost is celebrated at Whitsun, seven weeks after Easter.

The Divine Image

ll. 10–11: *Paradise Lost* 3: 44.

The Little Boy Lost

254 l. 8. The will o' the wisp traditionally lured men into swampy places, but Milton's adaptation of the idea to describe Satan (*Paradise Lost* ix. 634–42) is surely echoed here.

Night

255 ll. 42–3. Cf. Isa. 11: 6.

l. 45. *life's river*: Rev. 22: 1.

256 *Manuscript Lyrics between* Innocence *and* Experience. In Blake's notebook entries of this period there are many drafts for Experience lyrics, which have not been reproduced. Several of the poems that have been included were lightly deleted in the notebook, but had been brought to a sufficiently final form to justify printing here. These are: 'I saw a chapel', 'I asked a thief', 'I heard an angel', 'Why should I care', 'Oh lapwing',

'Are not the joys', 'Soft Snow', 'In a Myrtle Shade', and 'Day'. One group of poems was recopied by Blake elsewhere in his notebook and put under the heading 'Several Questions Answered': they are 'Eternity', 'The look of love', and 'The Question Answered'. He added to them the couplet, 'Soft deceit and idleness,/These are beauty's sweetest dress'.

258 *A Cradle Song.* Presumably devised as an *Experience* parallel to 'A Cradle Song', but rejected in favour of 'Infant Sorrow'.

'*Why should I care*'

259 l. 2. *chartered.* See note to 'London' 1 (below, p. 558).

'*Thou hast a lap full of seed*'

260 l. 2. *country*: possibly, by a device for which there are precedents, a bawdy equivocation for 'cunt'.

In a Myrtle Shade. The myrtle, a motif peculiar to the lyrics of this period, has strong connotations of licit and illicit love. The shrub is sacred to Venus, and Pliny refers to its use in conjugal rites, which continued (in Germany, for example) into the modern period. In Greek legend Myrrha, the incestuous mother of Adonis, was turned into a myrtle, and at her festivals married women wore myrtle crowns. There is a folk-tradition that dreaming of myrtle prognosticates lovers or second marriages. For an important Virgilian association see note to 'My Spectre around me' 42 (below, p. 560).

l. 3. *free love.* See note to 'Earth's Answer' 25 (below, p. 558).

262 *How to Know Love from Deceit.* This poem-title is deleted in the notebook.

263 *Merlin's Prophecy.* The tradition of Merlin the prophet and of the 'Prophecies of Merlin' are both due to the medieval historian Geoffrey of Monmouth. Part 5 of *The History of the Kings of Britain* consists of the prophecies, which are enigmatic predictions of calamity and unnatural phenomena in Britain. But Geoffrey's *History* was a rare book, especially in English, and Blake's unnatural events are anyway more in the spirit of the latter part of the group of sayings which Lear's Fool calls a prophecy by Merlin (*Lear* III. ii. 85–90).

265 *Lacedaemonian Instruction.* 'Lacedaemonian' means 'Spartan'. Blake characteristically associates Sparta's celebrated military culture with religion.

'*Her whole life is an epigram*'

266 l. 1. *smack-smooth*: perfectly smooth.

267 *Songs of Experience.* Some four years after he first issued the Innocence collection Blake started to put out copies in which these poems were combined with a second group entitled *Songs of Experience*, the two being headed with a joint title and the subtitle 'Showing the Contrary States of the Human Soul': for a joint 'motto' to the enlarged volume, never used,

see the preceding item. He continued to issue *Innocence* on its own, but it is virtually certain that *Experience* was only published as a companion to *Innocence*, never independently. This harmonizes with Blake's use elsewhere, at this period, of the terms 'innocence' and 'experience': the former was used rather frequently and deliberately (and affirmatively: see the annotations to Lavater (above p. 4) and *Visions of the Daughters of Albion* 165 (above, p. 201)) whereas 'experience' is not a common or a weighted term. When Blake used the latter in the title of his new collection of lyrics it was as an antonym to 'innocence' that carried no special significance (except that it may have had a pejorative force due to another writer's deployment of it: Milton uses 'experience' on several occasions with reference to the Fall).

Introduction

ll. 6–7. At Gen. 3: 8 the Lord God walks in the garden 'in the cool of the day' and calls Adam and Eve, but note that Milton has Eve dream of Satan doing almost the same (*Paradise Lost* v. 35–9).

l. 9. Milton's term for the night sky as seen by Adam and Eve (*Paradise Lost* iv. 724).

Earth's Answer

268 l. 25. *free love*. It is uncertain when in the 19th century this phrase entered English in the sense of an ideal of unconstrained choice of sexual partner, but Blake's use of it here almost amounts to that meaning.

Holy Thursday

269 l. 4. *usurous*: because of the extent to which the lending of money with interest would lie behind the wealth with which the charity schools are endowed. Less specifically, a gift that is usurous is not a gift at all.

271 *The Tiger*. The rhyme-scheme and metre, and some elements of the expression, in this song may be indebted to Crashaw's 'Hymn of the Church in Meditation of the Day of Judgement', a paraphrase of the famous liturgical poem *Dies Irae*.

Ah! Sunflower

273 l. 5. *pined away with desire*. Blake seems to be remembering the actual phrasing of the account of Clytia, who was deserted by Apollo and turned into a sunflower, as given in Lemprière's Classical Dictionary.

The Little Vagabond

274 l. 12. *bandy*. The connection between the disease of rickets (of which bandy-leggedness is a characteristic symptom) and lack of sunlight was not understood in Blake's day, but rickets was known to be linked to poor and overcrowded living-conditions.

London

l. 1. *chartered*: pejoratively, since charters deliver boroughs into the hands of corporations. For a significant parallel expression of hostility to charters see Paine's *Rights of Man* Part 1 (1791) (e.g. 'Every chartered town is an aristocratical monopoly in itself'). Blake is either not using the term literally, or is thinking of the streets of the City of London, the only part of London to have a corporation in his day (and even that, as a matter of strict history, was not established by a proper royal charter).

l. 10. *appalls*. There is a play on the old sense of 'make pale'.

Infant Sorrow

276 l. 6. *swaddling bands*. The age-old practice of swaddling children was becoming unpopular in advanced circles in Blake's day.

A Poison Tree

l. 14. *night . . . veiled the pole*. See Cowper's 'On the Death of Mrs Throckmorton's Bullfinch' (1789) 31.

l. 16. *outstretched beneath the tree*. Compare Milton's Adam after he has eaten the fruit (*Paradise Lost* x. 850-1).

278 *To Tirzah*. 'Tirzah' is a rare and obscure biblical name used once here by Blake, and then only, but often, in *The Four Zoas*, *Milton*, and *Jerusalem*. Blake was no doubt mainly struck by its use as the name of one of the daughters of Zelophehad (see note to *Milton* i. 419, below p. 568), but he would also have noticed its linking with that of Jerusalem at S. of S. 6: 4, and perhaps its appearance as the name of one of the conquered territories of the Israelites at Josh. 12: 24.

l. 3. *generation*. Blake's first use of a word that later became his standard term for the mortal world. Though he was echoing neo-Platonic practice this early use makes it clear that the ordinary sense of the word was in his mind too. He may also have remembered *Paradise Lost* i. 653.

279 l. 15. A formulation with a Pauline ring to it.

A Divine Image. Although never included in *Experience* this would-be counterpart to 'The Divine Image', unlike other poems rejected from the collection, was illustrated and etched.

280 *Manuscript Lyrics of the Felpham Years*. While the poems in this section may not have been composed entirely at Felpham itself (the first three items, in particular, are of uncertain date) they are predominantly the product of Blake's residence there.

'*A fairy stepped upon my knee*'. The functions attributed to fairies in this poem seem to derive in some measure from Pope's sylphs in *The Rape of the Lock*.

281 *On the Virginity . . . Joanna Southcott*. Joanna Southcott was a Devonshire girl who achieved great celebrity as a mystic. She claimed in 1802 to be the bride of God and mother-to-be of the Christ of the second coming.

282 *The Grey Monk*. Blake eventually used this title for a slightly shorter version of this poem than is given here. The original conception was even longer, but several verses were adopted (together with a couple also appearing here) to make a poem at the head of *Jerusalem* III (see below).

'*Mock on*'

284 l. 5. *every sand*. The sense of a single grain of sand must have been nearly obsolete in Blake's day.

285 l. 9. *Democritus*. The ancient Greek pioneer of atomic theory, for whom reality was composed of an infinity of uniform indivisible units, differentiated only by the equivalents of Locke's primary qualities.

l. 10. See note to 'To my friend Butts' 15 (above, p. 520).

'*My Spectre around me*'

286 l. 42. *infernal grove*. From the 'grove' described at *Paradise Lost* x. 547-52, which bears ashy fruit resembling 'the bait of Eve / Used by the tempter' (a phrase Blake would have noticed). Also remembered is the myrtle grove in Virgil's underworld (*Aeneid* vi. 440-76) which harbours the lovelorn, and where Aeneas last sees Dido.

286 '*O'er my sins*'. These verses obviously have a close connection with the preceding poem, though the relationship intended by Blake is not clear. They should perhaps be allotted to two speakers, as has been done conjecturally for the companion poem.

288 *The Mental Traveller*. While the title of this tantalizing and much-discussed poem refers primarily to the first-person speaker it perhaps has a secondary application to the protagonist of the poem: the male figure who passes through infancy, manhood, old age, and parenthood, till the 'maiden' of l. 56 brings about a reversal of the process (and if the latter is also the 'woman old' of l. 8 she follows, at the same pace, the opposite sequence: age–youth–age). It requires an attentive reading to see this pattern, however, since it is embellished with a number of other images of human development and transition in the sexual and social domains. Two senses of 'generation'—procreation, and the interval between old and young—are strongly operative, in a manner that recalls *Tiriel*. Indeed the special quality of 'The Mental Traveller' arises from its deployment in a ballad-like form of large, enduring human themes reminiscent of the earlier poem.

ll. 7-8. Ps. 126: 5.

Mary

293 l. 43. *face . . . human divine*: Cf. *Paradise Lost* iii. 44.

Lyrics from the Epic Poems

From the Preface to Milton

297 l. 3. *Lamb of God*. The phrase is particularly associated with John the Baptist who, uniquely in the Bible, uses it to refer to Jesus in John 1.

l. 12. *chariot of fire*. Elijah is taken up to Heaven in a 'chariot of fire' at 2 Kgs. 2: 11.

From Jerusalem (plate 27)

l. 2. *Primrose Hill*. In Blake's day Primrose Hill was used for ceremonies by a neo-Druidical organization calling itself The Order of the Welsh Bards, and Druids are prominent in the prose which goes before this poem (see above, p. 31).

298 l. 13. *Jew's Harp House*: a tea-garden in Marylebone Park (later part of Regent's Park). *Green Man*: a public house on the Marylebone Road (formerly New Road), just south of Marylebone Park and Willan's Farm (near the modern Great Portland Street tube station).

l. 15. *Willan's Farm*: also incorporated into Regent's Park.

l. 34. *Tyburn's brook*. Tyburn Hill (where Marble Arch now stands) was the site of public executions in London.

From Jerusalem (plate 52)

300 l. 1. *Charlemagne*: a Christian king, but in the mould of the Roman Emperors.

l. 7. *schools*: Medieval theology.

301 l. 21. *Titus! Constantine*: Roman emperors who were pro-Christian, but emperors none the less.

302 *Late Lyrics*. Blake continued to write verse in short, rhymed forms in the last twenty years of his life, but most of this material is satirical comment on art, overlapping with the texts in Section 2 above: only a few lyrics on other topics are reprinted here.

'If I e'er grow'

303 l. 4. *take the wall*: be accorded woman's privileges as conferred by sexual etiquette.

'You don't believe'

l. 7. *'Miracle'*. For Blake's attack on the idea of miracles as supernatural phenomena see above, p. 22.

l. 11. *'Only believe'*: Mark 5: 36.

'I rose up'

304 l. 15. *accuser of sins*. See note to *Vision of the Last Judgement* (above, p. 524).

'Why was Cupid a boy'

305 l. 18. Blake may be conjecturing that the Greek cult of male homosexual love was rooted in admiration of martial prowess.

'To Chloe's breast'

306 l. 2. *Myra*. This name, and Chloe, are probably those of imaginary women, but in the latter case Blake could have in mind the figure of

Myrrha: see note to 'In a Myrtle Shade' (above, p. 557).   *pocket-hole*: placket-hole, giving access to the genitalia.

THE LOS POEMS

309 *The Book of Urizen*. Blake had used the form of title 'The Book of...' once before to indicate an imitation of a biblical model (for *Thel* as an imitation of Job), and here he returns to the formula for three successive works: *Urizen*, *Ahania* (not reproduced), and *Los*. Each of these, moreover, has the appearance internally of a book of the Bible, with chapters and numbered verses. But Blake does not achieve anything remotely like a Bible-sized text, or even a Pentateuch-sized one (even if the imitations of Exodus, Numbers, Joshua, and Revelation in *Milton* and *Jerusalem* are counted into the project), and in *Urizen* there are signs of uncertainty of conception: it was originally entitled 'The First Book of Urizen', and the contents and their order vary from copy to copy more than in the case of any other Blake text.

Urizen is nevertheless a powerful experience. A *Europe*-like plasticity of time is taken to a point where all eternity and a single life seem to be coextensive. The poem modulates from a complex reworking of *Paradise Lost* into a strange psychomachia that is combined with an imitation of Genesis. There is nothing straightforward or doctrinaire about the poem's central subject, Urizen. He is a much more ambiguous figure than many critics recognize, and Blake expresses this ambiguity partly by giving him qualities of both Satan and God/Messiah in *Paradise Lost*, linking these figures ingeniously by way of allusions to their various conquests of chaos (see, for example, the note on 'petrific' below).

l. 1. The traditional word-order of the opening to an epic (starting with the subject of the poem, which is also the object of the first main verb) is preserved, as is the Miltonic genitive grammatical aspect, even though there is no main verb in Blake's opening, let alone one that takes a genitive.

l. 3. *place . . . north*. The headquarters of Milton's rebellious Satan in Heaven are in the north.

l. 6. *Dictate*: Milton's well-known word for the activity of his muse.

l. 14. *Brooding*: in *Paradise Lost* a repeated adjective (as here) denoting God's creativity.

l. 16. *ninefold darkness*. Milton's Satan lies for nine days in the abyss of Hell.

310 l. 33. *petrific*: Milton's memorable coinage for the powers of Death in solidifying chaos; but the context more resembles God's creation of the world out of chaos, which is an 'omnific' act (*Paradise Lost* vii. 217).

l. 35. *ten thousands of thunders*. Cf. Christ's armoury of 'ten thousand thunders' (*Paradise Lost* vi. 836).

l. 43. *globes of attraction*: Heavenly bodies held together by gravitational forces.

311 ll. 67–74. The passage has several echoes of Satan's crossing of chaos (*Paradise Lost* ii. 910 ff.), though 'self-balanced' recalls God the creator (it is Milton's epithet for the earth at vii. 242).

312 l. 117. *no light from the fires*. Cf. *Paradise Lost* i. 62–3 (though a traditional feature of descriptions of Hell).

313 l. 162. *hurtling*: perhaps 'clashing' (but see *Los* 134, below p. 329).

316 l. 260. As in the seventh day of creation in Genesis nothing new is formed in Blake's seventh age (though neither Urizen nor Los is very restful).

318 ll. 331–2. Enitharmon's tent is closely modelled on Isa. 54: 2, where the same imagery is associated with the fertility of the barren and unmarried.

321 ll. 438–45. Thiriel, Utha, Grodna, and Fuzon are, at one level, the elements of air, water, earth, and fire.

322 ll. 449–53. Urizen's feelings are comparable to those of God at the beginning of Gen. 3.

l. 459. *cities*. The appearance of cities at this point reflects their similar prominence in Gen. 11–14 and 18–19.

323 ll. 477–92. Blake's version of the destruction of Sodom and Gomorra and the transformation of Lot's wife in Gen. 19.

l. 494. See Gen. 2: 2.

324 ll. 509–24. The last four verses rapidly summarize the events of Exod. 1–14, climaxing in the passage of the Red Sea.

325 *The Book of Los*. Evidently, by virtue of its formal features, part of the same enterprise of biblical imitation as *Urizen*. It also expresses even more emphatically than that work the intimate association, amounting to a kind of interdependence, between Los and Urizen.

l. 13. *curled*: with curled hair; made-up to a degenerate extent.

326 *Chapter II*: a section full of memories of *Paradise Lost*, if not of exact echoes.

327 l. 100. *pliant*: probably with two senses: 'physically flexible' and 'well-adapted'.

328 l. 106. *polypus*: a jellyfish.

329 l. 149. *subtle particles*. See note to 'To my friend Butts' 15 (above, p. 520).

ll. 154–7. Accurately follows the actions of a smith in a forge.

330 l. 163. *redounding*: Surging up, abundant (almost an obsolete usage in Blake's day, though found on several occasions in the later writings).

l. 172. *four rivers*. According to Gen. 2: 10–4 four rivers were fed from the Garden of Eden.

331 *Milton*. Broadly speaking, each book relates a descent from the immortal to the mortal world which manifests itself to Blake as an incarnation of the immortal in his own body or in his immediate world in Felpham. In Book

I the descent and incarnation is performed by Milton; in Book II by Ololon, a group of immortals named after the river they dwell by and who become a single female Thel-like entity half-way through Book II (a kind of fusion that is common in the early history of Israel in the Old Testament: in particular, one name will often be used for a place, a ruler, and a tribe or nation). This pattern is much complicated, however, by chains or networks of repetitions of its material at almost every stage. The Milton figure, by his own account, is Satan, and is also plainly mirrored in Blake himself, who in turn is in some sense the bard of Book 1. Both have links with Los, and the latter engages in an incarnation-in-Blake, probably in Lambeth before Blake's departure for Felpham. Another such incarnation is that of the 'daughters of Beulah' in Blake's hand at the very outset. There are further examples of incarnation or spiritual occupation involving British geography and certain natural forms. The daughters of Beulah are also a repetition of the Ololon motif, which is further echoed in the six-fold Emanation of Milton, in Leutha, in Satan's seven angels, and in Los's consort Enitharmon.

Blake has sought to arrange these entities and episodes into a narrative sequence consisting chiefly of journeys and confrontations, but the degree of parallelism amongst them means that this cannot be done straightforwardly, and the result is one of the most obscure poems in English, *Jerusalem* not excluded (while it contains perhaps the most thrilling passages of blank verse that Blake wrote). Indeed *Milton* has no single story of a beginning–middle–end sort, and it is futile to search for one. One series of events in the poem is, however, worth emphasizing. It may be called the historical theme of *Milton*, and distinguishes this work from *Jerusalem*; for at this level *Milton* recounts certain alleged events in historical time—strange though they may be—involving Milton, Blake, Felpham, etc. in their ordinary literal senses (whereas in *Jerusalem* such literal materials are only the envelope of the narrative). This theme is very much concerned with the problem of how mankind is to achieve spiritual advancement without simply embarking on a new phase of religion, with its attendant sectarianism and narrowness (so *Milton* expresses Blake's old ambivalence about religion); it runs as follows:

Blake, living in Lambeth, has discerned the falseness of orthodox Christianity's claim to offer a new spiritual dispensation, and comes to Felpham inspired to express his vision. He composes a new account of Christian universal history (the bard's song of Book 1) which essentially rewrites the story of Satan as told in *Paradise Lost*. The gist of its subversiveness (as recognized by its audience at i. 311–12) is its arraigning of 'pity and love'. 'Pity', 'love', and 'mildness' are much reiterated terms in the vocabulary of the bard's song, which tells of their alliance with 'wrath', 'fury', and 'indignation', this alliance somehow occurring in the person of Satan (there are three rather different versions of what took place: at i. 33–7, 198–201, and 257–8). In this way the Father and Son of *Paradise Lost* are implicated by the bard in the rebellion in Heaven (where Satan's consort is called 'Leutha' by Blake, rather than 'Sin') and in the spiritual enslavement of mankind described at i. 43–176 (which is here due to the religion of the Old Testament, and not to the sinfulness of fallen Man). *Paradise Lost* had laid these to Satan's charge alone.

Milton hears the bard's song, and realizes that he has not only given a false account of universal history, but has renewed the reign of false religion in his own poetic creation. He must descend to Earth and atone with his death (evidently a parallel to Christ's resolve, reported in *Paradise Lost*, to die for mankind; but in Blake's version Christ's redemption has apparently not taken place, or has not yet become efficacious). His incarnation will be as part of the poet Blake, and the latter's poetic genius, or Los, at first opposing this development, is reconciled to it by the 'prophecy' that Milton will ascend again from Felpham. There is still the very important objection that Milton's work has tended to encourage the rise of deism, at the expense of true enthusiastic religion, but Los in a crucial speech (i. 719–92) urges that to be militant on behalf of the latter would be to repeat the calamity of the Reformation. In the fierce but organized physical life of the human individual and of human civilization lies the possibility of redemption from the rule of Satan. Book 1 closes with a long visionary description of the present physical world from its largest to its most minute elements, in which this too is displayed as a kind of human artefact.

In Book II Blake, already invested with the incarnated Milton, encounters a young girl in his garden who is seeking Milton. She provokes a stupendous manifestation of both Milton and the latter's 'Spectre', Satan. Satan represents the whole accumulated history of Man's spiritual life, in which contending religions have multiplied, and Milton declines to continue the process with an attack on him, but instead will himself die. Satan proclaims victory, but to his surprise and terror seems to be abetting the spiritual awakening of Man. The girl, Ololon, is still dismayed by the fact of the rationalistic religion that has held sway since Milton's time, but Milton affirms his power, having died into life as Blake, to transform mankind's spiritual being without imposing a new religion.

Evidently much is left out by a summary of this particular theme in *Milton*, especially from Book II and generally in relation to the Ololon figure. Indeed this whole domain of the poem—Ololon/Milton's Emanation/Leutha/Enitharmon/Satan's angels—is second only in importance to the historical theme. It too is 'historical' in so far as it springs from the facts of Milton's family life, and no doubt from Blake's relations with Catherine; but very general and permanent aspects of the human psyche and experience seem to be more at issue here. This side of the poem is better thought of as displaying these, rather than something unfolding in historical time.

The difficulties in *Milton* so far mentioned are greater than, but not essentially different from, those the reader encounters in some of Blake's shorter prophecies (in *Europe*, for example). But there is a new and important factor that makes *Milton* even less of a narrative poem than its predecessors: a change in Blake's approach to mythology. *Milton* continues the process of accretion and embellishment to a 'machinery' of Blake's own invention—agents and places bearing names never, or scarcely, heard before— which had started with 'Gwin, King of Norway'. But by Blake's standards there is relatively little fresh material of this sort, and alongside it appears a new kind of myth-making, which will become even more prominent in *Jerusalem*: myth-making which uses traditional names (in particular, from the Old Testament, and from British history and geography). The poet's contribution is to develop equivalences between these traditional names which have the effect of wrenching them from their familiar contexts. Stating these equivalences becomes a major concern for Blake, and hence the new kind of mythology brings with it a change of literary result: narrative about agents of whom we have probably never heard unless we have read certain other portions of Blake (into which we are nevertheless plunged as if they were well known) tends to be replaced by exposition of a framework of connections between elements we are familiar with, or know something of, in very different settings. Blake, in fact, starts to acquire and expound what is often attributed to him much too early in his career: a 'system'.

The Old Testament references in *Milton* have a surprisingly specific provenance, being drawn very largely from the story of the Israelites between the exodus from Egypt and the settling of the Promised Land (that is, the narrative portions of Exodus 14–40, Numbers, and Joshua— though due allowance must be made for the extent to which elements of this history are recalled throughout the Old Testament under the rubric of God's relationship with his people). This can be explained in part by regarding *Milton* as a sequel to *The Book of Urizen* (which seems to end with the crossing of the Red Sea): a project that Blake mentioned as long ago as *The Marriage of Heaven and Hell*—the 'Bible of Hell'—is now perhaps being more thoroughly pursued than hitherto because of a fuller loyalty to Scripture developing in Blake some time after *The Book of Los*. But something more seems to be required to explain Blake's emphasis on this particular phase of Jewish history (it continues to be the main source of Old Testament allusions in *Jerusalem*). This is the period when the Jewish religion is most under threat, when the Israelites are most frequently guilty of infidelities and heresies for which they are punished by Jehovah. The Jewish nation is also very clearly divided into its twelve tribes (as is especially emphasized in *Jerusalem*), and there is potential strife amongst them religiously and politically (e.g. Joshua 22). In *Milton* Blake does not so much reiterate biblical examples of the social and spiritual crises among the Jews as invent his own (e.g. i. 556–7, 751–7, 763–9), and it seems that, for this poem which so much concerns the relations of spiritual militancy and religious intolerance, he found the vicissitudes of post-Egyptian Israel peculiarly suggestive. This is also the story of the Israelites coming into their long-expected inheritance of the Promised Land, and under this aspect is compelling for the Blake of

Jerusalem, where the British (if only as a metaphor for humanity) are the chosen race, and Britain is the site of the true fulfilment of the Christian dispensation.

The text. Readers consulting certain other editions of *Milton* will notice passages not appearing in this version and/or differences of order. There are four copies of *Milton* in existence; two are identical and shorter than the others, and are generally followed here. The extra material in the longer copies is as follows:

i. 26. 28 lines in one copy, plus a further 76 in the other. (And all copies have a section of 35 lines which is widely regarded as an unsatisfactory later interpolation.)

i. 175. 21 lines in both copies.

i. 444. 51 lines in both copies.

ii. 96. 43 lines in both copies.

However, these longer versions also place i. 826–942 in the position they occupy here, whereas they appear after i. 1005 in the two shorter copies that I have otherwise followed: it seems the more natural order, but readers should be aware of the alternative.

1804. Though this date on the title-page probably represents the substantial completion of Blake's text the evidence is that nothing was engraved until at least four years later.

To justify . . . to men. See *Paradise Lost* i. 26.

Book the First

l. 1. *Beulah*. Isa. (62: 4) calls Jerusalem 'Beulah', i.e. married. It is Bunyan in *Pilgrim's Progress* who uses the term as a place-name. In Blake's day there was a rural hill in Norwood, on the southern edge of London, variously called Bewlay/Beaulieu/Beulah Hill, and some of his uses of the word seem to include an allusion to this site.

l. 10. *false tongue*: Cf. Ps. 120: 3.

l. 17. *mazes of providence*: combines the fallen angels of *Paradise Lost* ii. 558–61 with the heavenly dancing of v. 620–4.

l. 19. *sixfold Emanation*: Usually taken to refer to Milton's three wives and three daughters.

332 l. 40. *thousand years*: perhaps in allusion to the period of Satan's imprisonment as recorded in Rev. 19.

336 l. 149. *There is no other*. Echoes many claims to exclusivity by the Jewish deity in the Old Testament, but especially Exod. 20: 3 and Deut. 5: 7.

l. 174. *covering cherub*. This important phrase for Blake is used in the denunciation of Tyre at Ezek. 28: 16 (and in the preceding verses the 'cherub' is associated with Eden and 'the holy mountain of God').

l. 177 *eon*: best defined by this occurrence and two in *Jerusalem* (I. 752, II. 348).

339   l. 252. *Paradise Lost* ii. 760-1.

ll. 277-9. *ratified . . . decision*. The rapid, almost comic sequence of 'elections' that follows this parodically expressed event, apart from illuminating Blake's term 'elect', echoes the entrusting of Satan by Hell, and of Christ by Heaven, with missions to Earth in *Paradise Lost*.

340   l. 285. *Triple Elohim*: triple presumably with reference to the Christian doctrine of the Trinity (for Elohim see the note to the second *Jerusalem* preface (above, p. 521) on this term).   *Shaddai*: a common Hebrew term for God in the Old Testament, translated as 'The Almighty'.

l. 286. *Pahad*: another Old Testament term for God, translated as 'terror' or the equivalent.

l. 287. *loud he called*: presumably a reference to Christ's crying out on the cross 'with a loud voice' (according to three of the gospels).

341   l. 317: Jude 25.

343   l. 392. *Rock of Ages*. The phrase, used in the Authorized Version to gloss Isa. 26: 4, became very familiar as the opening of a late 18th-c. hymn.

l. 405. *tarsus*: the foot between heel and toe.

344   l. 419. The last five are several times listed as the daughters of Zelophehad in Numbers and Joshua. Their father is in fact dead, and they are conspicuous for their demand that they should inherit some of his wealth even though they have no brothers. On two different occasions (Num. 27, Josh. 17) and by two different patriarchs (Moses and Joshua) the problem is solved by granting the daughters a share of their uncles' inheritance.

l. 420. *Horeb*: the mountain of God; Sinai (l. 422) is equivalent to it, or part of it. Given *Paradise Lost* i. 7 the allusions are appropriate for Milton the dictating poet.

l. 424. Bashan, the kingdom of Og, was mountainous; the rest are all names of mountains in the Holy Land.

l. 428. The last three are all at various times the allied enemies of Israel. In Num. 20 the kingdom of Edom will not let the Israelites pass through its territory. Aram is Syria, a hostile neighbour associated in the Bible on several occasions with human sacrifice. In the course of *Milton* and *Jerusalem* Amalek and the Amalekites appear frequently as baneful agencies in a way that goes well beyond the historical actions ascribed to them in the Bible (rather in the manner that the Bible itself deploys the concept of Egypt). Blake is however building on the fact that Joshua's first victory was over the Amalekites, who are declared enemies of God 'from generation to generation' (Exod. 17: 8-16).

l. 439. *vehicular*: embodied, incarnate.

l. 449. *Arnon*: a resting-place for the Israelites on their way to the Promised Land. Crossing 'the brooks of Arnon' and 'the stream of the brooks that goeth down to the dwelling of Ar' is said at Num. 21: 14-15 to be comparable to the crossing of the Red Sea.

345 l. 451. *Mahanaim*: a frequently mentioned site in the Old Testament, perhaps most significant to Blake (from Josh. 13: 26, 30) as a point on the border between the transJordanian territories of Gad and Manasseh (see note to l. 557 below).

l. 455. *Succoth*: an Old Testament site associated with clay at 1 Kgs. 7: 46 and 2 Chron. 4: 17, but also the first stage on Israel's exodus from Egypt (Exod. 12: 37).

l. 458. *Beth Peor*: the place where Moses is buried, to be succeeded by Joshua. This event marks the close of Deuteronomy and the beginning of Joshua.

l. 462. *four Zoas*. Blake derives his idea of these four agencies from the four beasts of Rev. 4: 6-9 (collectively *zoa* in Greek). But 'zoa' is also an acceptable transliteration of the Greek word for life and existence.

l. 472. *Carmel*: a rocky promontory on the Palestinian coast.

l. 480. *Ephraim*: a son of Joseph and brother of Manasseh and Gad, and hence a tribal name. Alternatively, Mount Ephraim, the burial place of Joshua.

l. 483. *Cam*: the river on which Cambridge sits.

l. 484. *Rephaim*: an Old Testament people, sometimes said to be giants.

346 l. 494. *Where . . . his coming*: 2 Peter 3: 4.

ll. 502-3. *Hand, Hyle, Coban, Skofield*: Blake's disguises, of varying penetrability, for the Hunt brothers, editors of the *Examiner* (an allusion explained in the notes to *Public Address*, p. 537 above), Private Scolfield (the soldier with whom Blake had an altercation at Felpham, leading to a trial for high treason), and Hayley (Blake's Felpham patron—but Blake almost certainly has in mind also Greek *hyle*, 'matter', which, on a Gnostic view, the evil of creation produced). Coban could be a 'Cockburn' in reality, but he has not been identified. Although for the purposes of this edition Blake's various spellings of proper names deriving from his trial have been standardized the reader should be aware that, for whatever purpose, these names vary much more in spelling than any other group in Blake's mythology.

l. 503. *Reuben*: first-born of Jacob. For his gate see Ezek. 48: 31.

l. 507. *Manasseh*: the son of Joseph and brother of Ephraim, and also consequently a tribal name. Jointly Manasseh and Ephraim are particularly prominent in the episode of Jacob's blessing of them in Egypt (Gen. 48). See also ll. 769-70 below.

l. 509. *Hazor*: most prominent in the Old Testament as a city conquered by Joshua.

l. 521. *hand of clay*: with which Blake engraves and decorates his text.

l. 522. *Do . . . wilt*. Cf. Matt. 26: 39.

347 l. 537. *Og and Anak*: enemies of Israel of gigantic stature. For Anak see, in particular, Num. 13: 33.

l. 557. *Reuben, Gad*: two tribes that, together with half the tribe of Manasseh, have a very strong association in the early history of Israel as the only tribes who acquired lands east of Jordan before Israel crossed that river. In this sense they 'separated' from Israel. In Josh. 22 it is temporarily feared that they have also abandoned the God of Israel.

l. 558. *trough*: in which a smith cools hot metal with water.

348 ll. 577–9. Blake's inversion of Josh. 5: 15, perhaps with the deliverance of Peter from prison in mind (Acts 12: 8).

l. 581. *Those who Milton drove*: grammatically an inversion, with Milton as the object of the verb.

349 l. 621. *I am with you always*: Matt. 28: 20.

350 l. 625. *coming in the clouds*: Matt. 26: 64.

351 l. 658. *Sihon*: king of the Amorites, enemies of Israel and conquered during their advance; his territories were given to Reuben and Gad.

l. 680. The speech is now Rintrah's alone. George Whitefield and John Wesley are the two most famous figures in Methodism.

ll. 682–3: *took on . . . cross*: Phil. 2: 7–8.

352 l. 703. *Gwendolen and Conwenna*: both names derive from Geoffrey of Monmouth's British history.

353 l. 732. *winepresses*. The imagery of vintaging, which comes to form a triad with those of the forge and the loom in Blake's epics, is rooted in Rev. 14: 15–20.

l. 739. *seven eyes of God*. Cf. Zech. 4: 10.

l. 751. *Manazzoth*: presumably, since the other names here are those of tribes, Manasseh in another spelling. For yet another see l. 769 below.

ll. 763–4. Blake's episode rather than a biblical event, but an important instance of infidelity to Jehovah on the way to the Promised Land leads to a slaughter of Israelites by Amalekites and Canaanites in the wilderness (see Num. 14).

ll. 765–8: Obviously a reference to Gen. 37: 23–8, though the role of Amalekites is Blake's characteristic invention.

ll. 769–70. The prominent conjunction of Ephraim and Manasseh (see note to l. 507 above) in the Bible occurs in Egypt, but they are not credited with any gathering of the Israelites.

355 l. 805. *lilling*: Blake's coinage, or an error for 'lilting'.

l. 815. *serpent*: a large woodwind instrument.

l. 820. *Elias*: Elijah.

356 l. 869. Ps. 126: 5.

357 l. 875. *labour . . . for bread*: may refer to the inmates of the Royal Asylum for Female Orphans in Lambeth.

l. 886. *supper . . . bride*: Rev. 19: 7, 21: 9.

358 l. 907. *hem of their garments*: Matt. 9: 20, 14: 36.

ll. 909–11. See note to *Thel* 108 (above, p. 548).

359 l. 957. *lucky golden spinner*: the luck-bearing spider.

l. 965. *jubilee*: perhaps in the Miltonic sense of vocal celebration.

361 l. 1008. *Midsummer Night's Dream* v. i. 14–17.

363 l. 1075. *twenty-five cubits*: a distance which in the Bible is virtually confined to Ezek. 40, where it appears frequently among the dimensions of the Temple as shown to the prophet 'upon a very high mountain, which was as the frame of a city'.

364 l. 1126. *Zelophehad's daughters*. See note to l. 419 above.

l. 1131. *the Erythrean*: the Red Sea. Blake seems to accept the conjecture of Jacob Bryant (see note on p. 536 above) that this was the Indian ocean.

Book the Second

366 ll. 48–9. *coming . . . great glory*: Matt. 24: 30.

367 l. 77. *attending*: listening to (an archaism).

l. 82. *ever-during doors*. Cf. *Paradise Lost* vii. 206 (a section of the poem of peculiar interest to Blake).

l. 100. *daughter of Babylon*: in the Bible a formulation for Babylon itself. Blake seems to use it in a similar way, except that 'Babylon' is not the city but the figure from Rev. 17: 5, alluded to at l. 116.

369 l. 150. *River Storgé*. In Greek this would mean the river of parental love.

370 l. 177. *two fountains . . . life*: combines three biblical passages: Ezek. 47: 1–2, Zech. 14: 8, Rev. 22: 1.

l. 210. Probably a general reminiscence of *Paradise Lost* x. 304–5 and *passim*.

371 ll. 224–5. *fountain . . . streams*: combines Ezek. 47: 1–2 and Zech. 14: 8 with Exod. 17: 6.

l. 240. *Slumber nor sleep*: Ps. 121: 4.

372 l. 285. *wicker man of Scandinavia*. Multiple human sacrifices by burning in wicker effigies are only recorded of the Druids; but (contrary to *Jerusalem*) Blake seems at this stage not to wish to accuse the latter of cruelty, and transfers the anecdote to the Odin-worshipping Scandinavians.

373 ll. 289–90. *in him . . . churches*. In the following lines Blake seems to express this idea very directly, since the list of pagan deities is heavily indebted to Milton.

l. 294. *Baal*. His worship is strongly associated with Tyre in the Old Testament.   *Ashtaroth*. Blake uses Milton's occasional form of the name of the Sidonian goddess Astoreth (e.g. *Paradise Lost* i. 422). *Chemosh*: the Moabite national deity. Blake prefers the Bible spelling to Milton's 'Chemos'.

l. 299. *Dagon*. The idea that Dagon was a fish-god is a non-biblical conjecture familiar to Milton (see *Paradise Lost* i: 462–3).

l. 300. *Rimmon*: a deity strongly associated with Damascus in all traditions.

ll. 300–1. *Thammuz, Orus*: Milton's spellings of the Bible's Tammuz and the Egyptian god Horus.

l. 304. Blake is clearly following Milton's non-biblical interpretation of 'Belial' as a deity, and association of him with Sodom (*Paradise Lost* i. 502–3).

l. 307. *Rhea*: the Greek earth-goddess (see *Paradise Lost* i. 510–21).

ll. 310–16. The first twenty names are taken from biblical genealogies at Gen. 11 and Luke 3: many of them also appear elsewhere.

l. 311. *hermaphroditic*. Blake transfers from Milton's account of the pagan deities (especially Baal and Ashtaroth) the ideal of sexual indeterminacy and generalizes it to cover all three groups of 'heavens'.

l. 317. *dragon red*: Rev. 12: 3.

l. 320. *substances*. 'Substance' is the philosophical term used by Locke for the relatively undifferentiated matter, but possessing such Newtonian qualities as mass, volume, and velocity, which composes the universe.

374   l. 335. Cf. Num. 35: 7.

l. 340. *paved . . . stones*: Cf. Exod. 24: 10.

l. 350. *not . . . hands*: Heb. 9: 11.

375   l. 365. *one greater*: *Paradise Lost* i. 4.

l. 389. *Seven angels*: a recurrent motif throughout Revelation, but in association with the true deity. In this speech Blake has in effect transferred to the latter the fraudulent, Antichrist-like associations of Satan in Revelation.

376   ll. 402–3. *lake . . . fire*. Cf. Rev. 20: 10.

ll. 420–1. *Chaos, Sin, Death, Ancient Night*: ontologically the most intriguing of all the agents in *Paradise Lost*, being neither deity, angel, human, nor animal.

l. 426. *Legions*: an ancient British city of great renown, according to Geoffrey of Monmouth, where King Arthur was crowned.

l. 430. *basements*: foundations, sub-structure.

377   l. 464. *Bolingbroke*. The politician Henry St John, first Viscount Bolingbroke (1678–1751) is here because in his occasional philosophical writings he had enunciated a deistical Christianity.

378   l. 474. *Patmos*: where John is supposed to have received the vision recorded in Revelation.

ll. 476–7. *five . . . Philistia*: presumably the five lords of the Philistines (see Josh. 13: 3).

379   ll. 515–16. *curtains . . . rent*. Blake accepts the traditional association of the

veil of the temple (rent in Matt. 27: 51) with the curtains on the Old Testament ark of the covenant.

l. 538. *garment dipped in blood.* Cf. Rev. 19: 13.

l. 539. *Written within and without*: Ezek. 2: 10.

l. 540. *literal*: perhaps 'in letters'. The next paragraph is a kind of rephrasing, in English terms, of elements in Rev. 4, 7, 10, and elsewhere.

l. 545. *Jesus wept*: John 11: 35.

381 *Jerusalem*. In keeping with the new artistic emphases of Blake's later years *Jerusalem* is much longer than any other completed project, and much more formal in its organization: over 4,000 lines of verse arranged into four chapters, (the whole, together with prefaces, illustrations, etc., being etched on exactly 100 plates, 25 per chapter). It is for Blake an extraordinarily rigid arrangement, and sustained by him almost without hesitation; there are very few traces, especially by the standards of some of Blake's medium-length narratives, of changes of sequence, or supplementation of the plates to achieve his numerical targets (the only known case of a reordering of plates, in Ch. II, is noted below: essentially, in two of the five copies of *Jerusalem* ll. 594–883 are divided in two and inserted earlier in the chapter). Whatever else it has to offer, *Jerusalem* is the expression of a formidable energy and determination. It is not clear if the main numerical base (100 plates, 4 chapters) was chosen for reasons beyond the simplicity of the numbers themselves. Also intriguing is Blake's decision to call the sections of the poem 'chapters' rather than 'books', despite the obvious resemblances of the work as a whole to an epic. There may be some undiscovered model lying behind this usage, or it may be that Blake had come to see this work, with its very long lines generally filling his plates from side to side, as not unequivocally a work of verse.

The opening lines of *Jerusalem* offer clear hints about the organization of the text that follows and the interrelationship of its materials, which, if borne in mind, make *Jerusalem* a rather easier poem than *Milton* (in particular, the second aspect—the relationship between the main entities in the narrative—is not complicated in *Jerusalem* by the merging of mythological figures with Blake and his immediate world, which is so much the theme of *Milton*). It will be noticed that there is a considerable play with the notion of 'awaking' in these lines, and that there is an ambiguity about the status of l. 6. It seems at first to be the waking summons to the sleeping Blake, but l. 10's introduction of Albion implies that the 'sleeper' is Albion, that the awaking in question is the spiritual summons mentioned in l. 2 rather than the arousing of Blake to the task of transcribing his supernatural vision, and that the 'mild song' of *Jerusalem* proper (the poem is called a 'song' at its close) is already under way. In short, the text from l. 6 onwards is a vision of Blake's: and yet Blake wants to imply that he is on the same footing as, and analogous to, a figure *within* that vision—Albion. In the poem Los is the means by which this analogy is maintained: he is, quite simply, the manifestation of Blake within the visionary version of Britain that Blake has created, or which has been 'dictated' to him (just as 'Gwantok' and 'Kox', for example, are the real-life Guantock and Cock). All the main features of Blake's existence as he saw it—that of a married artist and representative of the eternal human imagination dwelling and working in the great city of London—have their correspondences in Los.

To be more accurate, *Jerusalem* is a sequence of five such visions by the poet Blake, in each of which Los is prominent, and in the first three of which he is the main actor. Los and Albion are quite separate— and Los's main initiatives are actions of foe and friend towards the latter—but they are also very alike: above all, the respective restorings of their calamitous relationships with their Emanations run completely in parallel. The sequence of visions gets under way properly at l. 100. Before that there is a second appearance, at l. 50, of the perfectly literal William Blake, working artist and visionary, who has spoken the very first lines. Each of the visions that are to follow is divided from the next by a resurfacing, so to speak, to this literal level—and scattered utterances in Blake's own voice start to crop up as the whole massive project moves to its climax.

First vision (I. 100–529): Los the artist
The first 240 or so lines concern Blake-as-Los's determination to remain the active exponent of imagination and seer of his society, despite the affronts he has endured and the possibility that his loyalty to his country is misplaced. This evidently has a connection with the neglect and humiliation (especially in the Scolfield affair) which Blake suffered at this period. In effect, Los must struggle to forgive his country (a lesson of peaceableness rather like that in *Milton*). Los' mastering of his Spectre is the artist's conquest of rancour and *amour propre*—a self-effacement which none the less calls forth its own heroic kind of self-assertion. Los will be in 'friendly contention' with Albion, a notion borne out in the design of the next two visions of his activity.

From l. 344 onwards are celebrated the glories of creativity once the energies of the self have been put at the service of the imagination. The building of Golgonooza is the equivalent of Blake's artistic activity—conceived as a parallel, spiritual London of outstanding majesty and beauty, and redeemed from all human cruelty.

Second vision (I. 529–II. 248): Los the foe of Albion's offspring
We return to Blake, deploring the disaster of modern European mental life, and sounding again the idea of the awaking of Albion. He then has a vision of a fierce, aggressive Los, who attacks first the sons of Albion (I. 633–II. 121), and then his daughters (II. 122–248). In the second, quite brief episode Los has to use a kind of long-range weapon, the Reuben-figure, since, as explained at the outset of the vision, 'he / Dare not approach the daughters openly' (I. 633–4).

The sons of Albion have to be thwarted in their attempt to achieve a complete casting-out of Albion's Emanation, Jerusalem. In a character-istic parallelism, Los is also motivated in this attack by the fact that his relationship with Enitharmon is bound up with that of Albion and

Jerusalem, and hence also at stake. In less mythological terms, such as Los uses in his speech at II. 102, what is occurring is a general distortion of relations between the sexes. Albion, indeed, at the close of Ch. I, feels the full horror of his alienation from Jerusalem (who is now closely linked, even fused, with Vala) and of the associated corruption of his spirit; so great is his remorse and despair that he 'dies', uttering the words 'hope is banished from me' which are later recorded by Blake as his last (II. 901). But, in an oscillation typical of his role in the poem, he is still alive in another manifestation, as the presiding spirit of a cruel rationalistic culture described at the beginning of Ch. II.

Third vision (II. 249–899): Los the friend of Albion
Blake's utterance resumes as that of the artist who has remained loyal to his visions at different times and in different places. He now sees Los's attempts to help Albion as a friend assisted by other friends. Los first turns to the twenty-eight English cities, who are represented as the concerned friends of a powerful but sick man dying at his home. The cities respond nobly but unavailingly.

Los then tries to work with the help of the Divine Vision, who explains that Albion must die to live again, an event that is reported in a different version by two refugees from the destruction of Albion. Los allies himself with the Divine Vision, and quests through Albion for the latter's enemies (this becomes the powerfully suggestive journey through London of ll. 812–41).

Fourth vision (II. 900–III. 1095): the world of the dead Albion
Blake inscribes, and depicts himself inscribing, Albion's last words, and there follows the vision of Albion's tomb or resting-place, of Jerusalem's departure, and of Erin, or Ireland, as the chief watcher and mourner at the tomb. Then almost the whole of Ch. III is given over to an account of life in Albion's world, with Jerusalem imprisoned and physically afflicted, and the malign forms of Albion and Jerusalem—Luvah and Vala—active in their stead. Certain spiritually enlightened individuals keep their faith in the Divine Vision and eventual redemption. Los and his daughters are among those using the materials and experiences of a fleshly, mortal world as a means of 'awaking again into Eternity.'

As in Ch. II Los is particularly the antagonist of the daughters of Albion, and most of the second half of this chapter (ll. 492–969) is a protracted vision of the cruel ascendancy of the feminine, astonishing in the profuseness of its images of violence and oppression, physical, psychical, and mental, as administered by the complex entity which is the daughters, Vala, and Rahab/Tirzah.

Fifth vision (III. 1096–end): The restoration of Jerusalem
This opens with an unusually long passage in Blake's own voice, leading to the picture of Jerusalem's ruined state, and her lament. The whole section clearly invokes those passages in the Bible, above all in Lamentations and Ezekiel, in which the captured city of Jerusalem is personified as a woman punished for transgressions (not that Blake accepts the imputation of the

sons of Albion), but the general effect is more important than verbal correspondences, which have not been cited in the notes.

The vision also concerns Enitharmon (who at least becomes Los's physical companion at IV. 362) and the daughters of Albion (who start to repent and make appeals to Los at IV. 381). There follows Los's vision of Jerusalem in her true beauty, and the beginnings of actual reconciliation between Los and Enitharmon at IV. 494, although their relationship is still fraught with suspicion and antagonism. The sense of oscillation between tokens of coming redemption and last-minute manifestations of the enemies of redemption is strong (see the remarks on Blake's use of Revelation below). Los is seen clearly as the visionary artist, rather as at the outset. Albion awakes, gives himself to Jesus, and calls back Jerusalem. She comes to him in the form of the whole spiritually awakened world.

Other schemes for *Jerusalem* have been proposed, but in my view this is the best. On the other hand, grasping the poem in the terms of this or any other simple overall plan is not the main challenge or satisfaction the reader should look for. Its considerable bulk, and great accumulation of detail, make such a response very natural; by the same token general schemes can too easily become a means of explaining the poem away, and of discounting it. Basically *Jerusalem* is worthwhile for the audacious, unflagging, and emotionally suggestive quality of the writing: it lives not by virtue of its total design, but for things such as Albion's confession (I. 927–1011), Los's quest through Albion (II. 800–74), the description and lament of Erin (II. 927–1058), the whole sequence from III. 131 to III. 899, the acts and utterances of Jerusalem, Vala, and the daughters of Albion at IV. 31–291, and the close of the poem, from IV. 426 to the end.

As far as the proper names in *Jerusalem* are concerned, there are no significant additions to Blake's own gallery of mythological figures, but there are more British names—both contemporary and historical—than hitherto, and more biblical ones. All the new appearances are explained in the notes, but some general comment on the Bible in *Jerusalem* is called for. For his Old Testament allusions Blake is still heavily indebted, as in *Milton*, to the early post-Egyptian history of Israel, and some relevant comment on this fact will be found in the headnote to *Milton*. As noted above, he supplements this with a general allusion, strong if unspecific, to the imagery of Jerusalem found in Ezekiel and Lamentations (and Ezekiel's vision of the temple is another definite element in the biblical background to the poem, though only directly echoed in two or three places). Of the New Testament books, Revelation can hardly be too much emphasized for its importance in *Jerusalem* (Blake may be argued to have completed his Bible of Hell with his last etched poem). Revelation is drawn on frequently in *Milton* also, but for *Jerusalem* it is the source of basic structural and conceptual motifs (the section of 'A Vision of the Last Judgement' based on Revelation (see above, p. 38) is a useful adjunct to reading *Jerusalem*). It is in Revelation above all that Jerusalem is seen simultaneously as a bride and a city (the crucial verse is Rev. 21: 2); her marriage and the entry of the saved into the city are indistinguishable crowning events of history. Blake associates the bride Jerusalem (no doubt with the encouragement of the Old Testament passages concerning her) with the 'woman clothed with the sun' of Rev. 12, who appears, gives birth to a child, and then flees into the wilderness (where she has 'a place prepared of God').

In this way Blake makes his own contribution to a characteristic structural pattern in Revelation, one which he adopts in his own text. John's vision is rich in figures who are antithetical to each other (on one side Antichrist-like, though the Antichrist is not named in Revelation). As a result the apocalyptic events seen by John tend to move towards their culmination (strongly so in ch. 12 if, like Blake, we take the 'woman clothed with the sun' to be Jerusalem), only to slip back in a counter-movement (which then turns out to be necessary to the whole process). Some of the more bewildering losses of narrative sequence in *Jerusalem* (as they seem) will strike the reader as familiar after a reading of Revelation. Finally, Revelation is a vision entrusted by God to one man, who is commanded to transmit to his fellows everything he has seen. Telling of his vision is itself in some way part of the coming millennium. The scheme adopted in analysing *Jerusalem* above—in terms of five visions of the poet Blake—will make it easy for the reader to discover how Revelation-like the poem is in this respect. Revelation 7 (where, for the last time, the tribes of Israel are numbered yet again) and 21: 12 (where the names of the tribes are seen on the gates of Jerusalem) may have been especially notable for Blake, who had already audaciously combined Britain and ancient Israel in *Milton*.

Μόνὸς δ 'Ιησοῦς: Jesus Alone.

Chapter IV

ll. 1–2. See note to *Urizen* 1 (above, p. 562).

l. 7. *I am . . . in me*: John 14: 20.

l. 18. *brother and friend*: combines Matt. 12: 49–50 and John 15: 14.

382   l. 28. *my own . . . myself.* Cf. Ezek. 29: 3.

383   l. 59. *Gwantok . . . Huttn*: a further sequence of names from Blake's own recent life. John Guantock, John Peachey, and William Brereton were judges and justices at Blake's trial; George Hutton was also involved on Scolfield's side; a 'Slayd', however, has not been identified. See note to *Milton* I. 502–3 (above, p. 569).

l. 61. *Kox, Kotope, Bowen*. Private Cock testified in support of Scolfield at the trial: the other two names have yet to be explained.

l. 67. The first time that twelve descendants of Albion are explicitly likened to the twelve tribes of Israel named after their patriarchs, but a natural development from the treatment of the tribes in *Milton*.

l. 68. *of beryl*: probably a reminiscence of Dan. 10: 5–6.

l. 74. *Mount Gilead*. Mount Gilead (as opposed to the mountainous region of that name occupied by Reuben, Gad, and Manasseh) is only mentioned once in the Bible (Judg. 7: 3).

l. 75. *Cordella and Ignoge*: according to Geoffrey of Monmouth, both royal princesses of ancient Britain (the former being Shakespeare's Cordelia).

l. 76. *covering cherub on Euphrates*. Euphrates, one of the Edenic rivers (Gen. 2: 14), is appropriate, given Blake's association of the covering cherub with Eden.

ll. 77–8. Gwineverra, Gonorill, Sabrina, Estrild, and Ragan are all women in the royal families of ancient Britain, as reported by Geoffrey (three are better known as Guinevere, Gonoril, and Regan). Gwinefred is Blake's Anglo-Saxon sounding coinage. In Mehetabel British and Old Testament traditions seem to converge: in context she should be Geoffrey's princess Methahel, but she bears a name from Gen. 36: 39.

385 ll. 134–42. For the Old Testament personal names see Gen. 10.

l. 146. *furnaces of affliction*: Isa. 48: 10.

386 l. 191. *Ranelagh, Strumbolo, Cromwell's Gardens*: all London pleasure-gardens in Blake's day.   *Chelsea . . . soldiers*: Chelsea Hospital for invalid soldiers.

387 l. 212. *Shinar*. See Gen. 11: 2.

l. 227. *abstain*: a rare transitive in English.

388 l. 241. *dry tree*: Isa. 56: 3.

391 l. 372. *Apollyon*. See Rev. 9: 11, and also the Valley of Humiliation episode in *Pilgrim's Progress*.

392 l. 384. *fatal tree*: the public gallows.

393 ll. 412–38. See Ezek, 1 and 10: 9–22. The emphasis on four gates at the cardinal points of the compass in the description of Golgonooza generally derives from Ezekiel's vision of the temple in chs. 40–7.

394 l. 467. *salamandrine*: capable of dwelling in fire.

397 l. 558. *valley of the son of Hinnom*. Hinnom only appears in the Old Testament (though quite frequently) in this particular formula. The valley was associated with human sacrifice, and is the equivalent of Gehenna (a biblical term for Hell).

398 l. 591. *to England and Scotland and Ireland*. In the next thirty-odd lines Blake makes explicit what has been adumbrated several times: the identity of Britain and the twelve tribes of Israel. For the tribal gates see note to iii. 986.

400 ll. 672–3. *unquenchable . . . worm*: Mark 9: 44.

401 l. 701. *cast ye Jerusalem forth*. Cf. 2 Kgs. 23: 27.

402 l. 722. *potter's field*: Matt. 27: 7.

403 l. 778. *Havilah*: the land drained by Pison, one of the rivers of Paradise.

406 ll. 878–9. *Nimrod . . . son*. Apparently a strange reversal of the biblical text. Blake transfers Nimrod's prowess in hunting (Gen. 10: 8–9) to Jehovah, and Nimrod's father Cush becomes the son of a huntsman.

407 l. 898. *ghast*: ghastly (a poeticism).

l. 903. *place . . . found*: Cf. Dan. 2: 35.

408 l. 937. *relapsing*: Blake's sense of 'falling' seems to be unique to him.

409 l. 968. *Jerusalem, Jerusalem*: Matt. 23: 37–9 (one of Blake's favourite chapters), Luke 13: 34–5.

410 l. 998. *Hesperia*: Italy.

l. 1009. *wast dead . . . alive*. See the parable of the prodigal son in Luke 15.

l. 1010. See Christopher Marlowe, *Dr Faustus* v. ii. 187.

ll. 1022–3. *one sparrow*. Cf. Matt. 10: 29.

411 l. 1026. Blake's version of John 1: 29 and its liturgical descendants. Cf. i. 1065.

Chapter II

412 l. 35. *fortuitous concourse*. For Democritus' atomic theory (giving rise to the phrase 'fortuitous concourse of atoms') see note to 'Mock on' 9 (above, p. 560).

l. 37. *stone of the brook*: as used by David to kill Goliath (1 Sam. 17: 40).

l. 40. Cf. Ps. 103: 16.

413 l. 80. *elevate*: elevated (a past participle).

414 l. 94. *neither . . . marriage*: Matt. 22: 30.

l. 120. *Mam Tor*: a hill in Derbyshire.

l. 124. *Zaretan, Stone of Bohan*: 1 Kgs. 7: 46, 2 Chron. 4: 17; Josh. 15: 6, 18: 17.

415 l. 137. *hide . . . dust*: Isa. 2: 10.

416 l. 164. *Heshbon*. The Moabite city of Heshbon was rebuilt by the tribe of Reuben, but eventually reoccupied by Moabites.

l. 166. *Gilead, Gilgal*. 1 Sam. 13: 7.

419 ll. 261–4. See the chronology of Blake's life in this edition for the significance of this sequence of place-names.

422 ll. 355–7. *Selsey . . . Chichester*: perhaps a reference to the removal of Selsey village to Chichester, as a result of coastal erosion, in 1075.

ll. 369 ff. A possible isolated case where Blake's change of plans about his text may have produced an awkwardness in the sequence of material (it is interesting that in two out of the five copies of *Jerusalem* ll. 594–674 of ch. II appear within this confusing part of the poem, at l. 400). The listing of the twenty-eight city-friends of Albion breaks off rather puzzlingly, not to be resumed until l. 566, and it may be that what comes in between is in some measure an interpolation. But most of it is important for theme and narrative, and there is no obvious way to reorganize the separate plates the text was etched on to make a more satisfactory sequence: so any change of plan here must date back to a hypothetical manuscript stage.

423 l. 391. *twenty-eight*: the twenty-four city-friends are combined with the four Zoas, a conjunction deriving from Rev. 4: 4–6 and subsequent passages.

424 l. 436. *Oshea . . . contend*: presumably based on the friendly confrontation at Josh. 14: 6–13 (Oshea is a variant of Joshua). The vocabulary of 'contending' is regularly used by Blake to denote friendly conflict. *Peor*: Num. 23: 28.

l. 438. *Balaam*: the gentile (Moabite) prophet who believed in the Jewish God and blessed Israel, contrary to the commands of King Balak.

425 l. 447. *We . . . English*: from the English folk rhyme that is supposed to be the chant of giants: 'Fee, fi, fo, fum . . .'.

428 l. 555. *leaves . . . Life*. Cf. Rev. 22: 2.

l. 565. The names that follow are all of cathedral towns and bishoprics.

430 ll. 634–7. Cf. Matt. 18: 2–14.

431 ll. 675 ff. In two of the five copies of *Jerusalem* the whole passage down to l. 883 appears earlier in ch. II, at l. 28. But it seems to belong much more naturally with the acts of Los as friend to Albion (his friendship for the latter being expressly mentioned at l. 802).

432 ll. 687–9. *reaction, action*. Blake plays on two quintessential terms of Newtonian dynamics. The political sense of 'reaction'—of reversal of, or resistance to, change—was also available in Blake's day.

l. 692. *Ephratah*: the Bible's ancient name for Bethlehem. See especially Ps. 132: 6.

l. 700. fear . . . flock: Luke 12: 32.

l. 703. *alone are escaped*. Cf. Job 1: 16, 19.

433 l. 721. Cf. 1 Cor. 4: 3–4.

l. 722. *If . . . die*. Cf. Ps. 104: 29.

l. 738. *Covered . . . foot*. Job 2: 7.

435 l. 792. Cf. Exod. 25: 18–22.

l. 794. This may refer to the second part of the second of the ten commandments (or 'rocky laws'). But see also III. 785.

436 ll. 808–10. The sequence of images accurately follows the process of brick-making.

l. 816. *kennels*: gutters in the street.

ll. 823–4. *Bethlehem . . . bread*. Bethelehem here is the London lunatic asylum at this date situated at Moorfields, rather than the biblical town. The Hebrew means 'house of bread'.

437 l. 839. *temple*: probably an ancient British temple hypothesized by Blake, rather than the Inn of Court of that name.

439 ll. 909–11. The main omissions from the canonical books of the Bible are Ruth, Chronicles, Job, The Song of Solomon, and the apostolic books other than Revelation. All are drawn on more or less frequently in Blake's poetry and prose, and Job was extremely important to him: the list corresponds exactly, however, to Swedenborg's prescription for the essential canon of the Bible, as approved by Blake in 1789 (see the chronology, p. xxiv above).   *five . . . Decalogue*: the first 5 books of the Bible (the Pentateuch), which culminate in the ten commandments and other basic doctrines of the Jewish religion.

441 l. 973. *Shiloh*: Joshua's capital and the seat of the tabernacle at that period.

442 l. 995. *chargeth . . . folly*. Cf. Job 4: 18.

   ll. 1015–16. Cf. Exod. 13: 21.

   l. 1023. *Beor*: Balaam's father.

443 l. 1037. *walks . . . fire*. Cf. Ezek. 28: 14.

   l. 1051. *Come, Lord Jesus*: Rev. 22: 20.   *if . . . died*. Cf. John 11: 21.

   l. 1062. *Ascending and descending*: Gen. 28: 12.

444 l. 1065. Cf. John 1: 29.

   ll. 1068–9. *let . . . sin*. Cf. Eph. 4: 26.

   Chapter III

445 l. 50. Cf. Matt 4: 1–3, Luke 4: 1–3.

   l. 55. *Agag*: The name/title borne by several kings of Amalek.

446 ll. 68–9. Recalls *Paradise Lost* ix. 823–5.

   ll. 76–7. *Joseph . . . Egypt*. Cf. Gen. 37: 28.

   l. 88. *The stars . . . fought*: Judg. 5: 20.

448 l. 130. Cf. Isa. 6: 8.

449 l. 163. *while*: meanwhile (a sense obsolete by Blake's day).

   l. 180. *Rosamond's bower*: the maze-like retreat allegedly built by Henry II at Woodstock to shelter his mistress Rosamond Clifford (where she was none the less successfully poisoned by Queen Elinor). Blake's knowledge of the legend probably comes from the 'Fair Rosamond' ballad in Percy's *Reliques*.

450 l. 199. *bonifying*: Blake's neologism.

452 l. 255. *four Zoas*. See note to *Milton* I. 462 (above, p. 569).

453 l. 297. Cf. *Exod*. 26: 7, 36: 35.

454 l. 315. *Mizraim*: Gen. 10: 6.

   l. 317. *Tesshina*. Although the context and Blake's practice at this period would suggest a known name the word has not been traced.

455   l. 357. Cf. Job 38: 31.

l. 364. *Lo . . . always*. Cf. Matt. 28: 20.

l. 365-6. *Only . . . sleepeth*. Combines John 11: 23-6 and Mark 5: 36-9.

l. 367. *behold Joseph and Mary*. The vision that follows in the next 27 lines is Blake's rewriting of Matt. 1: 18-24 in a vein that is not only anti-moralistic but rationalist.

ll. 372-3. *more pure . . . maker*: Job 4: 17.

456   l. 388. *without . . . price*: Isa. 55: 1.

l. 390. *none . . . sinneth not*. Of several biblical texts to this effect 1 Kgs. 8: 46 is the earliest.

ll. 397-9. *Euphrates, Gihon, Hiddekel, Pison*: the four Edenic rivers.

457   l. 406. *to the loathing . . . person*. Cf. Ezek. 16: 5.

l. 426. 'Seed of the woman' (because of Gen. 3: 15) implies descendants of Eve, and all the names that follow are those of biblical women (if 'Cainah' is taken to mean 'wife of Cain') forming a 'maternal line' to Mary, mother of Jesus, in the manner of the male-oriented genealogies at the beginnings of the gospels of Matthew and Luke. Blake also suggests, with considerable ingenuity, how this line of descent violates the idea of a licit, pure ancestry for Jesus, often with the direct authority of the Bible. Cainah, Ada and Zillah, and Naamah are respectively wife, daughters-in-law, and grand-daughter of Cain (the last of these, in one alternative Jewish tradition, is an evil woman who miscegenates with angels or, in yet another tradition, is the wife of Noah, unnamed in the Bible). Shuah's daughter, and her daughter Tamar, are both Canaanites and concubines of Judah: since Tamar is an ancestor of Jesus (Matt. 1: 3) his lineage is incestuous. And if the Rahab and Ruth of this genealogy are those written of at Josh. 2 and in Ruth, Jesus is also descended from a harlot of Jericho and a Moabite. Similarly Bathsheba the Hittite was the married mistress, and later wife, of King David, and mother of Solomon (Matt. 1: 6 in fact defines her as the ex-wife of Uriah). Jesus is descended from Solomon via Rehoboam, and according to 1 Kgs. 14: 21 his mother was Naamah, an Ammonite. The Bible only records of Zibeah that she was the mother of King Jehoash.

458   l. 433. *I know . . . last day*: John 11: 24.

l. 434. *in my flesh . . . God*: Job 19: 26.

l. 436. *I am . . . life*: John 11: 25.

l. 444. *a season*: Philem. 15 and John 16: 16.

459   l. 469. *Thor and Friga*: the Norse god of war and the principal wife of Odin. Here and in the following section Blake does more justice than hitherto to the allegations in the ancient sources that the Druids performed human sacrifices, though he still in some sense wishes to implicate Scandinavian forms of religion.

l. 471. *wheels filled with eyes*. Cf. Ezek. 10: 12 and Rev. 4: 8.

460 l. 499. *Druid knife*. Blake's sacrificial flint knife is perhaps an extrapolation from the celebrated golden knife with which the Druids were reputed to cut mistletoe: but see note to l. 641 below.

l. 500. *poetic vision*. Blake may have it in mind that knives and poison-cups feature in Elizabethan and Jacobean tragedies and in their descendants in the melodrama.

462 l. 542. *caves of Machpelah*. The cave of Machpelah was the family mausoleum of the great patriarchs Abraham, Isaac, and Jacob, and their wives.

l. 551. *poisonous blue*. Blake in effect suggests that woad was Prussian Blue (a characteristic ingredient of which, prussic acid, had been identified as a deadly poison in 1803).

466 l. 641. *knife of flint*. It is most significant that the Authorized Version glosses the 'sharp knives' used for circumcision at Josh. 5: 2–3 as 'knives of flint'.

ll. 644–9. Evidently closely imitative of Christ's crucifixion as described in the gospels of Matthew and Mark, with an additional allusion to Judg. 4: 2–3.

467 l. 691. *dove and raven*: Gen. 8: 7–8.

468 l. 700. *high Mona*: *Lycidas* 54.

l. 706. *blood of their covenant*. The primary use of the phrase is at Exod. 24: 8.

ll. 717–18. *Epicurean philosophy*. Underlying the Epicurean system was an atomistic physics inherited from Democritus.

l. 720. The rocks may be thought of as parents of men in a metaphysical system, such as the Epicurean, which holds that the inanimate and the animate (however intelligent) are constituted by the same materials. Also, Lucretius' account of Epicurus gives considerable emphasis to the phase of random and violent chaos which is supposed to precede the formation of the universe.

l. 728. Cf. Gen. 37: 31.

469 l. 762. *River Kanah*: formed the boundary between Ephraim and Manasseh.

470 l. 770. *Ebal, mount of cursing*. Cf. Deut. 27: 13–26.

l. 775. *Shechem*: modern Nablus.

ll. 784–5. Since the non-biblical deities in this list come in male–female pairs it may be deduced that Blake thinks of Molech and Chemosh likewise. See also IV. 401.

l. 790. *valley . . . Jebusite*: where Jerusalem (the Jebusite city) is located.

471 l. 805. *from Havilah to Shur*: Gen. 25: 18.

l. 818. *Uzzah*. See 2 Sam. 6: 3–7 (also 1 Chron. 13: 7–10).

l. 822. *Rehob in Hamath*. Cf. Num. 13: 21.

l. 828. *Meribah Kadesh*: Num. 13: 26, 20: 13–14. At Exod. 17: 7 the Authorized Version glosses Meribah as 'Chiding or Strife'.

472   l. 848. *Leah and Rachel*: the two contending wives of Jacob, notable for various subterfuges.

473   l. 877–9. *In the circumference*: since the tabernacle was set down at the centre of the Israelite camp, as described at Num. 1: 50–3 and 2.

l. 895. *key-bones*: collar-bone.

476   l. 968. *four . . . Scotland*: Aberdeen, Edinburgh, Glasgow, and St Andrews.

477   l. 980. *thus divided*. Blake follows the Bible (Num. 2) in grouping the twelve tribes in four camps named after four leading tribes and placed at the cardinal points of the compass (though the details are different in Blake's version). These four camps are associated with the four provinces of Ireland, and the complete list of twelve tribes is then associated with the constituent counties. For the latter the equivalent biblical event is the dividing of the Promised Land, as performed in Josh. 14–22.

l. 986. *Gates of Jerusalem*. The twelve gates of Jerusalem are referred to at Rev. 21: 12–13. On each is inscribed the name of one of the tribes. But the association of the tribes with each of the twelve gates of the city ultimately goes back to Ezek. 48.

l. 991. *four walls of Jerusalem*. The Jerusalem of Revelation also has four walls.

478   l. 1018. *Caffraria*: (or Kaffraria) the land of the Kaffirs (strictly, corresponding to the eastern part of Cape Province in South Africa).

l. 1019. *Zaara*: the Sahara.

ll. 1028–9. *Fénelon . . . Hervey*. François Fénelon, Jeanne Marie Guyon, St Teresa, George Whitefield, and James Hervey are instances of Catholic and Protestant religious teachers who in different ways were the advocates of enthusiastic and mystical forms of religion.

479   l. 1058. *Peleg and Joktan*: sons of Eber the descendant of Shem.

l. 1059. Voltaire ridiculed the idea of a benign God immanent in a world of suffering.

ll. 1065–6. A sequence that is meant to represent rulers in the ancient biblical, classical, and British worlds. The Bible does not record that Tubal was a ruler, though he is a direct descendant of Cain. Bladud and Belin are named as British kings in Geoffrey of Monmouth.

480   l. 1096. *the testimony of Jesus*: i.e. 'the spirit of prophecy' (Rev. 19: 10).

482   l. 1136. *Dinah*. For Dinah's story see Gen. 34.

l. 1139. *land of Cabul*. See 1 Kgs. 9: 13.

Chapter IV

483 l. 5. *potter . . . potsherds*: Isa. 30: 14.

484 l. 32. *arrows of the Almighty*: Job 6: 4.

l. 40. *mountain of blessing*. Cf. Deut. 11: 29.

l. 44. *Goshen*: a conquest of Joshua's.

485 l. 68. *Kishon*: a river in northern Palestine.

l. 69. *earth . . . down*. Cf. Job 1: 7.

486 l. 116. *I am . . . soul*. Cf. Ps. 22: 6.

489 l. 210. *zone*: girdle, belt.

490 l. 242. *Helle*: Hellas (Greece).

l. 257. *Bath Rabbim*: S. of S. 7: 4.

496 l. 448. *Holiness . . . Lord*: as with the frontlet worn by Aaron (Exod. 28: 36).

l. 463. Cf. Rev. 21: 2.

l. 476. *Javan*: a figure mentioned in Genesis, usually taken to be the ancestor of the Greeks. 'Isles of Javan' thus may be the Greek islands: but see l. 624 below.

498 l. 521. *not as . . . I will*: Matt. 26: 39, Mark 14: 36.

500 ll. 596–7. *Pharisaion, Grammateis, Presbyterion, Archiereus, Iereus, Saddusaion*: intended as transliterations of terms in the Greek New Testament denoting the following elements in the Jewish religious establishment, all implicated in the persecution of Jesus: Pharisees, scribes, elders, High Priest, priest, Sadducees.

l. 609. *dragon of the river*. Cf. Ezek. 29: 3.

501 l. 616. *Rabbath*: the chief Ammonite city.

l. 619. *scapulae . . . os humeri*: anatomical terms for shoulder-blades and upper-arm bone.

l. 637. *Anakim, Emim, Nephilim, Gibborim*. All giants, though the first two groups are giant enemies of Israel, and the latter two (in the Hebrew) the giants and their offspring of Gen. 6: 4.

505 ll. 752–3. *heavens . . . curtain*. Cf. Ps. 104: 2.

l. 754. *Hermes*: Hermes Trismegistus (see note (above, p. 551) to 'Africa' 18).

507 l. 813. See Gen. 30: 14.

l. 827. *communion of saints and angels*. 'The communion of saints' is an expression from the Apostles' creed. By supplementing it with angels Blake emphasizes that central to Christianity is the idea of a level of existence completely transcending the physical, which Man can also aspire to dwell in, such as is certainly played down, if not denied, by deism (though Bacon, Newton, and Locke have no relevant comment on the question).

508 l. 839. *damps*: poisonous gases.

512 l. 953. *fourfold*. The various quadruples in the passage that follows have their roots in those of Rev. 4–7.

513 l. 974. *non-ens*: non-being.

514 l. 992. *kingdoms . . . glory*. Cf. Matt. 4: 8.

FURTHER READING

Hazard Adams, *William Blake: a Reading of the Shorter Poems* (Seattle, 1963).

John Beer, *Blake's Humanism* (Manchester, 1968).

——, *Blake's Visionary Universe* (Manchester, 1969).

G. E. Bentley Jr., *Blake Records* (Oxford, 1969).

David Bindman, *Blake as an Artist* (Oxford, 1977).

Bernard Blackstone, *English Blake* (Cambridge, 1949).

Harold Bloom, *Blake's Apocalypse* (Ithaca, N.Y., 1963).

Anthony Blunt, *The Art of William Blake* (New York, 1959).

Jacob Bronowski, *A Man without a Mask* (London, 1944).

G. K. Chesterton, *William Blake* (London, 1910).

S. Foster Damon, *William Blake. His Philosophy and Symbols* (Boston, 1924).

——, *A Blake Dictionary* (Providence, R.I., 1965).

David V. Erdman, *Blake: Prophet against Empire* (Princeton, 1954).

David V. Erdman and John E. Grant (eds.), *Visionary Forms Dramatic* (Princeton, 1970).

Northrop Frye, *Fearful Symmetry* (Princeton, 1947).

Stanley Gardner, *Infinity on the Anvil* (Oxford, 1954).

D. G. Gillham, *Blake's Contrary States* (Cambridge, 1966).

Alexander Gilchrist, *Life of William Blake, 'Pictor Ignotus'* (London, 1863).

Robert Gleckner, *The Piper and the Bard* (Detroit, 1959).

E. D. Hirsch, *Innocence and Experience* (New Haven, 1964).

John Holloway, *Blake: the Lyric Poetry* (London, 1968).

Jack Lindsay, *William Blake: His Life and Work* (London, 1978).

Raymond Lister, *William Blake. An Introduction to the Man and his Work* (London, 1968).

Herschel M. Margoliouth, *William Blake* (New York, 1961).

John Middleton Murry, *William Blake* (London, 1933).

Martin K. Nurmi, *William Blake* (Kent, Ohio, 1976).

Morton D. Paley, *Energy and the Imagination* (Oxford, 1970).

Mark Plowman, *An Introduction to the Study of Blake* (London, 1927).

Kathleen Raine, *Blake and Tradition* (Princeton, 1968).

Alvin H. Rosenfeld (ed.), *William Blake: Essays for Foster S. Damon* (Providence, R.I., 1969).

Denis Saurat, *Blake and Milton* (New York, 1924).

Mark Schorer, *William Blake. The Politics of Vision* (New York, 1946).

Algernon Charles Swinburne, *William Blake. A Critical Essay* (London, 1868).

J. H. Wicksteed, *Blake's Innocence and Experience* (London, 1928).

Mona Wilson, *The Life of William Blake* (London, 1927).

INDEX OF TITLES AND FIRST LINES

POETRY

Abstinence sows sand all over   264
A fairy stepped upon my knee   280
A flower was offered to me   272
Africa   203
Ah! Sunflower   273
Ah! sunflower, weary of time   273
A little black thing among the snow   269
All the night in woe   241
America, a Prophecy   205
Ancient Proverb, An   262
And aged Tiriel stood before the gates of his beautiful palace   163
And did those feet in ancient time   297
And in melodious accents I   114
Angel, The   271
Anger and wrath my bosom rends   306
An old maid early, ere I knew   266
Answer to the Parson, An   265
Are not the joys of morning sweeter   261
Argument [to *The Marriage of Heaven and Hell*]   8
Asia   222
As I wandered the forest   261
Auguries of Innocence   27
Awake, awake my little boy   291
A woman scaly and a man all hairy   280

Birds, The   282
Blake's Apology for His Catalogue   108
Blossom, The   242
Book of Los, The   325
Book of Thel, The   175
Book of Urizen, The   309

Call that the 'public voice' which is their error   110
Can I see another's woe   244
Children of the future age   277
Chimney-Sweeper, The
    [*Innocence*]   252
    [*Experience*]   269
Clod and the Pebble, The   268
Come hither my boy; tell me what thou seest there   265
Come hither my sparrows   263
Come, kings, and listen to my song   127
Cradle Song, A
    [*Innocence*]   252

Cradle Song, A   258
Cruelty has a human heart   279
Crystal Cabinet, The   293
[Cynic's First Song, The]   232
[Cynic's Second Song, The]   234

Daughters of Beulah, muses who inspire the poet's song   331
Day   263
Dear mother, dear mother, the church is cold   274
Degrade first the arts if you'd mankind degrade   74
Divine Image, A   279
Divine Image, The   251
Dream, A   238

Each man is in his Spectre's power   300
Earth raised up her head   267
Earth's Answer   267
Echoing Green, The   249
England! awake! awake! awake!   301
Eno, aged mother   325
Enslaved, the daughters of Albion weep a trembling lamentation   196
Eternity   265
Europe, a Prophecy   214
Everlasting Gospel, The   45

Fair Elenor   124
Fairy, The   263
Father, father where are you going?   254
Five windows light the caverned man: through one he breathes the air   214
Fly, The   270
French Revolution, The   180
Fresh from the dewy hill, the merry year   230

Garden of Love, The   273
Golden Apollo, that through heaven wide   131
Golden Net, The   281
Great men and fools do often me inspire   306
Great things are done when men and mountains meet   304
Grey Monk, The   282
Grown old in love from seven till seven times seven   302
Gwin, King of Norway   127

Hail Matrimony, made of Love   234
Having given great offence by writing in prose   108
Hear the voice of the bard   267
Her whole life is an epigram, smack-smooth and neatly penned   266
He's a blockhead who wants a proof of what he can't perceive   302
He who binds to himself a joy   265
Holy Thursday
    [Innocence]   250
    [Experience]   269

How sweet I roamed from field to field   227
How sweet is the shepherd's sweet lot   243
How to Know Love from Deceit   262
Human Abstract, The   275

I am no Homer's hero you all know   302
I asked a thief to steal me a peach   257
I bless thee, O Father of Heaven and Earth, that ever I saw Flaxman's face   64
'I die, I die', the mother said   282
I dreamt a dream!—what can it mean?   271
I feared the fury of my wind   259
If I e'er grow to man's estate   303
If moral virtue was Christianity   45
If you have formed a circle to go into   306
If you play a game of chance, know before you begin   304
If you trap the moment before it's ripe   264
I have no name   244
I heard an angel singing   257
I laid me down upon a bank   256
I loved Theotormon   196
I love the jocund dance   228
I love to rise in a summer morn   246
Imitation of Pope; A Compliment to the Ladies   303
Imitation of Spenser, An   131
In a Myrtle Shade   260
In a wife I would desire   264
Infant Joy   244
Infant Sorrow   276
In futurity   239
In seed time learn, in harvest teach, in winter enjoy   10
Introduction [to Experience]   267
Introduction [to Innocence]   238
I rose up at the dawn of day   304
I saw a chapel all of gold   256
I saw a monk of Charlemagne   300
Is this a holy thing to see   269
I stood among my valleys of the south   34
I told my love, I told my love   256
I travelled through a land of men   288
I walked abroad in a snowy day   262
I wander through each chartered street   274
I was angry with my friend   276
I went to the Garden of Love   273
I will sing you a song of Los, the eternal prophet   203
I wonder whether the girls are mad   295

Jerusalem   381

Lacedaemonian Instruction   265
Lamb, The   243
Land of Dreams, The   291

Laughing Song   247
[Lawgiver's Song, The]   236
Leave, oh, leave me to my sorrows   236
Lily, The   273
Little Black Boy, The   248
Little Boy Found, The   254
Little Boy Lost, A   276
Little Boy Lost, The   254
Little fly   270
Little Girl Found, The   241
Little Girl Lost, A   277
Little Girl Lost, The   239
Little Mary Bell had a fairy in a nut   294
Little Vagabond, The   274
London   274
Long John Brown and Little Mary Bell   294
Love and harmony combine   228
Love seeketh not itself to please   268
Love to faults is always blind   262

Madman I have been called; fool they call thee   302
Mad Song   230
Mary   292
Memory, hither come   229
Mental Traveller, The   288
Merlin's Prophecy   263
Merry, merry sparrow   242
Milton   331
[Miss Gittipin's First Song]   234
[Miss Gittipin's Second Song]   236
Mock on, mock on, Voltaire, Rousseau   284
Morning   284
Motto [to Songs of Innocence and Experience]   266
My mother bore me in the southern wild   248
My mother groaned, my father wept   276
My pretty Rose Tree   272
My silks and fine array   227
My Spectre around me night and day   285

Night   254
Nought loves another as itself   276
Nurse's Song
    [Innocence]   250
    [Experience]   270

O Autumn, laden with fruit, and stained   122
[Obtuse Angle's Song]   235
O'er my sins thou sit and moan   286
Of the primeval priest's assumed power   309
Of the sleep of Ulro! and of the passage through   381

Oh for a voice like thunder, and a tongue 150
O holy virgin! clad in purest white 124
O lapwing, thou flyest around the heath 260
On Another's Sorrow 244
Once a dream did weave a shade 238
On the Virginity of the Virgin Mary and Joanna Southcott 281
O rose, thou art sick 270
O thou, who passest through our valleys in 121
O thou, with dewy locks, who lookest down 121
O Winter! bar thine adamantine doors 122

Phoebe dressed like beauty's queen 234
Piping down the valleys wild 238
Poison Tree, A 276
Prepare, prepare, the iron helm of war 151
Prologue, Intended for a Dramatic Piece of King Edward the Fourth 150
Proverbs of Hell 10

Question Answered, The 265

Rafael sublime, majestic, graceful, wise 103
Remove away that black'ning church 262
Riches 265
Rintrah roars and shakes his fires in the burdened air 8

Schoolboy, The 246
Shepherd, The 243
Sick Rose, The 270
Silent, silent night 259
Since all the riches of this world 306
Sleep, sleep, beauty bright 258
Smile, The 287
Soft Snow 262
Some look to see the sweet outlines 75
Some men created for destruction come 303
Song First by a Shepherd 232
Song Third by an Old Shepherd 232
Sound the flute 245
Spring 245
Sweet dreams, form a shade 252
Sweet Mary, the first time she ever was there 292

Terror in the house does roar 284
The angel that presided o'er my birth 302
The bell struck one, and shook the silent tower 124
The countless gold of a merry heart 265
The daughters of Mne Seraphim led round their sunny flocks 175
The dead brood over Europe; the cloud and vision descends over cheerful
    France 180
The fields from Islington to Marybone 297

The good are attracted by men's perceptions   266
The harvest shall flourish in wintry weather   263
The kings of Asia heard   222
The little boy lost in the lonely fen   254
The look of love alarms   265
The maiden caught me in the wild   293
The modest rose puts forth a thorn   273
The nameless shadowy female rose from out of the breast of Orc 214
There is a smile of love   287
The shadowy daughter of Urthona stood before red Orc   205
The sun arises in the east   263
The sun descending in the west   254
The sun does arise   249
The sword sung on the barren heath   264
The wild winds weep   230
This city and this country has brought forth many mayors   236
Thou fair-haired angel of the evening   123
Thou hast a lap full of seed   260
Three virgins at the break of day   281
Tiger, The   271
Tiger, tiger, burning bright   271
Tiriel   163
To a lovely myrtle bound   263
To Autumn   122
To be or not to be   235
To Chloe's breast young Cupid slyly stole   306
To find the western path   284
To God   306
To Mercy, Pity, Peace and Love   251
To Morning   124
To my Dearest Friend, John Flaxman   64
To my friend Butts I write   23
To my Myrtle   263
To Nobodaddy   261
To see a world in a grain of sand   27
To Spring   121
To Summer   121
To the Evening Star   123
To the Muses   126
To Tirzah   278
To Winter   122
'Twas on a Holy Thursday, their innocent faces clean   250

Venetian! all thy colouring is no more   78
Visions of the Daughters of Albion   196
Voice of the Ancient Bard, The   249

War Song to Englishmen, A   151
Welcome, stranger, to this place   232
Whate'er is born of mortal birth   278

Whate'er is done to her she cannot know   281
What is it men in women do require   265
When a man has married a wife he finds out whether   280
When early morn walks forth in sober grey   231
When my mother died I was very young   252
When nations grow old the arts grow cold   76
When old Corruption first begun   232
When silver snow decks Sylvio's clothes   232
When the green woods laugh with the voice of joy   247
When the voices of children are heard on the green
  [*Innocence*]   250
  [*Experience*]   270
Where thou dwellest, in what grave   282
Whether on Ida's shady brow   126
Why art thou silent and invisible   261
Why of the sheep do you not learn peace?   265
Why should I be bound to thee   260
Why should I care for the men of Thames   259
Why was Cupid a boy   305
William Bond   295
With happiness stretched across the hills   25
Wondrous the gods; more wondrous are the men   303

You don't believe—I won't attempt to make ye   303
Youth of delight come hither   249

PROSE

All religions are One   5

Berkeley, Bishop, *Siris*, annotations to   54
Boyd, Henry, *A Translation of the Inferno of Dante Alighieri*, annotations to   65

Contemplation   153
Couch of Death, The   152

Descriptive Catalogue of Pictures, A   81

Jerusalem, prefaces to
  (Chapter I)   72
  (Chapters II, III, IV)   31

King John, prologue to   150

Lavater, John Casper, *Aphorisms on Man*, annotations to   3

Marriage of Heaven and Hell, The   8
Milton, preface to   71

On Homer's Poetry   115
On Virgil   116

Reynolds, Sir Joshua, *Works*, annotations to vol. 1   73

Samson   154
Spurzheim, J. G., *Observations on . . . Insanity*, annotations to   45
Swedenborg, Emanuel, *Wisdom of Angels Concerning Divine Love and Divine
   Wisdom*, annotations to   5

There is No Natural Religion (1)   6
   ibid. (2)   7
Thornton, R. J., *The Lord's Prayer, Newly Translated*, annotations to   55

'*Vision of the Last Judgement*'   35
[*Vision of Pride, The*]   157

Watson, Richard, *An Apology for the Bible*, annotations to   20
'*Woe!*' cried the Muse   160
Wordsworth, William, *Poems* (1815), annotations to   116

RECIPIENTS OF LETTERS

Butts, Thomas   65, 67, 68, 70

Cumberland, George   61, 63

Trusler, Revd John   59, 60

INDEX OF NAMES AND MOTIFS

This index covers the following material: (1) all Blake's invented personal and topographical names, with the exception of a few minor items (2) words and phrases used regularly and idiosyncratically by Blake to express leading elements in his thought (3) all supernatural entities from religion and myth given significant mention in Blake's text or in the notes (4) legendary or quasi-historical figures and groups accorded an unfamiliar function by Blake.

The citations are, of course, restricted to the texts in this volume. For completeness they should be supplemented from the rest of Blake's surviving writings: in particular, *The Book of Ahania*, *The Four Zoas*, and the sections of 'Vision of the Last Judgement' not reproduced here.

Abomination of desolation, 23, 351, 379, 386, 482

Acis, 550

Adam, 9, 31, 38, 95, 203, 223, 339, 364, 373, 385, 394, 397, 415–16, 430, 446, 479–80, 482

Adona, 175

Adonis, 548, 557

Aeneas, 547, 548, 560

Ahania, 345, 395

Al-Ulro, 368

Albion (incl. Albion's Angel, Daughters of Albion, Sons of Albion), 19, 31, 39, 95, 148, 150–1, 196–7, 199, 202, 204, 206–7, 209–13, 217–20, 277, 298–9, 331, 334–48, 351–4, 356–7, 364, 366, 369–70, 373, 376–89, 391–3, 395–455, 457–68, 470–4, 476–500, 502–11, 514

Alla, 368, 501

Allamanda, 354, 358, 360–1, 363, 374, 422

Amalek/Amalekite, 344–5, 353, 382, 395, 425, 432, 457, 460, 471, 481, 487, 490, 492, 496, 506

Anak/Anakim, 347, 350, 367, 395, 442, 479, 501

Antamon, 203, 221, 361

Antichrist, 45, 50, 208, 500, 502, 577

Apollo, 82, 88, 93, 96, 131, 227, 511, 541, 542, 558

Apollyon, 391

Ariston, 203, 209, 552

Ark (of the Covenant), 40, 373, 379, 406, 470–1, 482, 490

Arthur, 94–5, 445, 461, 479, 482, 499, 536

Asher, 398–9, 478, 481

Ashtaroth, 373

Athena/Minerva, 93, 132, 547

Atlas, 95, 203, 552

Baal, 373

Babylon, 32, 116, 157, 298, 336, 351, 367–8, 374–5, 377–8, 402, 404, 406, 409, 413, 431, 454, 457, 480–2, 490, 501, 506

Bacchus, 88, 470

Behemoth, 505, 533

Belial, 373

Benjamin, 398–9, 459, 471, 477, 482, 502–3

Beulah, 27, 331, 333, 336–7, 339, 342–3, 346–8, 351, 356, 358, 360, 362, 365–73, 376–7, 381–3, 390–6, 400, 402–3, 408, 410–11, 416, 419, 422–3, 427–8, 430, 435, 438–46, 448, 452–54, 460, 472–3, 475, 478, 480, 483, 486, 493–5, 497, 501–2, 513

Bowen, 383, 385, 403, 416, 438, 476, 502–3

Bowlahoola, 352, 354–5, 358, 361, 363, 374, 422, 480, 482

Brahma, 203

Brereton, 383, 403, 416, 438, 476

Brittania/Brittanica, 39, 416, 508–9

Bromion, 196–9, 334, 353, 397, 476–8, 480

Cambel, 383, 385, 388, 406, 467, 475, 487–9, 491–2

Canaan/Canaanite, 49, 344–6, 353, 378, 382, 395, 397, 425, 432, 445, 451, 454, 457, 460, 462, 466, 471–2, 474, 481, 486, 490, 492, 495, 501, 505–6, 519

Cathedron, 340, 354, 358, 394, 452, 478, 480–2, 487, 491, 493, 497

Chemosh, 373, 470, 494, 501

Christ, 15, 18, 20–3, 40, 42–3, 50, 54–5, 62, 67, 71–2, 78, 109, 418, 502

Classes of Men, 15, 62, 72, 88, 90, 332, 337, 356, 358

Clytia, 558

Coban, 346, 352, 383, 385, 387–8, 402–3, 438, 416, 451, 475

Contraries, 32, 149, 369, 378–9, 389–90, 400, 408, 432, 446, 450, 462, 557

Conwenna, 352, 383, 405, 476, 501

Cordella, 383, 405, 475–6

Covering Cherub, 336, 352, 354, 372–3, 383, 500–1, 510

Cupid, 305–6

Dagon, 155, 373

Dan, 398–9, 468, 478, 481, 484

Daughters of Memory, 36, 71, 76, 82, 341

Devil, 9–10, 18–20, 47, 50, 52, 56, 87, 274, 294–5

Diana, 541

Diralada, 203

Divine Humanity, 40–1, 341, 403, 412, 424, 442, 457, 474, 511, 513

Druid etc., 31–2, 95, 98, 298, 335, 337, 373, 438, 443, 445, 449, 453, 457–60, 464, 468, 473, 483, 486, 489, 500–1, 506–7, 509, 512, 514

Elohim, 31, 340–1, 365, 447, 455, 479

Elynittria, 217, 338–40, 507

Enion, 346, 395, 433, 492, 497

Enitharmon, 26, 204, 208, 214–21, 319–21, 332–6, 339, 344–5, 349, 353, 355, 358, 364, 366, 370–1, 380, 390–1, 395, 399–401, 422, 429, 434, 452, 458–60, 490, 492, 495, 497–500, 506–7

Eno, 325

Entuthon/Entuthon Benython, 345, 358, 360, 363–4, 374, 382–4, 396, 483, 493, 499

Erin, 377, 388, 390–2, 426, 440, 443, 478, 482–3, 497, 499, 508

Esculapius, 88

Estrild, 383, 406, 476

Eternals, 309–11, 313, 318–20, 347, 407

Ethinthus, 220, 392, 451

Eve, 38, 446, 502, 547

Fates, 534

Friga, 459, 470

Fuzon, 322, 324

Gad, 347, 353, 398–9, 468, 481, 484, 486

Galatea, 550

Generation, 37–8, 41, 95, 278, 341, 366, 379, 385–6, 392–3, 399, 416, 429, 435, 443–4, 451, 469–70, 479, 501, 503, 514

Gog, Magog, 38, 481, 514

Golgonooza, 344, 347, 350, 353–4, 356, 358–70, 364, 366, 370–2, 374, 383, 389–90, 392–6, 444, 478–9, 483, 493, 497, 499, 514

Gonorill, 383, 476

Gordred, 127, 129–30

Grodna, 322

Gwantok, 383, 403, 416, 438, 475

Gwendolen, 352, 383, 385, 388, 405–6, 415, 449–50, 460, 462, 467, 475, 487–9, 491, 495

Gwin, 127–30

Gwinefred, 383, 476

Gwineverra, 383, 388, 476

Hand, 346, 352, 383, 385–8, 396, 401–3, 405, 414, 416, 430, 438, 451, 453, 469, 473–5, 481, 488, 491–2, 494, 502–3

Har, 164–7, 171–5, 177, 179, 203–4

Havilah, 403, 441, 446, 471

Hecate, 158, 534

Hela, 171–3

Helios, 548

Hercules, 82, 88, 96, 357

Hermes Trismegistus, 203, 505

Heuxos, 163–4

Heva, 164–7, 171–4, 204

Hoglah, 344, 370, 469

Huttn, 383, 385, 403, 416, 438

Hyle, 346, 352, 383, 385, 387–8, 396, 402–3, 405, 416, 430, 438, 451, 469, 475, 481, 488–90, 494, 503

Ignoge, 383, 391, 406, 475

Ijim, 168-70
Isis, 373
Issachar, 398, 477, 481, 484

Jehovah, 10, 45, 334-5, 338, 340-1, 352, 406, 428, 435, 442, 447, 455-7, 459-60, 471, 489, 513-14, 521, 566
Jerusalem, 31, 33-4, 36, 39, 157, 203, 223, 297-301, 331, 336-9, 346, 351, 357, 368, 370, 372, 375-9, 381-4, 386, 388-9, 391-2, 396, 398, 400-5, 407, 409-10, 412-13, 416, 418-19, 422-3, 425, 427-9, 431-2, 434, 436-43, 445-7, 451, 453-8, 468, 471-4, 476-8, 480-98, 500-1, 506, 511, 514
Jesus, 15, 18, 22, 31-7, 41, 43, 45-6, 48-56, 72-3, 88, 96, 203, 279, 331, 341, 350, 354, 365-6, 375, 379-80, 384, 413, 417-18, 422, 426-7, 443, 447, 458, 464, 473, 480, 482, 487, 499, 503-4, 507, 510-11
Joseph (Old Testament), 35, 398-9, 446, 468, 470, 477, 482, 486, 489, 502
Jove/Jupiter, 93, 131-2, 373, 528
Judah, 203, 353, 398-9, 477, 481, 484, 486, 502, 507, 510

Kotope, 383, 385, 387, 403, 438, 416, 468, 476
Kox, 383, 385, 387-8, 403, 416, 425, 476

Leutha, 196, 203, 220-1, 325, 337-40, 436, 494
Levi, 353, 398-9, 477, 481, 484, 486, 503, 507
Leviathan, 16, 505, 533
London Stone, 95, 298, 387, 430, 437, 449, 451, 465, 467, 481, 508
Los (incl. Sons of Los), 26-7, 203-4, 215, 222, 313-21, 326-7, 329-36, 344-5, 347, 349-66, 369-72, 374, 376-7, 380, 383-92, 394-401, 403, 414-17, 419-20, 422-3, 426-31, 434-8, 440, 444, 448-53, 458, 460-1, 466, 472-3, 476-83, 487, 490-1, 493-5, 497-500, 502-3, 505-10
Lotho, 169
Luban, 354, 356, 394
Lucifer, 46, 339, 447
Luvah, 176, 339, 345, 348, 355, 359-60, 364, 369, 371, 373, 385, 398, 403, 405-7, 410, 413-14, 416, 422-3, 433-9, 443, 445, 451, 453, 458-9, 461-2, 464-5, 479-80, 482, 487-8, 491-2, 502-3, 509, 511
Lyca, 239-42

Malah, 344, 370, 470
Manasseh, 346, 353, 469, 477, 484
Matha, 170
Mehetabel, 383, 475-6
Mercury, 112, 131-2
Merlin, 95, 263, 414, 416-17, 449, 482, 489, 499, 507
Michael, 334
Midas, 131
Milcah, 344, 370, 470
Mne Seraphim, 175
Mnemosyne, 82, 547
Mnetha, 164-7, 173-4
Molech, 43, 334, 339, 373, 447, 470-1, 494, 501
Mundane Egg/Shell, 344-5, 347, 349, 352, 356, 369, 371-4, 376-7, 394-5, 398, 431, 452, 460, 478-9, 482, 492, 506
Muse, 59, 76, 82, 126, 160, 331
Myratana, 163, 169, 171
Myrina, 546
Myrrha, 557, 562
Mystery, 36, 38, 275, 279, 351, 374, 482, 492, 507

Napthali, 398-9, 478, 481
Nobodaddy, 261

Oberon, 535
Ocalythron, 217, 507
Odin/Woden, 203, 357, 492, 546
Og, 43, 298, 347, 351, 367, 373-4, 395, 441-2, 479, 484, 501
Oithona, 550
Ololon, 348-50, 365-72, 377-9
Ona, 278
Oothoon, 196-7, 199, 201-3, 221, 340, 380, 423, 492
Orc, 203, 205, 207-8, 210-14, 216-17, 219, 221-3, 320-1, 348, 351, 363-4, 366, 395
Orcus, 170, 552
Or-Ulro, 368
Orus, 373

Osiris, 373
Ozoth, 362

Pahad, 340, 447
Palambron, 203, 216-17, 219, 332-40, 347, 350-3, 355, 363-4, 380, 397, 476-8, 480, 506-7
Pan, 131
Pandora, 159
Peachey, 383, 403, 416, 438, 475
Phaeton, 519
Pluto, 535
Poetic Genius, 5-6, 13, 341
Polyphemus, 550
Poseidon, 552
Prometheus, 518, 545
Proserpine, 535
Proteus, 547

Ragan, 383, 391, 405, 462, 476
Rahab, 32, 340, 344-6, 351, 356, 364, 366, 370, 372, 378, 415, 419, 431, 453, 457, 468-9, 473-4, 480, 482-3, 487-9, 495, 500
Rephaim, 345, 364, 422, 440, 462, 469, 471, 500-1, 506
Reuben, 346-7, 353, 391, 396-9, 414-17, 459-60, 462, 471, 473, 477-8, 481, 486, 489, 494-5, 502-3, 507
Rhadamanthus, 45
Rhea, 373
Rimmon, 373
Rintrah, 8-9, 203, 216-17, 219, 333-7, 347, 350, 352-3, 355, 363-4, 380, 397, 476-8, 480, 506-7

Sabrina, 383, 391, 405, 476
Satan/Satanic (incl. Satan the Accuser, Synagogue of Satan), 10, 15, 31-2, 38, 41-3, 48-9, 67, 75-6, 95, 99, 297-9, 301, 332-9, 341-2, 344, 347-8, 350-4, 356, 359-60, 364, 366, 371-8, 392, 394, 403, 412, 415, 419, 423, 430, 442-3, 446, 451-2, 454, 479, 503, 507, 562
Saturn, 373
Shaddai, 340, 428, 447, 471
Shadowy Female, 214, 351, 369
Sihon, 351, 374, 441-2, 501
Silenus, 88
Simeon, 353, 398-9, 477, 481, 507

Skofield, 346, 383, 385, 387, 391, 396, 401, 403, 406, 416, 425, 438, 451, 453, 468, 470, 476, 503
Slayd, 383, 403, 416, 438, 476
Sol, 520
Sotha, 203, 221, 361
States, 31-2, 37, 44, 88, 368, 411, 415, 417, 443, 474, 480, 557
Stone of Night, 207, 218

Thammuz, 373
Tharmas, 345, 354, 369, 377, 395, 423, 452, 459, 480, 492, 509, 511
Thel, 175-8
Theotormon, 26, 196-8, 201-3, 221, 334, 351, 353, 360-1, 397, 476-8, 480
Theseus, 546
Thiralatha, 221
Thiriel, 321
Thor, 357, 459, 470, 492
Tiriel, 163-74
Tirzah, 278, 340, 344-6, 351, 353, 356-7, 364, 370, 383, 415-16, 453, 468-9, 480, 500
Titania, 535
Titans, 95, 518

Udan Adan, 348, 350, 353, 358, 360, 382, 385, 394, 397, 499
Ulro, 336, 341, 343-4, 348-51, 354, 359-60, 363-4, 368-9, 371-3, 378-9, 381, 392-3, 408, 411, 416-17, 426, 428-9, 431, 447, 480, 483
Urizen, 199, 203-4, 208, 212, 216, 218, 220-1, 223, 309-10, 312-14, 317, 320-2, 324, 329-30, 344-6, 359-60, 369, 374, 377, 385, 395, 398, 416, 423, 437, 451-2, 459, 462, 464-5, 480, 499, 509, 511
Urthona, 19, 205, 216, 345, 369, 389, 419, 423, 434-5, 444, 452, 459, 480, 491, 509, 511
Utha, 322

Vala, 339, 346, 383-5, 391-2, 395, 400-8, 413-14, 416, 423, 427, 431-5, 437-9,

446–7, 452, 454, 458–65, 468, 474, 476–7, 483, 486–7, 489, 492, 495, 498, 500
Venus, 82, 93, 112, 470, 548, 557

Yuva, 163

Zazel, 164, 173
Zebulun, 398–9, 477, 481
Zoa, 345, 353, 355, 370, 376–7, 410, 416–17, 423, 430, 451–2, 459, 480, 482, 500
Zoroaster, 530